# Intelligent Systems Reference Library

## Volume 82

**Series editors**

Janusz Kacprzyk, Polish Academy of Sciences, Warsaw, Poland
e-mail: kacprzyk@ibspan.waw.pl

Lakhmi C. Jain, University of Canberra, Canberra, Australia, and
University of South Australia, Adelaide, Australia
e-mail: Lakhmi.Jain@unisa.edu.au

## About this Series

The aim of this series is to publish a Reference Library, including novel advances and developments in all aspects of Intelligent Systems in an easily accessible and well structured form. The series includes reference works, handbooks, compendia, textbooks, well-structured monographs, dictionaries, and encyclopedias. It contains well integrated knowledge and current information in the field of Intelligent Systems. The series covers the theory, applications, and design methods of Intelligent Systems. Virtually all disciplines such as engineering, computer science, avionics, business, e-commerce, environment, healthcare, physics and life science are included.

More information about this series at http://www.springer.com/series/8578

Guangquan Zhang · Jie Lu
Ya Gao

# Multi-Level Decision Making

## Models, Methods and Applications

 Springer

Guangquan Zhang
Faculty of Engineering and Information
    Technology
University of Technology Sydney
Sydney, NSW
Australia

Ya Gao
Faculty of Engineering and Information
    Technology
University of Technology Sydney
Sydney, NSW
Australia

Jie Lu
Faculty of Engineering and Information
    Technology
University of Technology Sydney
Sydney, NSW
Australia

ISSN 1868-4394              ISSN 1868-4408   (electronic)
Intelligent Systems Reference Library
ISBN 978-3-662-51634-8      ISBN 978-3-662-46059-7   (eBook)
DOI 10.1007/978-3-662-46059-7

Springer-Verlag GmbH Berlin Heidelberg is part of Springer Science+Business Media
(www.springer.com)

# Preface

Multi-level decision-making (MLDM) handles problems that require compromise between the objectives of two or more interacting entities which are arranged within a hierarchical structure with independent and perhaps conflicting objectives. Bi-level decision-making is a special and particularly popular case of MLDM, in which only two levels of decision entities are involved, each of which tries to optimize their individual objectives under certain constraints, and to act and react in a sequential manner. The MLDM problem appears naturally in critical resource management, production and transportation planning, and organizational policy making. There are two fundamental issues to address in dealing with an MLDM problem. One is how to model a multi-level decision problem, and the other is how to find an optimal solution to the problem.

This monograph presents the new developments in multi-level (in particular bi-level and tri-level) decision-making theory, technique and methodology in both modelling and solution issues. It especially presents how a decision support system (software) can support managers in reaching a solution to a multi-level decision problem in practice.

This monograph offers the following advantages:

- It focuses on one of the most complex and challenging decision-making structures, in which several levels of decision entities are involved in a hierarchical decision-making process in which each level may have more than one decision entity, and each entity may have more than one objective function and fuzzy parameters in the functions.
- It combines decision theories, methods, algorithms and applications effectively. We discuss in detail the models and solution algorithms of each issue of bi-level and tri-level decision-making, such as multi-leaders, multi-followers, multi-objectives, rule-based, and fuzzy parameters presented in this monograph, as well as the related case studies and/or software systems.
- It is designed as a unified whole in which the content in each chapter is related to the material that precedes it, and also to what will follow. A number of case-based examples, such as logistics, are discussed in various chapters for different

multi-level decision situations, as well as the use of decision support systems to obtain the desired solutions.

- It reflects the latest academic research progress and state-of-the-art development through the results of our own and other authors' recent publications in this field. It does not attempt to provide exhaustive coverage of every fact or research result that exists.

This monograph is principally based on our research developments over the past ten years, during which time we have produced more than 50 journal and conference publications in this field.

The monograph has 14 chapters, organized into five parts. The first part, from Chaps. 1 to 3, covers concepts of decision-making, decision support systems and bi-level decision-making in general. The second part of the monograph, from Chaps. 4 to 6, presents bi-level multi-follower and tri-level multi-follower decision-making, including related models, solution methods, algorithms and case studies. The third part, from Chaps. 7 to 9, focuses on uncertainty issues in multi-level decision-making and discusses fuzzy bi-level and fuzzy multi-objective bi-level decision models and solution algorithms. In Part IV, Chap. 10 deals with a non-programming multi-level decision issue and proposes the framework and methods of rule-set-based bi-level decision-making. The last part, from Chaps. 11 to 14, shows the development of bi-level and tri-level decision support systems and related applications by using the methods presented in the previous chapters. These applications include the power market, supply chain management and railway organization.

Our potential readers include organizational managers and practicing professionals, who can use the methods and software provided to solve their real decision problems; researchers in the areas of bi-level and multi-level decision-making and decision support systems; students at an advanced undergraduate or master's level in information systems, business administration, or the application of computer science programs.

We wish to thank the Australian Research Council (ARC), whose ARC discovery grant DP0557154 partially supported the work presented in this monograph; our co-workers, who have offered advice and have conducted some of the research results in this monograph with us; many researchers who have worked in multi-level decision-making, fuzzy set application, decision support systems and related areas over the past several decades, whose significant insights have been included in the monograph and whose well-known publications feature in the reference list; Dr. Yue Zheng, Mr. Jialin Han, Dr. Jun Ma, and the researchers and students in the Decision Systems and e-Service Intelligence (DeSI) laboratory, Centre for Quantum Computing and Intelligent Systems (QCIS), at the University of Technology Sydney (UTS) who suffered through several versions of the decision algorithms and decision support systems shown in this monograph; Sue Felix who proofread the main part of this monograph; and the editors and production staff at Springer, who helped us to ensure the monograph was as good as we were capable of making it.

Sydney, Australia, October 2014                                          Guangquan Zhang
                                                                                     Jie Lu
                                                                                    Ya Gao

# Contents

**Part I  Bi-level Decision Making**

**1  Decision Making and Decision Support Systems** .............  3
   1.1    Organizational Decision Making ......................  3
   1.2    Classification for Decision Problems and Techniques .......  4
          1.2.1    Decision Problem Classification ................  4
          1.2.2    Decision Support Technique Classification .........  5
   1.3    Main Decision Support Techniques ....................  6
          1.3.1    Mathematical Programming ...................  7
          1.3.2    Multi-criteria Decision Making .................  8
          1.3.3    Case-Based Reasoning .......................  10
          1.3.4    Data Warehouse and Data Mining ...............  11
          1.3.5    Decision Tree ...........................  12
          1.3.6    Fuzzy Sets and Systems .....................  13
   1.4    Decision Support Systems .........................  13
          1.4.1    Concepts .............................  14
          1.4.2    Characteristics .........................  15
          1.4.3    Components ...........................  15
   1.5    DSS Classification .............................  17
          1.5.1    Model-Driven DSS .......................  17
          1.5.2    Data-Driven DSS ........................  17
          1.5.3    Knowledge-Driven DSS or Intelligent DSS .........  18
          1.5.4    Group DSS ............................  19
          1.5.5    Web-Based DSS .........................  19
   1.6    DSS Software Illustration .........................  20
          1.6.1    Case 1: Decider .........................  20
          1.6.2    Case 2: A DSS for Ore Blending Cost Optimization
                of Blast Furnaces .........................  21
   1.7    Summary ................................  24

**2   Optimization Models**.............................................  25
   2.1   Concepts .........................................................  25
   2.2   Linear Programming ..............................................  28
   2.3   Non-linear Programming ..........................................  29
      2.3.1   Varieties of Non-linear Programming .............  29
      2.3.2   Theories and Optimality Conditions
         of Non-linear Programming ...................  30
      2.3.3   Methods for Solving Non-linear Programming
         Problems ....................................  31
   2.4   Multi-objective Programming .....................................  32
      2.4.1   Multi-objective Programming Model .............  32
      2.4.2   Multi-objective Linear Programming Methods ......  34
      2.4.3   A Case-Based Example ........................  36
   2.5   Goal Programming ...............................................  37
   2.6   Stackelberg Game Model.........................................  41
      2.6.1   Stackelberg Game and Bi-level Programming.......  42
      2.6.2   Stackelberg Game and Nash Game ..............  43
      2.6.3   Applications of Stackelberg Games ..............  43
   2.7   Particle Swarm Optimization ....................................  44
   2.8   Summary........................................................  46

**3   Bi-level Programming Models and Algorithms** ...............  47
   3.1   Bi-level Programming Model .....................................  47
   3.2   Solution Theories for Linear Bi-level Programming..........  49
   3.3   $K$th-Best Algorithm for Linear Bi-level Programming .......  53
   3.4   Kuhn-Tucker Approach for Linear Bi-level Programming ....  55
   3.5   Branch-and-Bound Algorithm for Linear Bi-level
      Programming ....................................................  57
   3.6   Penalty Function Method for Linear Bi-level Programming ...  59
   3.7   Multi-level Programming Model ..................................  61
   3.8   Summary........................................................  62

**Part II   Multi-level Multi-follower Decision Making**

**4   Bi-level Multi-follower Decision Making** ....................  65
   4.1   Problem Identification ...........................................  65
   4.2   Framework for Bi-level Multi-follower Decision Making .....  66
   4.3   Bi-level Multi-follower Decision Models .................  68
      4.3.1   BLMF Decision Entity-Relationship Diagram.......  68
      4.3.2   Linear BLMF Decision Models ...............  68
   4.4   Uncooperative Bi-level Multi-follower Decision Making .....  76
      4.4.1   Solution Concepts...........................  76
      4.4.2   Theoretical Properties ........................  77

|  |  | 4.4.3 | Uncooperative BLMF $K$th-Best Algorithm | 79 |
|  |  | 4.4.4 | Uncooperative BLMF Kuhn-Tucker Approach | 83 |
|  | 4.5 | Semi-cooperative Bi-level Multi-follower Decision Making | | 85 |
|  |  | 4.5.1 | Solution Concepts | 86 |
|  |  | 4.5.2 | Theoretical Properties | 87 |
|  |  | 4.5.3 | Semi-cooperative BLMF $K$th-Best Algorithm | 87 |
|  |  | 4.5.4 | Semi-cooperative BLMF Kuhn-Tucker Approach | 89 |
|  | 4.6 | Reference-Uncooperative Bi-level Multi-follower Decision Making | | 91 |
|  |  | 4.6.1 | Solution Concepts | 91 |
|  |  | 4.6.2 | Theoretical Properties | 93 |
|  |  | 4.6.3 | Reference-Uncooperative BLMF $K$th-Best Algorithm | 95 |
|  |  | 4.6.4 | Reference-Uncooperative BLMF Kuhn-Tucker Approach | 102 |
|  | 4.7 | Summary | | 104 |
| **5** | **Bi-level Multi-leader Decision Making** | | | **105** |
|  | 5.1 | Problem Identification | | 105 |
|  | 5.2 | Framework for Bi-level Multi-leader Decision Making | | 106 |
|  | 5.3 | Linear Bi-level Multi-leader Decision Models | | 107 |
|  | 5.4 | Concepts and Definitions | | 112 |
|  | 5.5 | Generalized Nash Equilibrium Solution | | 114 |
|  | 5.6 | BLML Particle Swarm Optimization Algorithm | | 115 |
|  | 5.7 | A Numerical Example | | 118 |
|  | 5.8 | Summary | | 120 |
| **6** | **Tri-level Multi-follower Decision Making** | | | **121** |
|  | 6.1 | Problem Identification | | 122 |
|  | 6.2 | Basic Tri-level Decision Models | | 123 |
|  | 6.3 | Tri-level Multi-follower Decision Framework | | 125 |
|  |  | 6.3.1 | TLMF Decision Concepts | 125 |
|  |  | 6.3.2 | TLMF Decision Problem Classification | 126 |
|  |  | 6.3.3 | TLMF Decision Framework | 127 |
|  |  | 6.3.4 | TLMF Decision Entity-Relationship Diagrams | 127 |
|  | 6.4 | Tri-level Multi-follower Decision Models | | 133 |
|  |  | 6.4.1 | General Model for TLMF Decision | 133 |
|  |  | 6.4.2 | Typical Standard Models for TLMF Decision | 134 |
|  |  | 6.4.3 | Hybrid TLMF Decision Models | 142 |
|  | 6.5 | Case Studies for TLMF Decision Modeling | | 144 |
|  |  | 6.5.1 | Case 1: S28 Model | 144 |
|  |  | 6.5.2 | Case 2: S27 Model | 147 |
|  |  | 6.5.3 | Case 3: S54 Model | 148 |
|  |  | 6.5.4 | Case 4: Hybrid of S41, S45 and S48 Models | 150 |

6.6     Tri-level Decision Solution Methods . . . . . . . . . . . . . . . . . .     151
        6.6.1    Solution Concepts. . . . . . . . . . . . . . . . . . . . . . . .     151
        6.6.2    Theoretical Properties . . . . . . . . . . . . . . . . . . . . .     153
        6.6.3    Tri-level $K$th-Best Algorithm . . . . . . . . . . . . . . . . .     155
        6.6.4    A Numerical Example . . . . . . . . . . . . . . . . . . . . .      157
6.7     Tri-level Multi-follower Decision Solution Methods . . . . . . . .     159
        6.7.1    Solution Concepts. . . . . . . . . . . . . . . . . . . . . . . .     159
        6.7.2    Theoretical Properties . . . . . . . . . . . . . . . . . . . . .     162
        6.7.3    TLMF $K$th-Best Algorithm. . . . . . . . . . . . . . . . . .       165
        6.7.4    A Numerical Example . . . . . . . . . . . . . . . . . . . . .      167
6.8     Summary. . . . . . . . . . . . . . . . . . . . . . . . . . . . . . . . . .       170

Part III    Fuzzy Multi-level Decision Making

7    **Fuzzy Bi-level Decision Making** . . . . . . . . . . . . . . . . . . . . .     175
        7.1     Problem Identification . . . . . . . . . . . . . . . . . . . . . . . .     175
        7.2     Fuzzy Sets and Systems. . . . . . . . . . . . . . . . . . . . . . . .     176
                7.2.1    Fuzzy Sets. . . . . . . . . . . . . . . . . . . . . . . . . .     177
                7.2.2    Fuzzy Numbers . . . . . . . . . . . . . . . . . . . . . .       178
        7.3     Fuzzy Bi-level Decision Models . . . . . . . . . . . . . . . . . . .     183
        7.4     Fuzzy Approximation $K$th-Best Algorithm . . . . . . . . . . . . .     188
                7.4.1    Property and Algorithm . . . . . . . . . . . . . . . . . .     188
                7.4.2    Illustrative Examples . . . . . . . . . . . . . . . . . . . .     197
        7.5     Fuzzy Multi-Follower Approximation $K$th-Best Algorithm . . . .     201
        7.6     Summary. . . . . . . . . . . . . . . . . . . . . . . . . . . . . . . . .       205

8    **Fuzzy Multi-objective Bi-level Decision Making** . . . . . . . . . . . . .     207
        8.1     Problem Identification . . . . . . . . . . . . . . . . . . . . . . . .     207
        8.2     Fuzzy Multi-objective Bi-level Decision Model . . . . . . . . . .     208
        8.3     Fuzzy Approximation Kuhn-Tucker Approach . . . . . . . . . . .     219
                8.3.1    Fuzzy Approximation Kuhn-Tucker Approach . . . . . .     220
                8.3.2    A Case-Based Example . . . . . . . . . . . . . . . . . .     220
        8.4     Summary. . . . . . . . . . . . . . . . . . . . . . . . . . . . . . . . .       228

9    **Fuzzy Multi-objective Bi-level Goal Programming** . . . . . . . . . . .     229
        9.1     Problem Identification . . . . . . . . . . . . . . . . . . . . . . . .     229
        9.2     Solution Concepts. . . . . . . . . . . . . . . . . . . . . . . . . . .     230
        9.3     Fuzzy Bi-level Goal-Programming Algorithm. . . . . . . . . . . .     241
        9.4     A Numerical Example and Experiments. . . . . . . . . . . . . . .     242
                9.4.1    A Numerical Example . . . . . . . . . . . . . . . . . . .     242
                9.4.2    Experiments and Evaluation . . . . . . . . . . . . . . . .     246
        9.5     Summary. . . . . . . . . . . . . . . . . . . . . . . . . . . . . . . . .       247

**Part IV    Rule-set-based Bi-level Decision Making**

**10   Rule-Set-Based Bi-level Decision Making** . . . . . . . . . . . . . . . . . .   251
   10.1   Problem Identification . . . . . . . . . . . . . . . . . . . . . . . . . . . . .   251
   10.2   Information Tables and Rule-Sets . . . . . . . . . . . . . . . . . . . . . .   252
       10.2.1   Information Tables . . . . . . . . . . . . . . . . . . . . . . . . . .   252
       10.2.2   Formulas and Rules . . . . . . . . . . . . . . . . . . . . . . . . .   253
       10.2.3   Decision Rule Set Function . . . . . . . . . . . . . . . . . . . .   255
       10.2.4   Rule Trees . . . . . . . . . . . . . . . . . . . . . . . . . . . . . . . .   256
       10.2.5   Rules Comparison . . . . . . . . . . . . . . . . . . . . . . . . . . .   258
   10.3   Rule-Set-Based Bi-level Decision Model . . . . . . . . . . . . . . . .   260
       10.3.1   Objectives . . . . . . . . . . . . . . . . . . . . . . . . . . . . . . . .   260
       10.3.2   Constraints . . . . . . . . . . . . . . . . . . . . . . . . . . . . . . .   261
       10.3.3   Rule-Set-Based Bi-level Decision Model . . . . . . . . . . .   261
   10.4   Rule-Set-Based Bi-level Decision Modeling Approach . . . . . . .   262
   10.5   Rule-Set-Based Bi-level Decision Solution Algorithms . . . . . .   264
       10.5.1   Concepts and Properties . . . . . . . . . . . . . . . . . . . . . .   264
       10.5.2   Rule-Based-Based Solution Algorithm . . . . . . . . . . . .   265
       10.5.3   Transformation-Based Solution Algorithm . . . . . . . . .   268
   10.6   A Case Study . . . . . . . . . . . . . . . . . . . . . . . . . . . . . . . . . . .   273
       10.6.1   Problem Modeling . . . . . . . . . . . . . . . . . . . . . . . . . .   273
       10.6.2   Solution . . . . . . . . . . . . . . . . . . . . . . . . . . . . . . . . .   276
   10.7   Experiments and Analysis . . . . . . . . . . . . . . . . . . . . . . . . . .   283
   10.8   Summary . . . . . . . . . . . . . . . . . . . . . . . . . . . . . . . . . . . . . .   286

**Part V    Multi-level Decision Support Systems and Applications**

**11   Fuzzy Bi-level and Tri-level Decision Support Systems** . . . . . . . . .   289
   11.1   A Fuzzy Bi-level Decision Support System . . . . . . . . . . . . . .   289
       11.1.1   System Configuration and Interfaces . . . . . . . . . . . . .   290
       11.1.2   System Structure . . . . . . . . . . . . . . . . . . . . . . . . . . .   291
       11.1.3   Linear Bi-level Decision Support Process . . . . . . . . . .   293
       11.1.4   Non-linear Bi-level Decision Support Process . . . . . . .   302
   11.2   A Tri-level Decision Support System . . . . . . . . . . . . . . . . . . .   306
       11.2.1   System Configuration and Tri-level Decision
              Support Process . . . . . . . . . . . . . . . . . . . . . . . . . . . .   306
       11.2.2   Detailed Operational Process
              and System Interface . . . . . . . . . . . . . . . . . . . . . . . . .   309
   11.3   Summary . . . . . . . . . . . . . . . . . . . . . . . . . . . . . . . . . . . . . .   314

**12  Bi-level Programming for Competitive Strategic Bidding
      Optimization in Electricity Markets**...................... 315
      12.1  Background ........................................ 315
      12.2  Bidding Strategy Analysis in Competitive
            Electricity Markets ............................... 316
            12.2.1  Strategic Pricing Model for Power Plants.......... 317
            12.2.2  Generation Output Dispatch Model
                    for Market Operator ....................... 318
      12.3  BLML Decision Model in Competitive Electricity Markets ... 319
      12.4  A Case Study...................................... 320
            12.4.1  Test Data............................... 320
            12.4.2  Experiment Results....................... 321
            12.4.3  Experiment Analysis ...................... 323
      12.5  Summary......................................... 324

**13  Bi-level Pricing and Replenishment in Supply Chains**.......... 325
      13.1  Background ....................................... 325
      13.2  Case Study 1: Hi-tech Product Pricing and Replenishment
            Strategy Making .................................. 326
            13.2.1  Problem Formulation....................... 326
            13.2.2  Experiments............................. 331
      13.3  Case Study 2: Hi-tech Product Pricing and Replenishment
            Strategy Making with Weekly Decline-Rates ............. 332
            13.3.1  Problem Formulation....................... 332
            13.3.2  Experiments............................. 334
      13.4  Summary......................................... 336

**14  Bi-level Decision Making in Railway Transportation
      Management**........................................... 337
      14.1  Case Study 1: Train Set Organization ................. 337
            14.1.1  Background ............................. 337
            14.1.2  Problem Formulation....................... 340
            14.1.3  Experiments............................. 342
      14.2  Case Study 2: Railway Wagon Flow Management ......... 344
            14.2.1  Background ............................. 344
            14.2.2  Problem Formulation....................... 345
            14.2.3  Experiments............................. 355
      14.3  Summary......................................... 356

**References**.............................................. 357

# Abbreviations

| | |
|---|---|
| AHP | Analytic Hierarchy Process |
| AID | Attribute Importance Degree |
| BLDM | Bi-level Decision-Making |
| BLDSS | Bi-level Decision Support Systems |
| BLMF | Bi-level Multi-follower |
| BLML | Bi-level Multi-leader |
| BLMLMF | Bi-level Multi-leader and Multi-follower |
| BLML-PSO | Bi-level Multi-leader Particle Swarm Optimization |
| BLP | Bi-level Programming |
| CBR | Case-based Reasoning |
| DBMS | Database Management System |
| DE | Decision Entities |
| DERD | Decision Entity-Relationship Diagrams |
| DSS | Decision Support Systems |
| EPEC | Equilibrium Problem with Equilibrium Constraints |
| FA-$K$th-Best | Fuzzy Approximation $K$th-Best |
| FBLDM | Fuzzy Bi-level Decision-Making |
| FBLDSS | Fuzzy Bi-level Decision Support System |
| FBLGP | Fuzzy Bi-level Goal Programming |
| FBLMF | Fuzzy Bi-level Multi-follower |
| FBLP | Fuzzy Bi-level Programming |
| FGDSS | Fuzzy Group Decision Support Systems |
| FMO-BLD | Fuzzy Multi-objective Bi-level Decision-Making |
| FMO-BLGP | Fuzzy Multi-objective Bi-level Goal Programming |
| FMO-BLMF | Fuzzy Multi-objective Bi-level Multi-follower |
| FMO-BLP | Fuzzy Multi-objective Bi-level Programming |
| GDM | Group Decision-Making |

| GDSS | Group Decision Support Systems |
|------|-------------------------------|
| GP | Goal Programming |
| GSS | Group Support Systems |
| IDSS | Intelligent Decision Support Systems |
| KBMS | Knowledge-base Management System |
| LP | Linear Programming |
| LTLP | Linear Tri-level Programming |
| MADM | Multi-attribute Decision-Making |
| MBMS | Model-base Management System |
| MCDM | Multi-criteria Decision-Making |
| MCGDM | Multi-criteria Group Decision-Making |
| MLDM | Multi-level Decision-Making |
| MLP | Multi-level Programming |
| MO-BLMF | Multi-objective Bi-level Multi-follower |
| MODM | Multi-objective Decision-Making |
| MODSS | Multi-objective Decision Support Systems |
| MOGP | Multi-objective Goal Programming |
| MOL-BLP | Multi-objective Linear Bi-level Programming |
| MOLP | Multi-objective Linear Programming |
| MOP | Multi-objective Programming |
| NES | Neighborhood Entity Set |
| NLP | Non-linear Programming |
| OLAP | Online Analytical Processing |
| PSO | Particle Swarm Optimization |
| RSBLD | Rule-set-based Bi-level Decision |
| RWFM | Railway Wagon Flow Management |
| TLDM | Tri-level Decision-Making |
| TLDSS | Tri-level Decision Support System |
| TLMF | Tri-level Multi-follower |
| TLMF-DERD | TLMF Decision Entity-Relationship Diagrams |
| TLP | Tri-level Programming |
| TOPSIS | Technique for the Order Preference by Similarity to Ideal Solution |
| TSO | Train Set Organization |
| UMP | Uniform Marginal Price |
| WDSS | Web-Based Decision Support Systems |

# Part I
# Bi-level Decision Making

# Chapter 1
# Decision Making and Decision Support Systems

This book addresses an important decision making area—multi-level decision-making. To help readers understand the following chapters of this book, this chapter presents fundamental concepts, models, and techniques of decision making and *decision support systems* (DSS), thus providing an introduction for the remaining chapters of this book.

This chapter is organized as follows. Section 1.1 discusses the features of organizational decision making. Section 1.2 gives the classification of decision problems and techniques. Section 1.3 introduces six popular decision support techniques: mathematical programming, multi-criteria decision-making, case-based reasoning, data mining decision trees, and fuzzy sets. Main concepts, characteristics and components of DSS are presented in Sect. 1.4. Section 1.5 discusses DSS classifications. Two kinds of DSS software, a general DSS tool and a specific DSS with application are illustrated in detail in Sect. 1.6. Finally, we give a summary in Sect. 1.7.

## 1.1 Organizational Decision Making

Organizational decision-making seeks to find the optimal or most satisfactory solution for a decision problem such as selecting the best from a set of product prototypes, making an optimized resource plan, choosing the most suitable supplier and determining a product's price.

Organizational decision problems are of various types, from daily operational decisions to long-term strategy business decisions, from internal single decisions to multi-level decisions or multi-organizational decisions. Decision makers can be at various levels according to their decision problems, such as a product distributor, a supermarket manager or a head of department. Different decision-making tasks may have different features and therefore are modeled in different forms or presented by different methods, and solved by different decision support techniques.

Organizational decisions can be made by an individual or a group of decision makers. The latter situation is called *group decision-making* (GDM). Individual

© Springer-Verlag Berlin Heidelberg 2015
G. Zhang et al., *Multi-Level Decision Making*,
Intelligent Systems Reference Library 82, DOI 10.1007/978-3-662-46059-7_1

decisions are often made at lower managerial levels and in small organizations, or for a short-term decision issue. Group decisions are usually made at high managerial levels and in large businesses, or for a long-term strategy issue. For the GDM problem, each group member may have their own understanding of the nature of the decision problem and the feasibility of particular solutions. Also, the decision makers in a decision group could be in the same organization, having the same objectives; in different organizations, and/or having different, even totally conflicting objectives; and could also be in the same organization but at different levels and with different objectives. The latter two situations will be the main focus of this book: multi-level decision-making.

*Multi-level decision-making* (MLDM) problems appear in many situations which require compromise between the objectives of two or more interacting entities and these entities are arranged within a hierarchical structure with independent and perhaps conflicting objectives. *Bi-level decision-making* (BLDM) is a special case of MLDM which addresses the problem in which two levels of decision makers are involved and each level tries to optimize their individual objectives under certain constraints, and to act and react in a sequential manner. The two levels of decision makers can be suppliers and buyers in a supply chain for a pricing problem or faculties and departments in a university on a budget arrangement. *Tri-level decision-making* (TLDM) is another special case of MLDM and has all the typical features of MLDM when the number of levels is greater than two.

## 1.2  Classification for Decision Problems and Techniques

### 1.2.1  Decision Problem Classification

In general, organizational decision problems can be classified based on their natures. A classical classification is based on a given problem's complexity degree, i.e., *structured, semi-structured* and *unstructured* (Turban et al. 2005). The last two are also called *ill-structured*. Different types of decision problems may require different modeling processes and different solution methods.

A *structured* decision problem can be described by classic mathematical models, such as linear programming or statistics methods. The procedure for obtaining an optimal solution is known by standard solution methods. For example, goal programming can be used to solve a linear programming model when the decision maker provides a goal for their decision objective. A typical structured decision example is that to select a supplier who has the lowest price of all the suppliers with the same quality/type of products, or determines a product plan which will bring the highest profit of all the possible product plans in a factory.

An *unstructured* decision problem is fuzzy, uncertain and vague, for which there is no standard solution method for obtaining an optimal solution, or where such an optimal solution does not exist. Human intuition is often the basis for decision making in an unstructured problem. Typical unstructured problems include planning new services to customers, hiring an executive for a big company, choosing a set of development projects for a long period, or making a set of policies for a social issue.

*Semi-structured* decision problems fall between structured and unstructured problems, having both structured and unstructured features, and reflecting most real-world situations. Solving *semi-structured* decision problems involves a combination of both standard optimization solution procedures and human judgment, and also needs the support of related intelligent information processing techniques and inference approaches.

In principle, computer-based decision support techniques can be more useful in structured and semi-structured decision problems than unstructured decision problems. In an unstructured decision problem only part of the problem can be assisted by computerized decision support techniques. For semi-structured decision problems, a computerized decision support technique can improve the quality of the information on which a decision is based, therefore increasing the decision maker's situation awareness to reach a better decision and improve decision efficiency.

Another classification of decision problems is based on levels of decision problems: *strategic planning, management control* and *operational control.*

*Strategic planning* refers to long-range goals and policies for resource allocation. Such decisions are at a high management level, normally unstructured, and have a higher degree of uncertainty.

*Management control* refers to the acquisition and efficient use of resources in the accomplishment of organizational goals, and related decisions are at middle management level.

*Operational control* decisions are about the efficient and effective execution of specific tasks, are normally structured, and are relatively easy to formulate by mathematical models and solve by computer-based tools.

Decision making can be also seen as a reasoning process, which can be *rational* or *irrational* (Simon 1993) and is based on explicit assumptions or tacit assumptions.

*Rational* decision making emphasizes fact collection and conducting research such as surveys, interviews and data analysis. A rational decision-making model involves a cognitive process (thinking through).

*Irrational* decision making makes assumptions and obtains results without accurate data and model analysis but is often driven by emotions.

These classifications can help to determine a suitable decision-making methodology and select a suitable support technique for a particular decision problem.

## *1.2.2 Decision Support Technique Classification*

Various decision support techniques including models, methods, algorithms and software tools have been widely developed by both academic and practical researchers. There are different classifications for these kinds of decision support techniques. Zachary (1987) proposed a cognition-based taxonomy for decision support techniques which has six basic classes as follows:

*Process models* are computational models that assist the projection of real-world complex processes, make some assumptions about the process and give a hypothetical decision. A typical model is the probabilistic model and the Markov chain is a canonical example of the probabilistic process model.

*Choice models* support the integration of decision criteria across alternatives to select the best alternative from a discrete set or continuous description space of decision alternatives. A typical choice model is the multi-criteria decision-making model.

*Information control* techniques provide support to information processing by helping with the representation, manipulation, access, and monitoring of bodies of data and knowledge. Typical techniques include data retrieval techniques, data warehouse approaches, and data mining techniques.

*Analysis and reasoning techniques* support the application of problem-specific expert reasoning procedures, such as linear program and other mathematical programming, goal-driven, process-driven and data-driven inference. Goal programming, evidential reasoning, case-based reasoning and sensitivity analysis are very successful techniques in the class of decision techniques.

*Representation aids* assist in the expression and manipulation of a specific representation of a decision problem. Typical techniques include natural language processing, graphic representation and graphic user interface, special human cognitive processing techniques, such as decision trees and decision tables, as well as techniques to capture the mental model used by expert decision makers and incorporate it into the interface as an aid to the novice user of a DSS.

*Human judgment amplifying/refining techniques* help in the quantification of heuristic judgments. Decision makers are able to solve problems heuristically or intuitively with results that are usually quite good but almost never truly optimal (Zachary 1986). Typical techniques in this class include human-aided optimization, and adaptive user modeling and prediction, as well as Bayesian updating.

The above cognition-based classification of decision support techniques provides a picture and guideline for decision technique selection for problem solving and DSS development. In practice, a DSS often integrates two or more of the techniques mentioned above to solve a complex organizational decision problem.

## 1.3 Main Decision Support Techniques

In this section, we will introduce six popular decision support techniques involved in modeling and executing the decision-making process.

## 1.3.1 Mathematical Programming

*Mathematical programming*, also called *optimization*, refers to the study of decision-making problems in which one seeks to minimize or maximize a function by systematically choosing the values of variables from an allowed set (a feasible set). A mathematical programming model includes three sets of elements: *decision variables*, *objective functions*, and *constraints* (*constraint conditions*), where *uncontrollable variables* or *parameters* are within the objective functions and the constraints. Many real-world decision problems can be modeled by mathematical programming models.

There are various types of mathematical programming models such as linear programming (Benayoun et al. 1971), multi-objective programming (Hwang and Masud 1979), and bi-level/multi-level programming (Bracken and McGill 1973; Candler and Norton 1977).

*Linear programming* is an important type of mathematical optimization in which there is only one objective function, and the objective function and constraints are expressions of linear relationships among decision variables. Linear programming is heavily used in various management activities, either to maximize the profit or minimize the cost of an organization. We use the example below to explain how to build a linear programming model and obtain a solution for a practical decision problem.

**Example 1.1** A company produces two kinds of product: $A$ and $B$. We know that producing one unit of $A$ returns \$40 profit, and $B$ returns \$70. However, the company has limitations on its labor (total 501 labor hours available per time slot; each $A$ needs 4 h and $B$ 3 h), machine time (total 401 machine hours available, each $A$ needs 2 h and $B$ 5 h), and marketing requirements (need to produce 10 units of $A$ and 20 units of $B$). The decision problem is to determine how many $A$ and $B$ should be produced to obtain the maximum profit. Within these settings and requirements, we can establish a linear programming model:

Suppose $x_1$ and $x_2$ are the respective units of products $A$ and $B$ to be produced,
Maximize total profit: $\max z = 40x_1 + 70x_2$
Labor constraint (hours): $4x_1 + 3x_2 \le 501$,
Machine constraint (hours): $2x_1 + 5x_2 \le 401$,
Marketing requirement for $x_1$ (units): $x_1 \ge 10$,
Marketing requirement for $x_2$ (units): $x_2 \ge 20$.

By using the simplex method, we obtain the following solution:

$$x_1 = 93, \quad x_2 = 43.$$

We therefore have the result $z = 6,730$. That is, by producing 93 units of product $A$ and 43 units of product $B$, the company can obtain a maximum profit of \$6,730.

We need to indicate that in order to apply the linear programming technique, the decision problem must be understood as one requiring an optimization problem, the objective function and constraints must be modeled in linear functions, and the data on the various variables in the objective function and constraints must be obtained to represent the coefficients of these functions.

When a decision problem has more than one objective function, it is called *multi-objective programming* or *multi-objective decision-making* (MODM). When the objective function and/or constraints are modeled by expressions of non-linear relationships between decision variables, we will use *non-linear programming* to obtain solutions.


## 1.3.2 Multi-criteria Decision Making

When we need to select the best option from a list of alternatives based on multiple criteria for a decision problem, it is often necessary to analyze each alternative in the light of its determination of each of these criteria. *Multi-criteria decision-making* (MCDM), also called *multi-attribute decision-making* (MADM), refers to making preferred decisions (e.g., evaluation, prioritization, and selection) in the presence of multiple and conflicting criteria over the alternatives available. An MCDM utility model combines all the criteria of a given alternative simultaneously through the use of a specific utility formula or utility function. Problems for MCDM may range from those in our daily life, such as the selection of a restaurant, to those affecting entire nations, such as the judicious use of money for the preservation of national security. However, even with this diversity, all MCDM problems share the following common characteristics (Hwang and Yoon 1981):

*Alternatives*: there are usually a limited number of predetermined alternatives, such as only three suppliers being available to a buyer in a supply chain.

*Multiple criteria*: there are a limited number of criteria. For example, price, quality of products, and speed of delivery are criteria for the buyer in selecting a supplier.

*Conflicting criteria*: multiple criteria are in conflict with one another; for example, the low price and high quality of products are in conflict.

*Incommensurable unit*: criteria may be measured in different units. For example, the unit price is the dollar and the unit of product quality is degree of satisfaction.

*Selection*: the solution to an MCDM problem is to select the best solution among previously specified finite alternatives.

Mathematically, an MCDM problem can be modeled as follows:

$$\begin{cases} \text{select from:} & \{A_1, A_2, \ldots, A_m\} \\ \text{s.t.} & C_1, C_2, \ldots, C_n \end{cases}$$

where $A = \{A_1, A_2, \ldots, A_m\}$ denotes $m$ alternatives and $C = \{C_1, C_2, \ldots, C_n\}$ represents $n$ criteria (also called attributes) for characterizing a decision situation.

The term *select* in the model is normally based on maximizing a multi-criteria value (or utility) function elicited from the decision makers. By mapping the alternatives onto a cardinal scale of value, the alternative with the highest cardinality is implicitly the best.

In many situations, the MCDM issue is conducted in a decision group in which several decision makers are involved. In such a situation, we need to aggregate all group members' opinions on all alternatives under all criteria, which is called *multi-criteria group decision-making* (MCGDM). Furthermore, these criteria may have different weights, and some experts' opinions may be more important than others. Criteria, in particular, can be in a hierarchy; that is, a criterion has several sub-criteria. Considering all these issues: decision group, decision makers' weights, criteria' weights, and multi-evaluation levels of criteria, Fig. 1.1 shows the working process of a more general MCGDM. The outcome of a MCDM or a MCGDM is a ranking of alternatives.

Multi-criteria decision making methods have been widely developed as reported by Hwang and Yoon (1981), Yager (2004) and other researchers. Some popular methods include the Technique for The Order Preference by Similarity to Ideal Solution (TOPSIS) (Hwang and Yoon 1981) and the Analytic Hierarchy Process (AHP) method (Saaty 1980). Example 1.2 will demonstrate an application of a MCDM method.

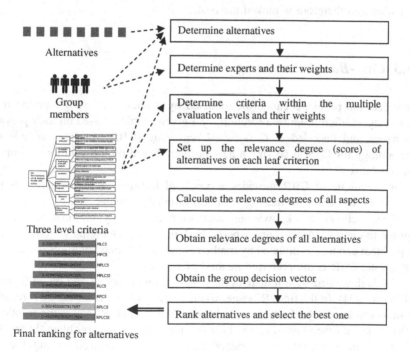

**Fig. 1.1** The working process of a general MCGDM

**Example 1.2** A logistic company plans to develop an e-business system. Companies $A$, $B$, $C$ have been selected as alternatives to take this job. Four criteria are used in the logistic company selection committee:

($C_1$) user friendly,
($C_2$) security,
($C_3$) easy to maintain, and
($C_4$) excellent in logistic decision support functions.

There are five members (e1, e2, e3, e4, e5) in the committee.

Each member gives their scores (1, 2, 3, 4, 5; 1-unsatisfactory, and 5-highly satisfactory) to each IT company's proposal under each criterion by a score matrix. For example, e1 gives the following scores for companies $A$, $B$, $C$ respectively under the four criteria:

$$\text{e1's score matrix}: \quad \begin{matrix} & C_1 & C_2 & C_3 & C_4 \\ A & \begin{pmatrix} 3 & 2 & 5 & 4 \\ B & 4 & 3 & 4 & 3 \\ C & 3 & 1 & 4 & 3 \end{pmatrix} \end{matrix}$$

Through aggregating the score matrix of all members using a MCDM method, such as TOPSIS, it is evident that Company $B$ has the higher value (cardinality) of evaluation and therefore is ranked the best.

### 1.3.3 Case-Based Reasoning

Many decision problems cannot be modeled by mathematical programming models. Managers often produce a solution for a given problem based on their previous experience and knowledge. *Case-based reasoning* (CBR) provides an effective methodology for DSS in solving a new problem based on the solutions of similar past problems.

The technique of CBR provides a powerful learning ability which uses past experiences as a basis for dealing with new similar problems. A CBR system can, therefore, facilitate the knowledge acquisition process by eliminating the time required to elicit solutions from experts. In dynamically changing situations where the problem cannot be modeled by mathematical models and solutions are not easy to generate, CBR is the preferred method of reasoning.

CBR is described by a four-R cycle: retrieve, reuse, revise and retain (Aamodt and Plaza 1994). In the first 'R' stage, when a new problem is input, CBR retrieves the most similar case(s) from the case base. In the second 'R' stage, the solution(s) of the retrieved case(s) is(are) reused for dealing with the new problem. In the third 'R' stage, the solution(s) of retrieved case(s) is (are) revised to suit the new problem, and in the fourth 'R' stage, the revised solution and the new problem are

**Fig. 1.2** CBR cycle

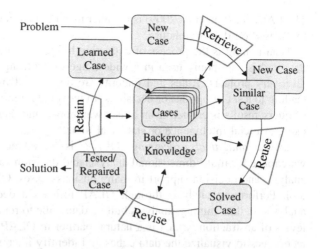

retained in the case base for future reuse. Figure 1.2 shows the CBR cycle. Clearly, CBR is naturally suitable for knowledge-based decision making. The success of a CBR system in decision making is mainly dependent on the suitability of the knowledge (cases and rules) and the correctness of reasoning.

To cope with all issues that arise when CBR is used in real-world applications and especially in rapidly changing environments, several additional phases have been accepted to extend the original four-R cycle of CBR. As a typical development, Reinartz et al. (2001) extended the standard four-R CBR cycle by two additional stages: *review* and *restore*. The review stage covers tasks to judge and monitor the current state of a CBR system and its knowledge containers, whereas the restore stage invokes mechanisms to change the system and its knowledge to improve performance.

CBR-based DSS have been developed for various applications to aid decision makers or to help a wide variety of decision making activities. Since the CBR method is a natural reasoning methodology for people, it is easily accepted and applied by decision makers. Normally people are good at using previous cases but not as good at recalling the right ones. CBR can assist decision makers to extend their memory by providing them with the correct cases to help them reason, yet still allowing them to do all the complex reasoning and decision making.

## 1.3.4 Data Warehouse and Data Mining

Data warehouse is a repository of an organization's electronically stored data. A data warehouse system involves not only data storage but also the techniques to retrieve and analyze data, to extract, transform and load data, and to manage the data dictionary. In particular, the data warehouse includes business intelligence tools to implement the above functions to better support business decision making.

The FACEST (Niu et al. 2009) is a higher management level's business intelligence tool based on data warehouse.

Data mining is the process of extracting hidden and undiscovered patterns from data and is commonly used in a wide range of profiling practices and knowledge discovery projects. Rules and patterns are discovered from data with the aim of leading to a set of options or decisions. In most data mining applications, a data file of query results is created from a data warehouse and then analyzed by a specialist using artificial intelligence or statistical tools.

*Online analytical processing* (OLAP) is an efficient way to access a data warehouse for multi-dimensional analysis and decision support. For the purpose of analysis and decision support in many business cases, OLAP provides a powerful tool. Performing analysis through OLAP follows a deductive approach to data analysis. In data mining, analysts build a data cube to represent the data at different levels of abstraction, which is a natural partner to OLAP. Analysts can use OLAP techniques to visualize the data cubes and identify interesting patterns, trends, and relationships among them. Thus, data mining becomes a powerful tool to analyze data and provides an effective way to support decision making. Recently, greater attention has been paid to data-driven DSS which, through its development, data mining and data analytics (especially streaming data analytics) can be successfully applied to support decision making in the current big data situation.

### 1.3.5 Decision Tree

A *decision tree* is a graphic description of a set of decision rules and their possible consequences. It can be used to create a plan to reach a goal of decision (Quinlan 1986). A decision tree, as a special form of tree structure, is a predictive model to map observations about an item with conclusions about the item's target value. Each interior node corresponds to a variable and an arc to a child node represents a possible value or splitting condition of that variable. As shown in Fig. 1.3, a leaf represents a predicted value (large) of the target variable (size) given the values of the variables, for example, "600, manufacturer" represented by the path from the root. From the example shown in Fig. 1.3, we can see that a decision tree can be used to support strategy making for medium and small businesses in different industries.

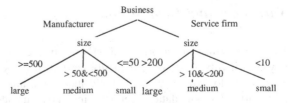

**Fig. 1.3** An example of decision tree

The decision tree approach, as a decision support tool, models a decision problem and its possible consequences in a tree-like graph. It is very suitable for a decision which involves possible chance event outcomes, resource costs, and utility. Decision trees are commonly used in decision analysis to help identify the strategy which is most likely to reach a goal. In applications, a decision tree or consequences can contain fuzzy numbers or linguistic terms and are therefore called fuzzy decision trees (Janikow 1998).

### 1.3.6 Fuzzy Sets and Systems

Whatever decision techniques are used, a critical issue we need to deal with is uncertainty. Decision environments and data sources often have various uncertain factors, resulting in uncertain relations among decision objectives and decision entities. In the meantime, data itself is with uncertainty as well. For example, an individual's preference for alternatives and judgment for criteria are often expressed by linguistic terms, such as 'low' and 'high', which are uncertain expressions. Precise mathematical and inference approaches are not efficient enough to tackle such uncertainty.

Various uncertain information processing techniques have therefore been developed by using fuzzy sets, fuzzy numbers and fuzzy logic in decision making activities. Related research has been reported by Zadeh (1975), Zimmermann (1986), Kacprzyk and Yager (1985), and many other researchers. Research results include new methodologies and algorithms of fuzzy multi-objective decision making, fuzzy multi-criteria decision making, fuzzy case-based reasoning, fuzzy decision trees, fuzzy data retrieval and fuzzy association rules. Various applications of fuzzy decision making have been developed as well. For example, a web-based fuzzy group decision support system was developed for the application of team situation awareness support (Lu et al. 2008b) and a fuzzy hierarchical criteria group decision-making method was developed for new product comprehensive evaluation (Lu et al. 2011).

## 1.4 Decision Support Systems

The DSS field has grown rapidly, drawing technology from many disciplines and pursuing applications in a variety of domains. This section will introduce the concepts, characteristics, types and components of a DSS as a software tool, which will be used in subsequent chapters, such as Chap. 11 on bi-level DSS tools.

## 1.4.1 Concepts

Decision support systems are a specific class of computerized information systems that support organizational or individual decision-making activities. A properly-designed DSS is an interactive software-based system which supports decision makers in compiling useful information from raw data, meta-data, documents, knowledge, and/or models to solve related decision problems. It enables complex models to be used in real decision problem solving, including long-term planning and emergency situations which need to be responded to in a very short time. As many organizational decision problems involve large data sets which are stored in different databases, data warehouses, and possibly even at websites outside the organization, the new DSS technology can quickly search and transmit the required distributed data to help in decision making.

Since the term DSS was proposed in the early 1970s, its definition has been the subject of great interest. Gorry and Scott-Morton (1971) defined DSS as interactive computer-based systems which help decision makers to utilize data and models to solve ill-structured problems. The classic definition of DSS, provided by Keen and Scott-Morton (1978), is that DSS couples the intellectual resources of individuals with the capabilities of the computer to improve the quality of decisions. In this book, a DSS is described as a computer-based information system which supports decision makers confronting structured and semi-structured decision problems through direct interaction with data, models, algorithms, and other related techniques (Lu et al. 2007).

From the application point of view, a DSS can be seen as an approach for supporting decision making. It uses an interactive, flexible and adaptable system especially developed for supporting the solution for a specific decision problem. It uses data, provides a user-friendly interface, and can incorporate decision makers' own insights. In addition, a DSS can use various decision support techniques, including an optimization model, case-based reasoning, and data warehousing which are built by an interactive and iterative process, or a combination of some of these techniques.

Some DSS are developed for a specific decision problem and can only be used for a particular organization/situation. We call these specific DSS. For example, an ore blending cost optimization model-based multi-role DSS for blast furnaces in iron and steel enterprises (Zhang et al. 2011b) is a specific DSS. This DSS is constructed by pre-processing the data about materials and elements, abstracting the related optimal operations models and introducing the reasoning mechanism of an expert system. This multi-role DSS includes a non-linear ore blending model for blast furnaces, related matching algorithms, a database, a model base and a knowledge base. This DSS achieved good economic benefits when it was used in an iron and steel company (see Sect. 1.6.2).

In contrast, some DSS are developed for general decision problems, such as Decider (Ma et al. 2010), a fuzzy MCGDM-based DSS. This is a general DSS tool which can be used, in general, by any organization for solving any fuzzy MCDM or fuzzy MCGDM problem.

A DSS is intended to support rather than replace decision making; to be an adjunct to decision makers to extend their capabilities but not to replace their judgments. Sometimes, there may not be an optimal solution for decision problems that arise in ill-structured situations and uncertain environments. The decision therefore must evolve through the interaction of decision makers with resources such as data analysis models to obtain the most satisfactory solution for the circumstances.

## *1.4.2 Characteristics*

Several ideal characteristics of DSS have been discussed by Turban and Aronson (1998) and many other researchers. We list nine main DSS characteristics below:

- dealing with decision problems;
- supporting decision of managers at different levels;
- supporting decision groups and individual decision-makers;
- supporting a variety of decision styles and processes;
- adaptability and flexibility in carrying out a decision support task and approach of the users;
- interactive and extremely user-friendly to allow non-technical decision makers to interact fully with the DSS and a web-based interface;
- combining the use of models and analytic techniques;
- combining the use of artificial intelligence;
- accessing a wide variety of data sources.

With these characteristics, a DSS can improve decision makers' effectiveness, efficiency and productivity.

## *1.4.3 Components*

Classically, there are three main components in a DSS: a data management component which could be a *database management system* (DBMS), a model component which could be a model base and its *model-base management system* (MBMS), and a user interface. With the development of DSS, a knowledge base and related reasoning component, called *knowledge-base management system* (KBMS) becomes a component. Since the web is widely used, it is possible to have a web-based user interface for a DSS. Figure 1.4 shows these components and their relationships under a DSS framework. However, any given DSS may be composed of only some of these components.

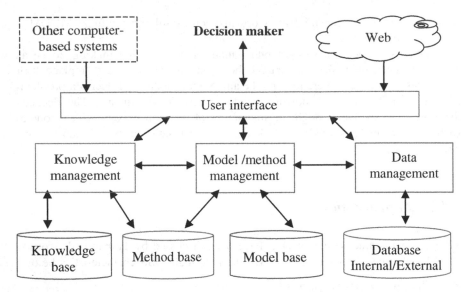

**Fig. 1.4** The main components of a DSS

Now we present a more detailed explanation of the main components.

The *data management component* in a DSS mainly includes data sets, databases or data warehouses, which contain relevant internal or external data and is managed by a DBMS or related software. The component can be connected to the corporate intranet, an extranet, and/or the Internet through the web.

The *model/method management component*, as another important component of a DSS, includes various quantitative decision models (such as a bi-level decision model) and methods (such as the $K$th-Best algorithm) that provide the analytical and solution generation capabilities of a DSS. This component is connected to a model base and/or method base. The model base provides specific decision problem models of reality, such as a product planning MODM model, a new product evaluation MCDM model, and bi-level supply chain decision model. A method base is related to the model base, which consists of a set of methods/algorithms used to solve decision problems described by models in the model base. For example, a fuzzy multi-objective group decision support system, called FMODSS (Lu et al. 2007) has a model base which contains multiple models.

The *Knowledge management component* can support any of the other components such as the user interface, methods, or data management, or it can act as an independent component. It can be interconnected with the organization's knowledge base, which may consist of a rule base, a fact base, and a case base to conduct inference for possible solutions.

The *user interface* is designed for users to communicate with the DSS. A decision problem and possible solutions are all derived through the user interface. The user interface in an active DSS provides effective intensive interaction channels between computers and decision makers.

In summary, a DSS is an interactive computer-based information system which can help decision makers to utilize data, models, methods, knowledge and communication to solve decision problems. In practice, a DSS may have a number of special components, such as text-base, multi-media database, and prediction and varying functions.

## 1.5 DSS Classification

There is no universally accepted taxonomy of DSS. Not every DSS even fits neatly into one category of classification and may be a mix of two or more architectures in one. In this section, we detail five main types of DSS and also discuss other classifications reported in the literature.

### 1.5.1 Model-Driven DSS

This type of DSS emphasizes the access to various models, such as the linear programming model, to generate solutions to assist decision makers. A model-driven DSS has a model base and has the ability to apply models in the model base. *Multi-objective DSS* (MODSS) is a typical model-driven DSS in which several multi-objective decision-making models are stored a model base of the DSS (Lu et al. 2007). It allows decision makers to analyze multiple objectives, and uses a variety of multi-objective decision making models and related methods to derive efficient solutions. *Bi-level DSS* (BLDSS) is another typical model-driven DSS which will be discussed in the chapters that follow. The working principle for a model-driven DSS is that it incorporates users' input in various phases of modeling and achieving solutions.

In a model-driven DSS, decision makers can interact in various stages of model development, management, and problem solving. A BLDSS, for example, is intended to provide necessary computerized assistance in bi-level decision-making models to decision makers to solve their problems. Decision makers are encouraged to explore any support available in an interactive fashion with the aim of further defining the nature of the problems. The challenges of a model-driven DSS are decision problem modeling and solution finding.

### 1.5.2 Data-Driven DSS

Many decision problems are led by findings obtained from data mining and data analytics. For example, a data pattern or data association rules can be found through data mining. This pattern generates a decision problem and therefore a DSS is established to solve it. In general, this type of DSS collects and provides real-time access to a large database or data warehouse to support decision making. Such a database or data warehouse can have internal or external data sets. A data-driven DSS can also provide queries and management reports according to decision makers' requirements. The more advanced data-driven DSS is combined with OLAP and data mining (such as spatial data mining, correlation mining, link mining, and web mining) to support more complex decision-making in a dynamic online environment. More recently, big data analytics has become a vital competitive force in business. A data-driven DSS in a big data environment tends to provides better decisions through using big data analytics since it enables decision makers to strategize on the basis of large data-based evidence.

## 1.5.3 Knowledge-Driven DSS or Intelligent DSS

A knowledge-based DSS is a DSS that uses the methods and techniques of artificial intelligence and/or computational intelligence to improve the support to decision making. The core components of a knowledge-driven DSS are knowledge base and inference mechanisms (Turban and Watkins 1986; Gregor and Benbasat 1999). A knowledge base is a special kind of database for knowledge storage and management, providing the means for the computerized collection, organization, and retrieval of knowledge. In principle, expert systems, CBR systems, neural networks and fuzzy logic-based DSS are all particular types of knowledge-based DSS. Also, they are practical applications of artificial intelligence and computational intelligence techniques which combine knowledge of a particular domain with inference capability to enable the system to reach a decision making level of performance. As they are applications of intelligence technologies they are also called *intelligent decision support systems* (IDSS) (Angern and Luthi 1990). Ideally, an IDSS should be like a human consultant, supporting decision makers to better understand their problems and generate better solutions (Bui and Lee 1999).

Combined with classical decision making methods and intelligent technology, IDSS are capable of delivering more reliable decision support tools for users. Moreover, IDSS have the potential to facilitate effective and swift decision making in the selection of appropriate applications that best match an organization's strategy. In particular, an IDSS' ability to provide an unprecedented level of automated guidance on the analysis of a class of decisions can enable decision makers with few skills in decision analysis to conduct effective decision analysis in that domain.

### 1.5.4 Group DSS

This type of DSS supports multiple decision-makers to work collaboratively in a group to solve a decision problem (Gray 1987). As communication is very important in this situation, it is also called communication-driven DSS. A *group DSS* (GDSS) is characterized as a DSS that combines the capabilities of distributed database technologies, computer communication technologies and group decision technologies to support the identification, analysis, formulation, evaluation, and solution of problems by a group in a user-friendly computing environment. Due to the complexity of group decision making, specific decision making models and methods are needed in a GDSS to establish a systematic means of supporting effective and efficient group decision making.

A GDSS typically offers a wide range of capabilities, including computerized support for interactive modeling, idea generation, knowledge sharing, group preference aggregation mechanisms, and optimal group solution generation. Importantly, a GDSS is used in decision groups, not in general group meetings, and to support decision making, not only to create alternatives. This is one of the foundational differences between GDSS and *group support systems* (GSS). Basically, there are two types of GDSS. One is online DSS, that is, all decision makers can access the GDSS and participate in a decision process with communication in a decision room, such as the *fuzzy GDSS* (FGDSS) discussed in Lu et al. (2007). Another is offline GDSS. All decision makers' options are collected through survey, and are entered into a GDSS, and an aggregated group decision is obtained. The GDSS tool, Decider (Ma et al. 2010), is in this type. The focus in GDSS research has been primarily on group decision-making models, methods, interaction and communication, with a strong emphasis on the achievement of consensus.

### 1.5.5 Web-Based DSS

The implementation of a *web-based DSS* (WDSS) has been popular since the mid-1990s when Internet technology began to develop rapidly around the world (Shim et al. 2002). A WDSS uses a web browser to access the Internet or Intranet. It can be model-driven, data-driven, knowledge-driven, communication-driven or hybrid. Recent developments in e-Commerce, e-Business, e-Government and e-Service provide a fertile ground for this new type of DSS application.

The types of DSS discussed above can all be combined with each other to form a hybrid DSS such as a knowledge-based GDSS or an intelligent WDSS. There are many other classifications of DSS in the literature. We introduce some related results below.

Using the relationship with the user as the criterion, there are passive and active DSS (Jelassi et al, 1987). A *passive* DSS is a system that aids the process of decision making but cannot bring out explicit decision suggestions or solutions. An *active*

DSS can bring out such decision suggestions or solutions. Many traditional DSS are passive in their operations. As information technology has recently adopted a more active role in corporate strategy, DSS could also undertake a more active stance by identifying gaps in existing operations and suggesting ways to strengthen the standing of a business.

Another classification for DSS is created by Power (2002) using the mode of assistance as the criterion. It includes *communication-driven DSS*, *data-driven DSS*, *document-driven DSS*, *knowledge-driven DSS*, and *model-driven DSS*.

Using scope as the criterion, Power (2004) differentiates *enterprise-wide DSS* and *desktop DSS*. An *enterprise-wide DSS* is linked to large data warehouses and serves many managers in the company. A *desktop, single-user DSS* is a small system that runs on an individual manager's PC.

More classifications of DSS are available, such as text-oriented DSS, database-oriented DSS, spreadsheet-oriented DSS, solver-oriented DSS, rule-oriented DSS, and compound DSS (Holsapple and Whinston 1996). Also, many hybrid DSS have been developed for more specific and complex decision problems. More recently, mobile-based DSS has been proposed to support mobile users' decision making and have been applied in health field (Kuntagod and Mukherjee 2011; Amailef and Lu 2013).

## 1.6 DSS Software Illustration

We will introduce two DSS software in this section. One is a general DSS software kit called Decider (Ma et al. 2010). Another is a specific DSS which is only used for ore blending (Zhang et al. 2011b).

### 1.6.1 Case 1: Decider

The software Decider is a fuzzy multi-criteria group decision support system. It is designed to handle MCDM problems in group decision making. It allows (1) multiple criteria with multi-level hierarchy, (2) the involvement of several decision makers in group decision making, and (3) the use of linguistic terms, that is, decision makers' opinions can be expressed in linguistic terms and fused by uncertain information process techniques such as fuzzy numbers and fuzzy aggregation operators. Decider can deal with both subjective and objective information. That is, some criteria can be evaluated by machine assessment and fused with human opinions. Decider is a fuzzy group DSS.

The decision methodology/working process of Decider is shown in Table 1.1. Figures 1.5, 1.6 and 1.7 show the interfaces of Decider. Decider can be used for different multi-criteria group decision problems as long as it has a set of alternatives, a set of criteria and one or more decision makers. The outcome of Decider is

**Table 1.1** Working process of decider

| Step | Process |
| --- | --- |
| Step 1 | Identify alternatives |
| Step 2 | Identify hierarchies of criteria and evaluators as well as their weights |
| Step 3 | Identify information sources and their connection with criteria |
| Step 4 | Collect information from information sources |
| Step 5 | Evaluators evaluate collected information to generate initial decision matrix for each alternative |
| Step 6 | Apply the fuzzification method to assessments in initial decision matrix |
| Step 7 | Apply the fuzzy aggregation method to obtain overall assessments of each alternative |
| Step 8 | Generate ranking for each alternative by the fuzzy aggregation method and ranking strategy |

**Fig. 1.5** Interface of fabric-hand based hierarchical criteria in decider

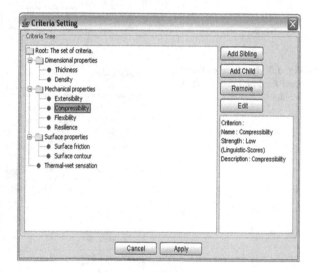

the ranking of the set of alternatives (see Fig. 1.7). For example, it can be used to rank a set of new product prototypes with a set of criteria by a group of designers, or a set of grant applications with a set of criteria by a group of committee members (Ma et al. 2010).

## 1.6.2  Case 2: A DSS for Ore Blending Cost Optimization of Blast Furnaces

In iron and steel enterprises, it is difficult to obtain the lowest-cost optimal solution to an ore blending problem for blast furnaces because of the complexity of materials

**Fig. 1.6** Interface of expert input by linguistic terms for fabric-hand in decider

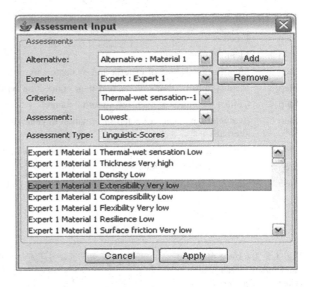

**Fig. 1.7** Interface of fabric-hand final ranking results in decider

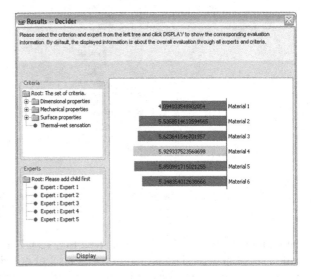

and burden of workflow. An ore blending DSS has been developed to solve the problem. This DSS integrates a database, a model base and a knowledge base. It has made economic gains since it was implemented in Xiangtan Iron and Steel Group Co. Ltd., China. Table 1.2 shows the workflow of the ore blending DSS, which has five steps.

The ore blending DSS provides a friendly human-computer interactive interface. Users can manually input various process parameters, such as market price, internal price, supply at corresponding periods, upper and lower bounds, and element

**Table 1.2** Working process of the ore blending DSS

| Step | Process |
|------|---------|
| Step 1 | A decision maker inputs general information, such as the element proportion of each kind of coal and ore |
| Step 2 | A field expert enters all kinds of rules about ore blending and formulas of technical parameters |
| Step 3 | A user inputs an example through the human-computer interactive interface, which includes the type, upper and lower bounds, price of materials and proportion constraints of each element in the finished product |
| Step 4 | A decision modeling expert provides an appropriate model for the reasoning machine from the model base. The reasoning machine fetches data from the database, acquires rules from the knowledge base and carries out forward reasoning. The result is returned to the analyst via explanation |
| Step 5 | An analyst analyzes the computing results to determine whether they are reasonable, and submits the conclusion to the user via the human-computer interface |

percentage content for each material. The ore blending DSS consists of two kinds of computing: optimization computing and experienced computing. For the former, an expert chooses an appropriate model to obtain the optimal solution. For the latter, a field expert and an analyst collaboratively conduct the inverse operation for the

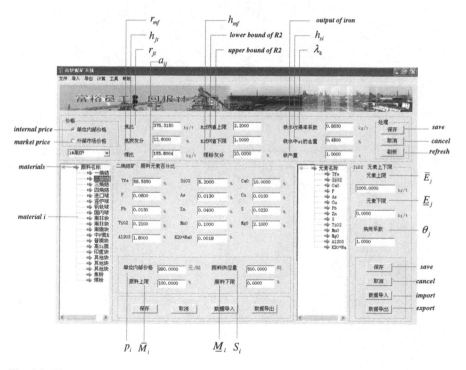

**Fig. 1.8** The main interface of the ore blending DSS

optimal solution and validate whether it satisfies the corresponding constraints and meets the technological requirements.

When a user clicks the optimization computing button, the system will load data in the database and parameters in the memory workspace, and will perform optimization computing in the background according to the selected model. When the computing is finished, the results will be transferred back to the system for display. The main interface for the ore blending cost optimization system is shown in Fig. 1.8 (Zhang et al. 2011).

## 1.7 Summary

This book aims to help readers understand what a multi-level decision-making problem is, how to model it and what it can support to achieve a solution. Furthermore, it will help readers to know how to develop and use a multi-level DSS tool to handle complex multi-level decision-making problems. This chapter presents fundamental concepts about decision making, decision models, and DSS, as well as DSS software illustration. It provides the necessary fundamental techniques and preliminary knowledge for the following chapters of this book.

# Chapter 2
# Optimization Models

To model and solve a bi-level or multi-level optimization problem, we have to first understand basic single-level optimization models and related solution methods. This chapter introduces related concepts, models and solution methods of basic single-level optimization including linear programming, non-linear programming, multi-objective programming, goal programming, Stackelberg game theory, and particle swarm optimization. These knowledge will be used in the rest of the book.

This chapter is organized as follows. Section 2.1 introduces basic single-level optimization concepts and models. Section 2.2 presents the solution method of linear programming. Section 2.3 addresses non-linear programming by its definition, classification, theories, and solution methods. Section 2.4 gives the models and solution methods of multi-objective programming. Section 2.5 introduces goal programming and its solution process. In Sect. 2.6, we present the principles, theorems and applications of Stackelberg game theory. Particle swarm optimization, which will be used as a solution method for non-linear optimization problem, is then introduced in Sect. 2.7. Section 2.8 presents a summary.

## 2.1 Concepts

The core of the decision process is to formulate an identified decision problem and then find an optimal solution. Many decision models have been developed and different types of decision models require different kinds of decision-making methods to obtain solutions. Popular decision models include (1) *Analytic Hierarchy Process* (AHP), which allows consideration of both the qualitative and quantitative aspects of a decision problem and reduces a complex decision to a series of pairwise comparisons; (2) Grid Analysis, also known as decision matrix analysis or multi-attribute utility theory, in which the decision matrices are the most effective, and multiple alternatives and criteria will be taken into account in the decision process; (3) Decision Tree, which is a graph of decisions and their possible consequences, is used to create a plan to reach a goal; and (4) Optimization model which is a more sophisticated approach to solve decision problems and is the main focus of this book.

© Springer-Verlag Berlin Heidelberg 2015
G. Zhang et al., *Multi-Level Decision Making*,
Intelligent Systems Reference Library 82, DOI 10.1007/978-3-662-46059-7_2

Optimization, also called mathematical programming, refers to the study of decision problems in which one seeks to minimize (min) or maximize (max) a function by systematically choosing the values of variables within their allowed sets. Many real-world decision problems can be modeled by an optimization framework. To model a decision problem as an optimization model, we need, in principle, three sets of basic variables: *decision variables, result variables* and *uncontrollable variables (or parameters)*.

*Decision Variables* describe alternative courses of action and are determined by related decision makers. For example, for a product planning problem, the number of products to be produced is a decision variable.

*Result Variables* are outputs and are often described by objective functions, such as profit (max) and cost (min). The outputs are determined by decision makers, the factors that cannot be controlled by decision makers, and the relationships among the variables.

*Uncontrollable Variables* (or Parameters) are the factors that affect the result variables but are not under the control of decision makers. These factors can be fixed, in which case they are called parameters, or they can vary. These factors are uncontrollable because they are determined by elements of the system environment. Some of these variables limit decision makers and therefore form what are called the constraints of the problem. For example, each product's cost of production should be less than the total profit, and each product should meet marketing requirements and so on in a product planning problem.

There are many types of optimization models such as linear programming, non-linear programming, multi-objective programming, and bi-level programming.

*Linear Programming* (LP) is an important type of optimization in which the objective function and constraints are all linear. Linear programming problems include specialized algorithms for their solutions and for other types of optimization problems by solving linear programming problems as sub-problems. Linear programming is heavily used in various management activities, either to maximize the profit or minimize the cost. It is also the key technique of other optimization problems.

Now, we re-consider Example 1.1 discussed in Chap. 1 to explain how to build a model for an LP practical decision problem.

*Example 2.1* A company produces two kinds of products: *A* and *B*. We know that the profit of one unit of A and B is \$40 and \$70, respectively. However, the company has limitations in its labor (a total of 501 labor hours available per time slot; each A needs 4 h and B 3 h), machine (a total of 401 machine hours available, each A needs 2 h and B 5 h), and marketing requirements (the need to produce 10 units of A and 20 units of B respectively). The decision problem is how many A and B should be produced to obtain the maximum profit. Using these settings and requirements, we can establish a linear programming model:

Decision variables:

$$x_1 = \text{units of } A_1 \text{ to be produced};$$
$$x_2 = \text{units of } A_2 \text{ to be produced}.$$

Result variable (objective function):
Maximize total profit: $40x_1 + 70x_2$,
Labor constraint (hours): $4x_1 + 3x_2 \le 501$,
Machine constraint (hours): $2x_1 + 5x_2 \le 401$,
Marketing requirement for $x_1$ (units): $x_1 \ge 10$,
Marketing requirement for $x_2$ (units): $x_2 \ge 20$.

This is a linear programming problem and can be modeled by linear programming (see Sect. 2.2).

*Non-linear Programming* (NLP) is the process of solving a programming problem subject to certain constraints, over a set of unknown real variables, along with an objective function to be maximized or minimized, as with linear programming, but where some of the constraints or the objective function are nonlinear. For example,

$$\min_{x_1, x_2} \quad 40x_1^2 + 70x_2^3$$
$$\text{s.t.} \quad x_1^2 + 20x_2 \le 100,$$
$$2x_1 + 3\sqrt{x_2} \le 140,$$
$$x_1 \ge 10, x_2 \ge 20.$$

*Multi-objective Programming* (MOP) is the process of simultaneously optimizing two or more conflicting objectives subject to certain constraints. MOP problems can be found in a variety of fields, such as product and process design, aircraft design, automobile design, or wherever optimal decisions need to be made in the presence of trade-offs between two or more conflicting objectives. Maximizing profit and minimizing the cost of a product; maximizing performance and minimizing the fuel consumption of a vehicle; and minimizing weight while maximizing the strength of a particular component are all examples of multi-objective optimization problems.

In general, a multi-objective programming problem should not have a single solution that simultaneously minimizes or maximizes each objective to its fullest. In each case an objective must have reached a point such that, when attempting to optimize the objective further, other objectives suffer as a result. Finding such a solution, and quantifying how much better this solution is compared to other solutions, is the goal when setting up and solving a multi-objective optimization problem. For example,

$$\min_{x_1,x_2} \begin{pmatrix} 40x_1 + 70x_2 \\ 50x_1 + 60x_2 \end{pmatrix}$$

$$\text{s.t.} \quad 10x_1 + 20x_2 \leq 109,$$
$$20x_1 + 30x_2 \leq 419.$$

*Bi-level programming* (BLP) and *multi-level programming* (MLP) are complex optimization situations where one optimization problem is embedded in another one. A bi-level programming problem is a multi-level programming problem having two levels. Below is an example of bi-level programming. More detail will be presented in Chap. 3.

$$\min_{x_1} \quad 40x_1 + 70x_2$$

$$\text{s.t.} \quad 10x_1 + 20x_2 \leq 119,$$
$$20x_1 + 30x_2 \leq 409,$$
$$\min_{x_2} \quad 50x_1 + 60x_2$$
$$\text{s.t.} \quad 10x_1 + 8x_2 \leq 109,$$
$$x_1 \geq 10, x_2 \geq 2.$$

We can see that optimization is an ideal model for decision making. The single limitation is that it works only if the problem is structured and, for the most part, deterministic. An optimization model defines the required input data, the desired output, and the mathematical relationships in a precise manner.

## 2.2 Linear Programming

*Linear programming* is a mathematical approach to determining a means to achieve the best outcome (such as maximum profit or minimum cost) in a given mathematical model. This model is defined by an objective function and one or more constraints which have linear formats. A typical example would be taking the limitations of materials and labor described by linear inequalities, and then determining the "best" production levels for the maximal profit defined by a linear formula, under those limitations.

LP problem can be written as:

$$\max_{x} \quad f(x) = cx$$
$$\text{s.t.} \quad A_x \leq b, \tag{2.1}$$

where $x$ represents the vector of decision variables, $c$ and $b$ are vectors of known coefficients, and $A$ is a known matrix of coefficients. The expression $f(x)$ to be maximized (in other cases, it may be minimized) is called the objective function. The equations $Ax \leq b$ are the constraints which specify a convex polytope over which the objective function is to be optimized. Both $f(x)$ and $Ax$ have linear formats.

Linear programming has a tremendous number of application fields. It has been used extensively in business and engineering, in the areas of transportation, energy, telecommunications, and manufacturing. It has been proved to be useful in modeling diverse types of problems in planning, routing, scheduling, assignment, and design.

Just as with standard maximization problems, the method most frequently used to solve LP problems is the simplex method (Charnes and Cooper 1957). This method provides us with a systematic way of examining the vertices of the feasible region to determine the optimal value of the objective function. As is well-known, the simplex method has proven remarkably efficient in practice.

## 2.3 Non-linear Programming

*Non-linear programming* is the process of solving a problem of equalities and inequalities, collectively termed constraints, over a set of unknown real variables, along with an objective function to be maximized or minimized, where some of the constraints or the objective function are non-linear. Formally, an NLP problem can be written as:

$$\min_{x} \quad f(x) \tag{2.2a}$$

$$\text{s.t.} \quad h(x) = 0, \tag{2.2b}$$

$$g(x) \geq 0, \tag{2.2c}$$

where $x \in R^n, f : R^n \to R, h : R^n \to R^p, g : R^n \to R^q$. A point $x$ that satisfies the constraints given by (2.2b) and (2.2c) is called a feasible solution to problem (2.2a)–(2.2c). A collection of all such feasible solutions forms the feasible region. NLP is then use to search a feasible solution $\bar{x}$ such that $f(\bar{x}) \leq f(x)$ for any feasible solution $x$. $\bar{x}$ is called an optimal solution to the problem (2.2a–2.2c). In special cases when the objective function of (2.2a) and constraints (2.2b) and (2.2c) all have linear forms, the problem (2.2a–2.2c) reduces to a linear programming problem (2.1).

### 2.3.1 Varieties of Non-linear Programming

Based on the mathematical characteristics of the objective function (2.2a) and the constraints (2.2b) and (2.2c), NLP can be in many different formats. For an objective function or a constraint, the format can be linear, sum of squares of linear functions, quadratic functions, sum of squares of non-linear functions, sparse non-linear functions, or non-smooth non-linear functions.

Based on combinations of the above formats of the objective and constraints, an NLP problem can be a specific type (such as linear objective function, but the constraint is a quadratic function) and thus have particular properties.

## 2.3.2  Theories and Optimality Conditions of Non-linear Programming

In this section, we introduce the most important and widely used theories and optimality conditions of NLP. We first denote the feasible region of problem (2.2a–2.2c) by $S$. The following definitions and results can be found in Bazaraa et al. (2013).

**Definition 2.1** A point $x^* \in S$ is called a *relative* or *local minimum* of $f(x)$ over $S$ if there is an $\varepsilon > 0$ such that $f(x) \leq f(x^*)$ for all $x \in S$ within a distance $\varepsilon$ of $x^*$. If $f(x) > f(x^*)$ for all $x \in S$, $x \neq x^*$ within a distance $\varepsilon$ of $x^*$, then $x^*$ is called a *strict relative minimum* of $f(x)$ over $S$.

**Definition 2.2** A point $x^* \in X$ is called a *global minimum* of $f(x)$ over $S$ if $f(x) \geq f(x^*)$ for all $x \in S$. If $f(x) > f(x^*)$ for all $x \in S$, $x \neq x^*$, then $x^*$ is called a *strict global minimum* of $f(x)$ over $S$.

For situations where constraints are absent, the following two theorems hold.

**Theorem 2.1** *Let $f : R^n \to R$ be twice continuously differentiable throughout a neighborhood of $x^*$. If $f$ has a relative minimum at $x^*$, then it necessarily follows that*

1. *The gradient vector $\nabla f(x^*) = 0$.*
2. *$F(x^*)$ is positive semi-definite, where $F(x^*)$ is the Hessian matrix of $f(x)$ at $x^*$.*

**Theorem 2.2** *Let $f : R^n \to R$ be twice continuously differentiable throughout a neighborhood of $x^*$. Then a sufficient condition for $f(x)$ to have a strict relative minimum at $x^*$, where $\nabla f(x^*) = 0$ holds, is that $F(x^*)$ is positive definite.*

For NLP problems involving only equality constraints, the following definition and theories hold.

**Definition 2.3** A point $x^*$ satisfying the constraints $h(x^*) = 0$ is called a *regular point* of the constraints if the gradient vectors $\nabla h_1(x^*), \ldots, \nabla h_m(x^*)$ are linearly independent.

**Theorem 2.3** *At a regular point $x^*$ of the surfaces $S = \{x|h(x) = 0\}$, the tangent plane is equal to $T = \{y|\nabla h(x)y = 0\}$.*

**Theorem 2.4** *Suppose that $x^*$ is a local minimum of $f(x)$ subject to $h(x) = 0$ as well as a regular point of these constraints. There then exists a vector $\lambda \in R^m$ such that $\nabla f(x^*) - \lambda \nabla h(x^*) = 0$.*

The following definitions and theories are used for the general NLP problem (2.2a–2.2c).

**Definition 2.4** Let $x^*$ be a point satisfying the constraints $h(x^*) = 0$ and $g(x^*) \geq 0$; and let $J$ be the set of indices $j$ such that $g_j(x^*) = 0$. Then $x^*$ is called a *regular point* of these constraints if the gradient vectors $\nabla h_i(x^*)(1 \leq i \leq m)$, $\nabla g_j(x^*)(j \in J)$ are linear independent.

**Theorem 2.5** (Kuhn-Tucker Conditions) *Let $x^*$ be a relative minimum for the problem (2.2a–2.2c) and suppose that $x^*$ is a regular point for the constraints. Then there exists a vector $\lambda \in R^m$ and a vector $\mu \in R^q$ such that*

$$\nabla f(x^*) - \lambda \nabla h(x^*) - \mu \nabla g(x^*) = 0, \tag{2.3a}$$

$$\mu g(x^*) = 0, \tag{2.3b}$$

$$\mu \geq 0, \tag{2.3c}$$

$$h(x^*) = 0, g(x^*) \geq 0. \tag{2.3d}$$

### 2.3.3 Methods for Solving Non-linear Programming Problems

For an NLP problem in which the objective function and constraints have linear forms, the problem becomes an LP problem which can be solved using the well-known simple algorithm.

If the objective function of an NLP problem is convex (for the minimization problem), or concave (for the maximization problem), and the constraint set is convex, then the programming problem is called a *convex programming* problem and general methods from convex optimization can be used.

Several methods are available for solving non-convex problems. One method is to use special formulations of LP problems. Another involves the use of the branch-and-bound technique, where the programming problem is divided into subclasses to be solved with convex (minimization problem) or linear approximations that form a lower bound on the overall cost within the subdivision. With subsequent divisions, an actual solution will be obtained at some point whose cost is equal to the best lower bound obtained for any of the approximate solutions. This solution is optimal, although possibly not unique. The method may also be terminated early, with the assurance that the best feasible solution is within a tolerance of the best point found; such points are called $\varepsilon$-optimal solution. Terminating to $\varepsilon$-optimal solution is typically necessary to ensure finite termination. This is especially useful for large, difficult problems, and problems with uncertain costs or values where the uncertainty can be estimated with appropriate reliability estimation.

Under differentiability and constraint qualifications, the Kuhn–Tucker conditions provide the necessary conditions for a solution to be optimal. Under convexity, these conditions are also sufficient.

The most popular methods for NLP problems include Zoutendijk's feasible direction method, the gradient projection method, the penalty method, and the Lagrangian method (Bazaraa et al. 2013).

The above-mentioned methods depend on certain mathematical properties of the NLP problems to be solved. Sometimes, these properties are difficult to satisfy. In such situations, these methods become invalid. Heuristics-based methods such as *Genetic Algorithms* (Tang et al. 2011), *Particle Swarm Optimization* (Nezhad et al. 2013), on the other hand, do not have this limitation and are thus another direction for NLP problems.

## 2.4 Multi-objective Programming

The main characteristics of *Multi-objective Programming* (MOP) are that decision makers need to achieve multiple objectives simultaneously while these multiple objectives are non-commensurable and conflict with each other.

### 2.4.1 Multi-objective Programming Model

An MOP model considers a vector of variables, objective functions, and constraints. It attempts to maximize (or minimize) the objective functions. Since this problem rarely has a unique solution, we expect to choose a solution from among the set of feasible solutions, which will be explained later in this section. Generally, a MOP problem can be formulated as follows:

$$\max_{x} \quad f(x)$$
$$\text{s.t.} \quad x \in X = \{x | g(x) \leq 0\} \tag{2.4}$$

where $f(x)$ represents $k$ conflicting objective functions, $g(x) \leq 0$ represents $m$ constraints, and $x \in R^n$ is a $n$-dimensional vector of decision variables.

*Multi-objective linear programming* (MOLP) is one of the most important forms of MOP problems, which are specified by linear objective functions subject to a set of linear constraints. The standard form of a MOLP problem can be written as follows:

$$\max_{x} \quad Cx$$
$$\text{s.t.} \quad x \in X = \{x | Ax \leq b\} \tag{2.5}$$

where $C$ is a $k \times n$ objective function matrix, $A$ is an $m \times n$ constraint matrix, $b$ is a $m$-dimensional vector, and $x$ is a $n$-dimensional vector of decision variable.

We have the following notion for a complete optimal solution.

**Definition 2.5** (Sakawa 1993) $x^*$ is said to be a complete optimal solution, if and only if there exists a $x^* \in X$ such that $f_i(x^*) \geq f_i(x)(i = 1, \ldots, k))$ for all $x \in X$.

Also, ideal solution, superior solution, or utopia point are equivalent terms indicating a complete optimal solution (Lu et al. 2007).

In general, a complete optimal solution that simultaneously maximizes (or minimizes) all objective functions does not always exist when the objective functions conflict with each other. Thus, a concept of Pareto optimal solution is introduced into MOLP.

**Definition 2.6** (Sakawa 1993) $x^*$ is said to be a Pareto optimal solution, if and only if there does not exist another $x \in X$ such that $f_i(x) \geq f_i(x^*)$ for all $i$ and $f_i(x) \neq f_i(x^*)$ for at least one $i$.

The Pareto optimal solution is also called a non-dominated solution, non-inferior solution, efficient solution, and non-dominate solution.

In addition to the Pareto optimal solution, the following weak Pareto optimal solution is defined as a slightly weaker solution concept than the Pareto optimal solution.

**Definition 2.7** (Sakawa 1993) $x^*$ is said to be a weak Pareto optimal solution, if and only if there does not exist another $x \in X$ such that $f_i(x) > f_i(x^*), i = 1, \ldots, k$.

Here, let $X^{CO}$, $X^P$ and $X^{WP}$ denote complete optimal, Pareto optimal, and weak Pareto optimal solution sets, respectively. Then from above definitions, we can easily obtain the following relations:

$$X^{CO} \subseteq X^P \subseteq X^{WP}. \tag{2.6}$$

A satisfactory solution belongs to a reduced subset of the feasible set that exceeds all of the aspiration levels of each objective. A set of satisfactory solutions is composed of acceptable alternatives. Satisfactory solutions do not need to be non-dominated, and a preferred solution is a non-dominated solution selected as the final choice through decision makers' involvement in the information processing stage.

The rest of this chapter focuses mainly on MOLP, the linear form of MOP.

## 2.4.2 Multi-objective Linear Programming Methods

The methods for solving MOLP problems have been well developed and classified into four classes by Hwang and Masud (1979) and Lai and Hwang (1994). We list them in Table 2.1.

As shown in Table 2.1, the first class of MOLP methods basically does not require any more information nor interaction with decision makers once the objective functions and constraints have been defined. The solution to a MOLP problem is presented on the basis of assumptions made about decision makers' preferences.

The second class of MOLP methods assumes that decision makers have a set of goals to achieve and that these goals will be established before formulation of a mathematical programming model. The multi-objective goal programming (MOGP) assumes that decision makers can specify goals for the objective functions. The key idea behind goal programming is to minimize deviation from the goals or aspiration levels set by decision makers. In most cases, therefore, MOGP seems to yield a satisfactory solution rather than an optimal one. More details about MOGP problem will be discussed later.

**Table 2.1** A classification of MOLP methods

| | Stage at which information is needed | Type of information | Typical methods |
|---|---|---|---|
| 1 | No articulation of preference information | | • Global criteria method (Hwang and Masud 1979, Salukvadze 1974) |
| 2 | A priori articulation of preference information | Cardinal | • Weighting method (Hwang and Masud 1979, Sakawa 1993) |
| | | Ordinal and cardinal | • (Multi-objective) goal programming (GP) (Ignizio 1976) |
| 3 | Progressive articulation of preference information (interactive method) | Explicit trade-off | • Efficient solution via goal programming (ESGP) (Ignizio 1981) |
| | | | • Interactive multiple objective linear programming (IMOLP) (Quaddus and Holzman 1986) |
| | | | • Interactive sequential goal programming (ISGP) (Hwang and Masud 1979) |
| | | | • ZW method (Zionts and Wallenius 1983) |
| | | Implicit trade-off | • STEP method (STEM) (Benayoun et al. 1971) |
| | | | • STEUER (Steuer 1977) |
| 4 | A posteriori articulation of preference information (non-dominated solutions generation method) | Implicit/ explicit trade-off | • Parametric method (Hwang and Masud 1979) |
| | | | • Constraint method (Hwang and Masud 1979; Sakawa 1993) |

The third class of MOLP, interactive methods, requires more involvement and interaction with decision makers in the solving process. The interaction takes place through decision makers' computer interface at each iteration. Trade-off or preference information from decision makers at each iteration is used to determine a new solution, therefore decision makers actually gain insights into the problem. Interactive programming was first initiated by Geoffrion et al. (1972) and further developed by many researchers. The STEP method (Benayoun et al. 1971) in particular is known to be one of the first interactive MOLP techniques, to which there have been a number of modifications and extensions. The interactive MOGP method was also proposed (Dyer 1972), which attempts to provide a link between MOGP and interactive methods.

Lastly, the purpose of the fourth class is to determine a subset of the complete set of non-dominated solutions to a MOLP problem. It deals strictly with constraints and does not consider the decision makers' preferences. The desired outcome is to narrow the possible courses of actions and select the preferred course of action more easily.

Interaction is one of the most important features for solving MOLP problems. There are three types of interaction in the MOLP problem solving process: pre-interaction (before the solution process), pro-interaction (during the solution process), and post-interaction (after the solution process). The seven MOLP methods selected from Table 2.1, ESGP, IMOLP, ISGP, MOGP, STEM, STEUER, and ZW, have differences in the interaction processes with decision makers. The MOGP, IMOLP and ISGP methods involve pre-interaction with users prior to the solution process through the collection of weights, goals, and priorities of objectives from users. The STEM method engages in pro-interaction during the solution process. Its principle is to require decision makers to nominate the amounts to be sacrificed of satisfactory objectives until all objectives become satisfactory. It first displays a solution and the ideal value of each objective. It then asks decision makers to accept or reject this solution. If it is accepted, the solution is taken as the final satisfactory solution. However, decision makers often make further searches so that more alternative solutions can be generated. If the current solution is rejected, a relaxation process starts. Decision makers will accept a certain level of relaxation of a satisfactory objective to allow the improvement of unsatisfactory objectives. When the relaxation fails, the system enables decision makers to continue re-entering a set of relaxation values and a new solution is then found. If decision makers accept this solution, it becomes the final satisfactory solution. Otherwise the system repeats the above process. Post-interaction is used in all seven methods. After a set of candidate solutions has been generated, decision makers are required to choose the most satisfactory solution.

Now, we give details of the weighting method for solving MOLP problems.

The key idea of the weighting method is to transform the multiple objective functions in the MOLP problem (2.5) into a weighted single objective function, which is described as follows:

$$\begin{aligned} \max_{x} \quad & wCx \\ \text{s.t.} \quad & x \in X = \{x|Ax \le b\} \end{aligned} \tag{2.7}$$

where $w = (w_1, w_2, \ldots, w_k) \ge 0$ is a vector of weighting coefficients assigned to the objective functions.

*Example 2.2* Let us consider the following example of a MOLP problem.

$$\begin{aligned} \max_{x_1, x_2} \quad & f(x) = \begin{pmatrix} f_1(x) \\ f_2(x) \end{pmatrix} = \begin{pmatrix} 2x_1 + x_2 \\ -x_1 + 2x_2 \end{pmatrix} \\ \text{s.t.} \quad & -x_1 + 3x_2 \le 21, \\ & x_1 + 3x_2 \le 27, \\ & 4x_1 + 3x_2 \le 45, \\ & 3x_1 + x_2 \le 30. \end{aligned} \tag{2.8}$$

Let $X$ denote the feasible region of problem (2.8). When $w_1 = 0.5, w_2 = 0.5$, the weighting problem is formulated as

$$\begin{aligned} \max \quad & wf(x) = 0.5x_1 + 1.5x_2 \\ \text{s.t.} \quad & (x_1, x_2) \in X. \end{aligned}$$

The optimal solution is $(x_1^*, x_2^*) = (3, 8)$, and the optimal objective function value is $f^*(x) = (f_1^*(x), f_2^*(x)) = (14, 13)$.

When $w_1 = 1, w_2 = 0$, the optimal solution is $(x_1^*, x_2^*) = (9, 3)$, and the optimal objective function value is $f^*(x) = (f_1^*(x), f_2^*(x)) = (21, -3)$.

When $w_1 = 0, w_2 = 1$, the optimal solution is $(x_1^*, x_2^*) = (0, 7)$, and the optimal objective function value is $f^*(x) = (f_1^*(x), f_2^*(x)) = (7, 14)$.

## 2.4.3 A Case-Based Example

*Example 2.3* A manufacturing company has six types of milling machine, lathe, grinder, jig saw, drill press, and band saw, whose capacities are to be devoted to producing three products $x_1$, $x_2$, and $x_3$. Decision makers have three objectives: maximizing profit, quality, and worker satisfaction. It is assumed that the parameters and objectives of the MOLP problem are defined precisely in this example. For instance, to produce one unit of $x_1$ requires 12 h of machine milling, as listed in Table 2.2 (Lai 1995).

**Table 2.2**  Production planning data

| Machine | Product $x_1$ (unit) | Product $x_2$ (unit) | Product $x_3$ (unit) | Machine (available hours) |
|---|---|---|---|---|
| Milling machine | 12 | 17 | 0 | 1,400 |
| Lathe | 3 | 9 | 8 | 1,000 |
| Grinder | 10 | 13 | 15 | 1,750 |
| Jig saw | 6 | 0 | 16 | 1,325 |
| Drill press | 0 | 12 | 7 | 900 |
| Band saw | 9.5 | 9.5 | 4 | 1,075 |
| Profit | 50 | 100 | 17.5 | |
| Quality | 92 | 75 | 50 | |
| Worker satisfaction | 25 | 100 | 75 | |

This problem can be described by a MOLP model as follows:

$$\max_{x_1,x_2,x_3} \quad f(x) = \begin{pmatrix} 50x_1 + 100x_2 + 17.5x_3 \\ 92x_1 + 75x_2 + 50x_3 \\ 25x_1 + 100x_2 + 75x_3 \end{pmatrix}$$

$$
\begin{aligned}
\text{s.t.} \quad & 12x_1 + 17x_2 \leq 1400, \\
& 3x_1 + 9x_2 + 8x_3 \leq 1000, \\
& 10x_1 + 13x_2 + 15x_3 \leq 1750, \\
& 6x_1 + 16x_3 \leq 1325, \\
& 12x_2 + 7x_3 \leq 900, \\
& 9.5x_1 + 9.5x_2 + 4x_3 \leq 1075, \\
& x_1, x_2, x_3 \geq 0.
\end{aligned}
\tag{2.9}
$$

We can see that this is a typical multi-objective programming problem.

## 2.5  Goal Programming

*Goal programming* (GP), originally proposed by Charnes and Cooper (1957), is a great strategy to deal with multi-objective optimization problems by setting multiple goals, as we mentioned before. In some decision situations, a decision maker may have more than one objective, with the improvement on one objective to be achieved only at the expense of others. For example, a coordinator of a multi-division firm considers three objectives in making an aggregate production plan: to maximize the net profit, to maximize the quality of products, and to maximize worker satisfaction (Example 2.3). The three objectives could be in conflict with each other, but must be considered simultaneously. Any improvement in one objective may be achieved only at the expense of other objectives.

Goal programming takes a 'satisfactory solution' strategy. It requests a decision maker to set a goal or a target for the objective (a set of goals for a MOLP) that the person wishes to attain. A preferred solution is then defined to minimize the deviation from the goal. Therefore, goal programming would appear to yield a satisfactory solution rather than an optimal one. Now we give a formal description of the method adopted by goal programming.

Suppose that a MOLP problem is defined as follows:

$$\max_{x} \quad f(x) = (\alpha_1 x, \alpha_2 x, \ldots, \alpha_k x)$$
$$\text{s.t.} \quad Ax \leq b. \tag{2.10}$$

For problem (2.10), there are a total of $k$ objectives $(\alpha_1 x, \alpha_2 x, \ldots, \alpha_k x)$ to achieve. We give goals $g_i (i = 1, 2, \ldots, k)$ for the $i$th objective. Our effort is now focused on making each objective $\alpha_i x$ as close to its goal $g_i (i = 1, 2, \ldots, k)$, as possible. The problem (2.10) is then transformed as follows:

$$\min_{x, v_1^-, v_1^+, \ldots, v_k^-, v_k^+} \quad v_1^- + v_1^+ + \cdots + v_k^- + v_k^+$$
$$\text{s.t.} \quad \alpha_1 x + v_1^- - v_1^+ = g_1,$$
$$\alpha_2 x + v_2^- - v_2^+ = g_2,$$
$$\vdots$$
$$\alpha_k x + v_k^- - v_k^+ = g_k,$$
$$v_1^-, v_1^+, \ldots, v_k^-, v_k^+ \geq 0,$$
$$Ax \leq b. \tag{2.11}$$

To give a more clear understanding of the idea adopted, $v_i^-$ and $v_i^+ (i = 1, \ldots, k)$ can be defined as follows:

$$v_1^+ = \frac{1}{2(|\alpha_1 x - g_1| + (\alpha_1 x - g_1))},$$
$$v_1^- = \frac{1}{2(|\alpha_1 x - g_1| - (\alpha_1 x - g_1))},$$
$$\vdots$$
$$v_k^+ = \frac{1}{2(|\alpha_k x - g_k| + (\alpha_k x - g_k))},$$
$$v_k^- = \frac{1}{2(|\alpha_k x - g_1| - (\alpha_k x - g_k))}.$$

In the above formula, $v_i^-$ and $v_i^+$, are deviation variables representing the under-achievement and over-achievement of the $i$th goal $g_i$, for the $i$th objective $\alpha_i x (i = 1, \ldots, k)$, respectively.

The problem defined by (2.11) is a standard linear programming problem which can be solved by the simplex method.

There are some variants of goal programming. The initial goal programming formulations order the deviations between objectives and goals into a number of priority levels. The minimization of the deviation at a higher priority level are more important than the deviations at lower priority levels. This is called lexicographicor pre-emptive goal programming (Amador and Romero 1989). When clear priority ordering of the goals to be achieved exists, lexicographic goal programming can be used.

Weighted or non pre-emptive goal programming can be used if a decision maker is more interested in making direct comparisons of the objectives. In this situation, all the deviations between objectives and goals are multiplied by weights, which reflect the relative importance of the objectives. We add these weighted deviations together as a single sum to form the objective function. This process is defined by the following formula:

$$\min_{x, v_1^-, v_1^+, \ldots, v_k^-, v_k^+} \quad v = w_1^- v_1^- + w_1^+ v_1^+ + \cdots + w_k^- v_k^- + w_k^+ v_k^+$$

$$\text{s.t.} \quad \alpha_1 x + v_1^- - v_1^+ = g_1,$$
$$\alpha_2 x + v_2^- - v_2^+ = g_2,$$
$$\vdots$$
$$\alpha_k x + v_k^- - v_k^+ = g_k,$$
$$v_1^-, v_1^+, \ldots, v_k^-, v_k^+ \geq 0,$$
$$Ax \leq b,$$

where $w_i^-$ and $w_i^+ (i = 1, 2, \ldots, k)$ are non-negative constants representing the relative importance to be assigned to the positive and negative deviations for each of the relevant goals.

Based on goal programming as previously introduced and the MOLP model, MOGP requires that goals are set for each objective, following which a preferred solution is defined as one which minimizes the deviations from those goals.

We assume that the goals $g = (g_1, \ldots, g_k)$ are specified for objective functions $f(x) = (f_1(x), \ldots, f_k(x))$ by decision makers, and a decision variable $x^* \in X$ in the MOLP problem is sought so that the objective functions $f^*(x) = (f_1^*(x), \ldots, f_k^*(x))$ are as close as possible to the goals $g = (g_1, \ldots, g_k)$.

The deviation between $f^*(x) = (f_1^*(x), \ldots, f_k^*(x))$ and $g = (g_1, \ldots, g_k)$ is usually defined as a deviation function $D(f(x), g)$. The MOGP can then be defined as an optimization problem:

$$\min_{x \in X} \quad D(f(x), g)$$
$$\text{s.t.} \quad x \in X = \{x \in R^n | Ax \leq b\}, \tag{2.12}$$

that is, find an $x^* \in X$, which minimizes $D(f(x), g)$ or

$$x^* = \arg \min_{x \in X} D(f(x), g). \tag{2.13}$$

Normally, the deviation function $D(f(x), g)$ is a maximum of deviation of individual goals,

$$D(f(x), g) = \max\{D_1(f_1(x), g_1), \ldots, D_k(f_k(x), g_k)\}. \tag{2.14}$$

From (2.12) and (2.14), the min–max approach is applied to the GP problem:

$$\min_{x \in X} \max\{D_1(f_1(x), g_1), \ldots, D_k(f_k(x), g_k)\}. \tag{2.15}$$

By introducing an auxiliary variable $\gamma$, (2.15) can then be transformed into the following linear programming problem:

$$\min_{x} \quad \gamma$$
$$\text{s.t.} \quad D_1(f_1(x), g_1) \leq \gamma,$$
$$D_2(f_2(x), g_2) \leq \gamma,$$
$$\vdots$$
$$D_k(f_k(x), g_k) \leq \gamma,$$
$$Ax \leq b. \tag{2.16}$$

*Example 2.4* Let us consider the following example of a MOLP problem:

$$\max_{x_1, x_2} \quad f(x) = \begin{pmatrix} 2x_1 + x_2 \\ -x_1 + 2x_2 \end{pmatrix}$$
$$\text{s.t.} \quad -x_1 + 3x_2 \leq 21,$$
$$x_1 + 3x_2 \leq 27,$$
$$4x_1 + 3x_2 \leq 45,$$
$$3x_1 + x_2 \leq 30,$$
$$x_1, x_2 \geq 0.$$

Suppose the goals are specified as $g = (10, 10)$ for the two objective functions. The original MOLP problem can be converted as the following LP problem with the auxiliary variable $\gamma$:

$$\min_{x_1,x_2} \quad \gamma$$

$$\text{s.t.} \quad 2x_1 + x_2 - 10 \leq \gamma,$$
$$-x_1 + 2x_2 - 10 \leq \gamma,$$
$$-x_1 + 3x_2 \leq 21,$$
$$x_1 + 3x_2 \leq 27,$$
$$4x_1 + 3x_2 \leq 45,$$
$$3x_1 + x_2 \leq 30,$$
$$x_1, x_2 \geq 0.$$

The optimal solution then is $(x_1^*, x_2^*) = (2, 6)$, and the optimal objective function values are $f^*(x) = (f_1^*(x), f_2^*(x)) = (10, 10)$.

When the goals are specified as $g = (15, 15)$, the optimal solution is $(x_1^*, x_2^*) = (1.865, 7.622)$, and the optimal objective function values are $f^*(x) = (f_1^*(x), f_2^*(x)) = (11.351, 13.378)$. We learn from the optimal objective function values that the goals are not achieved. The reason is that the goals specified are beyond the feasible constraint area. The point of $(x_1^*, x_2^*) = (1.865, 7.622)$ is on the boundary of the feasible constraint area.

Goal programming has the advantages of being simple and easy to use. It can handle relatively large numbers of variables, constraints and objectives, which accounts for the large number of goal programming applications in many diverse fields, such as business management, transportation planning, and resource optimization. A limitation of goal programming is that setting the goals for some of the objectives may not be straight forward. In-depth field knowledge might be required to solve a decision problem, and experiments sometimes need to be carried out to set suitable goals.

## 2.6 Stackelberg Game Model

The Stackelberg game model, which is also called a leader-follower game, was first proposed by Heinrich von Stackelberg in 1952 (Stackelberg 1952). It is based on economic monopolization phenomena. In a Stackelberg game, one player acts as a leader and the rest as followers. The problem is then to find an optimal strategy for the leader, assuming that the followers react in a rational way which will optimize their objective functions, given the leader's actions.

Stackelberg used a hierarchical model to describe a market situation in which decision makers try to optimize their decisions based on individually different objectives but are affected by a certain hierarchy.

## *2.6.1 Stackelberg Game and Bi-level Programming*

The Stackelberg leadership model considers the case of a single leader and follower. Let $X$ and $Y$ be the strategy sets for the leader and follower respectively. Denote their objective function by $F(x,y)$ and $f(x,y)$ respectively. Knowing the selection $x$ of the leader, the follower can select his best strategy $y(x)$ such that his objective function $f(x,y)$ is maximized, i.e.,

$$y(x) \in \Phi(x) = \arg\max_{y \in Y} f(x,y). \tag{2.17}$$

The leader then obtains the best strategy $x \in X$ as

$$x \in \arg\max_{x \in X} \{F(x,y) | y \in \Phi(x)\}. \tag{2.18}$$

Formulae (2.17) and (2.18) can be combined to express the Stackelberg game as follows:

$$\max_{x} \quad F(x,y)$$
$$\text{s.t.} \quad x \in X,$$
$$y \in \arg\max_{y \in Y} f(x,y).$$

Bi-level programming (see Chap. 3) is more general than Stackelberg game in the sense that the strategy sets (also called the admissible sets) depend on both $x$ and $y$. This leads to a general bi-level programming (Candler and Norton 1977) as follows:

$$\max_{x} \quad F(x,y)$$
$$\text{s.t.} \quad G(x,y) \leq 0, \tag{2.19}$$
$$y \in \arg\max\{f(x,y) | g(x,y) \leq 0\}.$$

Bi-level programming problem (2.19) is a generalization of several well-known optimization problem (Dempe 2002). For example, if $F(x,y) = -f(x,y)$, then it is a classical min–max problem; if $F(x,y) = f(x,y)$, we have a realization of the decomposition approach to optimization problem; if the dependence of both the leader's and the follower's problem on $y$ is dropped, the problem is reduced to a bi-criteria optimization problem.

### 2.6.2 Stackelberg Game and Nash Game

The Stackelberg game can be considered as an extension of the well-known Nash game (Nash 1951). In the Nash game, we assume that there are $k$ players, and the $i$th player has a strategy set $X_i$, and his objective function is $f_i(x)$ for $i = 1, 2, \ldots, k$, where $x = (x_1, x_2, \ldots, x_k)$. Each player chooses a strategy based on the choices of the other players and there is no hierarchy. The unstructured problem is modeled as follows: for $i = 1, 2, \ldots, k$, we have $\max\limits_{x_i \in X_i} f_i(x)$.

This is a Nash game in which all players aim to maximize their corresponding objective functions.

In contrast, there is a hierarchy between the leader and followers in the Stackelberg game. The leader is aware of the choices of the followers, thus the leader, being in a superior position with regard to everyone else, can achieve the best objective while forcing the followers to respond to this choice of strategy by solving the Stackelberg game. Without loss of generality, we now assume that the first player is the leader, and the rest of the players are followers. Let $X_{1-} = X_2 \times X_3 \times \cdots \times X_k, f_{1-}(x) = (f_2(x), \ldots, f_k(x))$, and $x_{1-} = (x_2, \ldots, x_k) \in X_{1-}$. The above Nash game is accordingly transformed into a Stackelberg game, which is given as follows:

$$\max_{x_1 \in X_1} \quad f_1(x)$$

$$\text{s.t.} \quad x_{1-} \in \text{argmax}\{f_{1-}(x) | x_{1-} \in X_{1-}\}.$$

This is a Stackelberg game or leader-follower game.

### 2.6.3 Applications of Stackelberg Games

The investigation of Stackelberg games is strongly motivated by real world applications, and Stackelberg games techniques have been applied with remarkable success in many domains, such as transportation network design, production planning and logistics.

Stackelberg games have been applied to the network design problem (Ben-Ayed 1988) arising in transportation systems. In the accompanying formulation, a central planner controls investment costs at the system level, while operational costs depend on traffic flow, which is determined by the individual user's route selection. Because users are assumed to make decisions to maximize their peculiar utility functions, their choices do not necessarily coincide with the choices that are optimal for the system. Nevertheless, the central planner can influence users' choices by improving certain links, making some relatively more attractive than others. In deciding on these improvements, the central planner tries to influence users' preferences in such a way that total costs are minimized. The partition of the control variables between the upper and lower levels naturally leads to a bi-level formulation.

Moreover, a fuzzy Stackelberg game model was set up to control traffic flow in a disaster area after an earthquake (Feng and Wen 2005). When a severe earthquake occurs, roadway systems usually suffer various degrees of damage, reducing their capacity and causing traffic congestion. Maintaining viable traffic functions to facilitate the saving of more lives is a crucial mission task following an earthquake. The aim of the commander of the government Emergency-Response Centre at county and city level (the upper level) is to allow traffic to pass through disaster areas to the extent that is possible given the roadway's capacity, while road users (at the lower level) always choose the shortest route to affect emergency rescues. To solve this decision problem, the bi-level technique has been used post-earthquake to provide an efficient traffic control strategy for recovery from chaos.

A Stackelberg game has been formulated for a newsboy problem. The decision makers are the manufacturer and retailers. The former acts as a leader who controls the product price, and the retailers as the followers who decide the quantity of newspapers to order. The relationship between the manufacturer and retailers is a sequential non-cooperative game. The manufacturer first decides the product price, and the retailers then decides the quantity. The manufacturer tries to determine product price and maximize his profit after considering the retailers' behavior. The retailers' decision is to optimize the order quantity so as to maximize his profit at a given product price. Clearly, this newsboy problem can be modeled as a Stackelberg game.

In addition, Stackelberg games are frequently utilized in many other real-world cases, such as resource allocation, network investigation, and engineering. These applications have provided stimulating environments for the development of Stackelberg games.

## 2.7  Particle Swarm Optimization

In the computational intelligence area, *particle swarm optimization* (PSO) is a computational method that optimizes a problem by iteratively trying to improve a candidate solution with regard to a given measure of quality. PSO is a heuristic algorithm proposed by Kennedy and Eberhart (1995), Shi and Eberhart (1998).

Inspired by the social behavior of animals, such as fish schooling and bird flocking, PSO is a kind of population-based algorithm. The population of PSO is called a swarm, and each individual in the swarm is called a particle. The similarity between PSO and other evolutionary algorithms lies in the fact that an individual in the community is moved to a good area according to its fitness for the environment. Unlike other evolutionary computation methods, however, each particle in PSO has an adaptable velocity (position change), according to which it moves in the search space (Parsopoulos and Vrahatis 2002). Moreover, each particle has a memory, remembering the best position it has ever visited in the search space (Kennedy and Eberhart 1995). Thus, its movement is an aggregated acceleration towards its best previously visited position and towards the best particle of a topological neighborhood.

Suppose the current search space for PSO is $n$-dimensional, then the $i$th particle of the swarm can be represented by an $n$-dimensional vector, $x_i = (x_{i1}, \ldots, x_{in})$. The velocity (position change) of this particle can thus be represented by another $n$-dimensional vector $v_i = (v_{i1}, \ldots, v_{in})$. The best previously visited position of the $i$th particle is denoted as $p_i = (p_{i1}, \ldots, p_{in})$. Defining $g$ as the index of the best particle in the swarm (i.e., the $g$th particle is the best), and letting the superscripts denote the iteration number, the swarm is manipulated according to the following two equations (Eberhart et al. 1996):

$$v_{id}^{k+1} = wv_{id}^k + cr_1^k\left(p_{id} - x_{id}^k\right) + cr_2^k\left(p_{gd}^k - x_{id}^k\right),$$
$$x_{id}^{k+1} = x_{id}^k + v_{id}^{K+1},$$

where $d = 1, \ldots, n$ denotes the $d$-dimensional vector, $i = 1, 2, \ldots, N$ denotes the $i$th particle, $N$ is the size of the swarm, $w$ is the inertia weight, $c$ is a positive constant, called the acceleration constant, $r_1, r_2$ are random numbers, uniformly distributed in [0,1], and $k$ determines the iteration number.

Like many other global optimization methods, whether deterministic or evolutionary, PSO suffers from the problem of local optima. The existence of many local optimal solutions makes it difficult for PSO to detect the global optimal solution. In some cases, sub-optimal solutions are acceptable although not desirable, while in others, a global optimal solution is indispensable. The development of robust and efficient methods for avoiding local solutions is the subject of current PSO research.

Stretching technique (Parsopoulos and Vrahatis 2002) has been shown through simulation experiments and it provide an effective way for the PSO method to escape local optimal solution.

The idea behind the function of Stretching is to perform a two-stage transformation of the original objective function $F(x)$. The two-stage transformation can be applied immediately after a local optimization solution $\bar{x}$ of the function $F(x)$ has been detected. This transformation has been proposed by Parsopoulos and Vrahatis (2002) and is defined as follows:

$$G(x) = F(x) + \gamma_1 |x - \bar{x}|(\text{sign}(F(x) - F(\bar{x})) + 1), \tag{2.20}$$

$$H(x) = G(x) + \gamma_2 \frac{\text{sign}(F(x) - F(\bar{x})) + 1}{\tanh(\mu(G(x) - G(\bar{x})))}, \tag{2.21}$$

where $\gamma_1, \gamma_2$ and $\mu$ are arbitrary chosen positive constants, and $\text{sign}(x)$ is defined by:

$$\text{sign}(x) = \begin{cases} 1, & \text{if } x < 0; \\ 0, & \text{if } x = 0; \\ -1, & \text{if } x < 0. \end{cases}$$

The first transformation stage, defined in (2.20), elevates the function $F(x)$ and eliminates all the local optimization solutions that are less optimal than the result of $F(\bar{x})$. The second stage, defined by (2.21), stretches the neighborhood of $\bar{x}$ upwards, since it assigns higher function values to those points. Neither stage changes the local optimal solutions which can produce more optimal results than $\bar{x}$. Thus, the location of the global solution can be left unchanged.

Because PSO requires only primitive mathematical operators and is computationally inexpensive in terms of both memory requirements and speed (Parsopoulos and Vrahatis 2002), it has good convergence performance and has been successfully applied in many fields such as neural network training (Zhang et al. 2007), integral programming (Kitayama and Yasuda 2006), multi-objective optimization (Ho et al. 2006), and decision making (Nenortaitė 2007).

## 2.8 Summary

This chapter addresses the basic concepts and models of optimization theory: linear programming, non-linear programming, goal programming, multi-objective programming, Stackelberg games, and particle swarm optimization are introduced. These concepts, models and solution techniques will be used in the rest of this book.

# Chapter 3
# Bi-level Programming Models and Algorithms

This chapter introduces basic definitions, theorems, models and algorithms for bi-level programming (bi-level decision-making) and also basic models of multi-level programming, which will be used in the remaining chapters of this book.

Bi-level programming is a special situation of multi-level programming in which there are only two levels of optimization, that is, the levels of decision (optimization) $n = 2$. Bi-level and multi-level programming techniques are developed for solving decentralized decision-making problems with decision entities (also called decision makers) in a hierarchical system. The decision entity at the upper level is termed the *leader*, and at the lower level is termed the *follower* (Bard 1998). Each decision entity (leader or follower) tries to optimize their own objectives by considering the objective of the other level only partially or not at all, but the decision of each level affects the strategy of other level decision makers. The hierarchical optimization structure appears naturally in critical resource management and policy making, including tourism resource planning, water resource management, financial planning, healthcare planning, land-use planning, production planning, transportation planning, and power market pricing.

This chapter is organized as follows. Section 3.1 introduces the basic concepts and models for linear and non-linear bi-level programming. Section 3.2 presents the solution theories and some basic results for linear bi-level programming. The $K$th-Best algorithm, the Kuhn-Tucker (Karush-Kuhn-Tucker, or KKT) approach, the Branch-and-bound algorithm and the penalty function method are introduced in Sects. 3.3–3.6 respectively to show how to solve a linear bi-level programming problem. In Sect. 3.7, based on the bi-level programming concept, we present a multi-level programming model and in particular, a tri-level programming model (the levels of decision $n = 3$). Lastly, a summary is provided in Sect. 3.8.

## 3.1 Bi-level Programming Model

*Bi-level programming* (BLP) is closely related to the economics problem addressed by Stackelberg (1952) through its development of strategic game theory. The original formulation for bi-level programming appeared in 1973, in a paper authored by Bracken and McGill (1973), although it was Candler and Norton (1977) who

© Springer-Verlag Berlin Heidelberg 2015
G. Zhang et al., *Multi-Level Decision Making*,
Intelligent Systems Reference Library 82, DOI 10.1007/978-3-662-46059-7_3

first used the designation "bi-level" programming. However, it was not until the early 1980s that these problems started to receive the attention they deserved.

Bi-level programming supposes that the leader has control over the variable $x \in X \subset R^n$ and that the follower has control over the variable $y \in Y \subset R^m$. The leader goes first and selects an $x$ in an attempt to optimize (maximize or minimize) their objective function $F(x, y(x))$ subject to some additional constraints. The notation $y(x)$ stresses the fact that the leader's problem is implicit in the $y$ variable. Although for the sake of simplicity this notation will not be maintained in general, it should be remembered that $y$ is always a function of $x$. The follower observes the leader's actions and reacts by selecting a $y$ to optimize their objective function $f(x, y)$, subject to a set of constraints in the $y$ variable for the value of $x$ chosen.

A large part of the research on bi-level programming has centered on its linear version, i.e., the linear bi-level programming, or so-called linear bi-level optimization or linear bi-level decision-making, in which all formulas of the objective functions and the constraints from both the leader and follower are all linear functions.

The general form of linear bi-level programming is defined in Definition 3.1.

**Definition 3.1** (Bard 1998) For $x \in X \subset R^n$, $y \in Y \subset R^m$, $F : X \times Y \rightarrow R$, and $f : X \times Y \rightarrow R$, a linear bi-level programming problem is given as follows:

$$\min_{x \in X} \quad F(x, y) = c_1 x + d_1 y \tag{3.1a}$$

$$\text{s.t.} \quad A_1 x + B_1 y \leq b_1, \tag{3.1b}$$

$$\min_{y \in Y} \quad f(x, y) = c_2 x + d_2 y \tag{3.1c}$$

$$\text{s.t.} \quad A_2 x + B_2 y \leq b_2, \tag{3.1d}$$

where $c_1, c_2 \in R^n$, $d_1, d_2 \in R^m$, $b_1 \in R^p$, $b_2 \in R^q$, $A_1 \in R^{p \times n}$, $B_1 \in R^{p \times m}$, $A_2 \in R^{q \times n}$, $B_2 \in R^{q \times m}$.

Although the definitions of linear bi-level programming vary considerably from one reference to another, most recent publications tend to agree on Definition 3.1 as the general form. Nevertheless, there are many attempts to describe a more general form of bi-level programming to cover both linear and non-linear bi-level programming problems. That is, the objective functions and constraints can be in any form of functions. A general bi-level programming problem is written as follows:

$$\min_{x \in X} \quad F(x, y)$$
$$\text{s.t.} \quad G(x, y) \leq 0,$$
$$\min_{y \in Y} \quad f(x, y)$$
$$\text{s.t.} \quad g(x, y) \leq 0,$$

where $F$, $f : R^m \times R^n \to R$, $G : R^m \times R^n \to R^p$ and $g : R^m \times R^n \to R^q$ are continuous and twice differentiable functions.

This chapter mainly focuses on linear bi-level programming. We will first give related solution theories and then provide some popular solution methods.

## 3.2  Solution Theories for Linear Bi-level Programming

**Definition 3.2** (Basener 2006) A topological space is compact if every open cover of the entire space has a finite subcover. For example, $[a, b]$ is compact in $R$.

**Definition 3.3** (Bard 1998)

(a)  Constraint region of the linear bi-level programming problem (3.1a–3.1d):

$$S = \{(x, y) | x \in X, y \in Y, A_1 x + B_1 y \le b_1, A_2 x + B_2 y \le b_2\}.$$

(b)  Feasible set for the follower for each fixed $x \in X$:

$$S(x) = \{y \in Y | B_2 y \le b_2 - A_2 x\}.$$

(c)  Projection of $S$ onto the leader's decision space:

$$S(X) = \{x \in X | \exists y \in Y, A_1 x + B_1 y \le b_1, A_2 x + B_2 y \le b_2\}.$$

(d)  Follower's rational reaction set for $x \in S(X)$:

$$P(x) = \{y \in Y | y \in \arg \min[f(x, \hat{y}) | \hat{y} \in S(x)]\},$$

where $\arg \min[f(x, \hat{y}) | \hat{y} \in S(x)] = \{y \in S(x) | f(x, y) \le f(x, \hat{y}), \hat{y} \in S(x)\}$.

(e)  Inducible region or feasible region of the leader:

$$IR = \{(x, y) | (x, y) \in S, y \in P(x)\}.$$

To ensure that the problem (3.1a–3.1d) has an optimal solution, Bard (1998) gave the following assumptions. These assumptions serve as an introduction to an existence theorem of solutions.

**Assumption 3.1**

1.  $S$ is nonempty and compact.
2.  For decisions taken by the leader, the follower has some room to respond, i.e., $P(x) \ne \emptyset$.
3.  $P(x)$ is a point-to-point map with respect to $x$.

**Assumption 3.2** *IR* is nonempty.

The rational reaction set $P(x)$ defines the response while the inducible region *IR* represents the set over which the leader may optimize its objective. Thus in terms of the above notations, problem (3.1a–3.1d) can be written as:

$$\min \quad \{F(x,y)|(x,y) \in IR\}. \tag{3.2}$$

**Theorem 3.1** *If assumptions 3.1 and 3.2 are satisfied, there exists an optimal solution for a linear bi-level programming problem (3.1a–3.1d).*

*Proof* It follows immediately from Theorem 6 (Mersha and Dempe 2006).  □

**Theorem 3.2** *The inducible region IR can be written equivalently as a piecewise linear equality constraint comprised of support hyper-planes of S.*

*Proof* Let us begin by writing the inducible region of Definition 3.3(e) explicitly as follows:

$$IR = \{(x,y)|(x,y) \in S, d_2y = \min[d_2\tilde{y}|B_2\tilde{y} \le b_2 - A_2x, \tilde{y} \ge 0]\}.$$

Now define:

$$Q(x) = \min\{d_2y|B_2y \le b_2 - A_2x, y \ge 0\} \tag{3.3}$$

For each value of $x \in S(X)$, the resulting feasible region of problem (3.1a–3.1d) is non-empty and compact. Thus, $Q(x)$ which is a linear programming problem parameterized in $x$, always has an optimal solution. From duality theory, we get

$$\max \quad \{u(A_2x - b)|uB_2 \ge -d_2, u \ge 0\}, \tag{3.4}$$

which has the same optimal value as (3.3). Let $u^1, \ldots, u^s$ be a listing of all the vertices of the constraint region of (3.4) which is given by $U = \{u|uB_2 \ge -d_2, u \ge 0\}$. Because we know that an optimal solution to (3.4) occurs at a vertex of $U$, we get an equivalent problem

$$\max \quad \{u(A_2x - b_2)|u \in \{u^1, \ldots, u^s\}\} \tag{3.5}$$

which demonstrates that $Q(x)$ is a piecewise linear function. Rewriting *IR* as

$$IR = \{(x,y) \in S|Q(x) - d_2y = 0\} \tag{3.6}$$

yields the desired result.                                                □

**Corollary 3.1** *The linear bi-level programming problem (3.1a–3.1d) is equivalent to minimizing $F(x,y)$ over a feasible region comprised of a piecewise linear equality constraint.*

*Proof* By the problem (3.2) and Theorem 3.2, we have the desired result.

The function $Q(x)$ defined by the problem (3.3) is convex and continuous. In general, because we aim to minimize the linear function $F(x,y) = c_1 x + d_1 y$ over $IR$, and because $F(x,y)$ is bounded from below by the finite optimal value, i.e., $\min\{c_1 x + d_1 y | (x,y) \in S\}$, the following can be concluded. ☐

**Corollary 3.2** *An optimal solution to the linear bi-level programming problem* (3.1a–3.1d) *occurs at a vertex of IR.*

*Proof* Since $F(x,y) = c_1 x + d_1 y$ is linear, if an optimal solution exists, it must occur at a vertex of $IR$. The proof is completed. ☐

**Theorem 3.3** *The optimal solution* $(x^*, y^*)$ *of the linear bi-level programming problem* (3.1a–3.1d) *occurs at a vertex of S.*

*Proof* Let $(x^1, y^1), \ldots, (x^r, y^r)$ be the distinct vertices of $S$. Since any point in $S$ can be written as a convex combination of these vertices, let $(x^*, y^*) = \sum_{i=1}^{\bar{r}} \alpha_i (x^i, y^i)$, where $\sum_{i=1}^{\bar{r}} \alpha_i = 1$, $\alpha_i > 0$, $i = 1, \ldots, \bar{r}$ and $\bar{r} \leq r$. It must be shown that $\bar{r} = 1$. To see this let us write the constraints of problem (3.1a–3.1d) at $(x^*, y^*)$ in their piecewise linear form (3.6).

$$
\begin{aligned}
0 = Q(x^*) - d_2 y^* &= Q\left(\sum_i \alpha_i x^i\right) - d_2\left(\sum_i \alpha_i y^i\right) \\
&\leq \sum_i \alpha_i Q(x^i) - \sum_i \alpha_i d_2 y^i \quad \text{by convexity of } Q(x) \\
&= \sum_i \alpha_i \left(Q(x^i) - d_2 y^i\right).
\end{aligned}
$$

But by the definition of $Q(x^i)$, it follows that

$$
Q(x^i) = \min_{y \in S(x^i)} d_2 y \leq d_2 y^i.
$$

Therefore, $Q(x^i) - d_2 y^i \leq 0$, $i = 1, \ldots, \bar{r}$. Noting that $\alpha_i > 0$, the equality in the preceding expression must hold or else a contradiction would result in the sequence above. Consequently, $Q(x^i) - d_2 y^i = 0$ for all $i$. This implies that $(x^i, y^i) \in IR$, $i = 1, \ldots, \bar{r}$, and $(x^*, y^*)$ can be written as a convex combination of points in $IR$. Because $(x^*, y^*)$ is a vertex of $IR$, a contradiction results unless $\bar{r} = 1$. ☐

**Corollary 3.3** *If* $(x^*, y^*)$ *is an extreme point of IR, it is an extreme point of S.*

*Proof* Let $(x^*, y^*)$ be an extreme point of $IR$ and assume that it is not an extreme point of $S$. Let $(x^1, y^1), \ldots, (x^r, y^r)$ be the distinct vertices of $S$. Since any point in $S$ can be written as a convex combination of these vertices, let $(x^*, y^*) = \sum_{i=1}^{\bar{r}} \alpha_i (x^i, y^i)$, where $\sum_{i=1}^{\bar{r}} \alpha_i = 1$, $\alpha_i > 0$, $i = 1, \ldots, \bar{r}$ and $\bar{r} \leq r$. It must be shown that $\bar{r} = 1$. To see this let us write the constraints to problem (3.1a–3.1d) at $(x^*, y^*)$ in their piecewise linear form (3.6).

$$0 = Q(x^*) - d_2 y^* = Q\left(\sum_i \alpha_i x^i\right) - d_2\left(\sum_i \alpha_i y^i\right)$$

$$\leq \sum_i \alpha_i Q(x^i) - \sum_i \alpha_i d_2 y^i \quad \text{by convexity of } Q(x)$$

$$= \sum_i \alpha_i \left(Q(x^i) - d_2 y^i\right).$$

But by the definition of $Q(x^i)$, we find that

$$Q(x^i) = \min_{y \in S(x^i)} d_2 y \leq d_2 y^i.$$

Therefore, $Q(x^i) - d_2 y^i \leq 0$, $i = 1, \ldots, \bar{r}$. Noting that $\alpha_i > 0$, the equality in the preceding expression must hold or else a contradiction would result in the sequence above. Consequently, $Q(x^i) - d_2 y^i = 0$ for all $i$. This implies that $(x^i, y^i) \in IR$, $i = 1, \ldots, \bar{r}$ and $(x^*, y^*)$ can be written as a convex combination of points in $IR$. Because $(x^*, y^*)$ is an extreme point of $IR$, a contradiction results unless $\bar{r} = 1$. This means that $(x^*, y^*)$ is an extreme point of $S$. The proof is completed. $\qquad \square$

In searching for a way to solve the linear bi-level programming problems, it would be helpful to have an explicit representation of $IR$ rather than the implicit representation given by formula (3.6). This can be achieved by replacing the follower's problem (3.1c, 3.1d) with its KKT conditions and appending the resultant system to the leader's problem. Let $u \in R^q$, and $v \in R^m$ be the dual variables associated with constraints (3.1d) and $y \geq 0$, respectively. We have the following result.

**Proposition 3.1** (Bard 1998) *A necessary condition that $(x^*, y^*)$ solves the linear bi-level programming problem (3.1a–3.1d) is that there exist (row) vectors $u^*$, and $v^*$ such that $(x^*, y^*, u^*, v^*)$ solves:*

$$\min_{x,y,u,v} \quad c_1 x + d_1 y \tag{3.7a}$$

$$\text{s.t.} \quad A_1 x + B_1 y \leq b_1, \tag{3.7b}$$

$$u B_2 - v = -d_2, \tag{3.7c}$$

$$u(b_2 - A_2 x - B_2 y) + v y = 0, \tag{3.7d}$$

$$A_2 x + B_2 y \leq b_2, \tag{3.7e}$$

$$x \geq 0, y \geq 0, u \geq 0, v \geq 0. \tag{3.7f}$$

## 3.3 *Kth*-Best Algorithm for Linear Bi-level Programming

Theorem 3.3 and Corollary 3.3 have provided a theoretical foundation for the *Kth*-Best algorithm. This means that by searching extreme points on the constraint region $S$, we can efficiently find an optimal solution for a linear BLP problem. The basic idea of the *Kth*-Best algorithm is that according to the upper level objective function value, we place all the extreme points on $S$ in ascending order and select the first extreme point to check whether it is on the inducible region *IR*. If it is, the current extreme point is the optimal solution. If not, we select the next one and check.

Let $(x_{[1]}, y_{[1]}), \ldots, (x_{[N]}, y_{[N]})$ denote the $N$ ordered extreme points to the following linear programming problem

$$\min \quad \{c_1 x + d_1 y | (x, y) \in S\}, \tag{3.8}$$

such that $c_1 x_{[i]} + d_1 y_{[i]} \le c_1 x_{[i+1]} + d_1 y_{[i+1]}, i = 1, \ldots, N - 1.$

Let $\tilde{y}$ denote the optimal solution to the following problem:

$$\min \quad \{f(x_{[i]}, y) | y \in S(x_{[i]})\}. \tag{3.9}$$

We only need to find the smallest $i$ $(i \in \{1, \ldots, N\})$ such that $y_{[i]} = \tilde{y}$. Let us write the problem (3.9) as follows:

$$\min_y \quad f(x, y)$$
$$\text{s.t.} \quad y \in S(x),$$
$$x = x_{[i]}.$$

From Definition 3.3(a) and (c), the above problem is equivalent to:

$$\min_y \quad f(x, y) = c_2 x + d_2 y \tag{3.10a}$$

$$\text{s.t.} \quad A_1 x + B_1 y \le b_1, \tag{3.10b}$$

$$A_2 x + B_2 y \le b_2, \tag{3.10c}$$

$$x = x_{[i]}. \tag{3.10d}$$

The idea of the *Kth*-Best algorithm is equivalent to selecting one ordered extreme point $(x_{[i]}, y_{[i]})$ and then solving the problem (3.10a–3.10d) to obtain an optimal solution $\tilde{y}$. If $\tilde{y} = y_{[i]}$, $(x_{[i]}, y_{[i]})$ is a globally optimal solution to the problem (3.1a–3.1d). Otherwise, check the next extreme point.

This can be accomplished with the following procedure.

---

**Algorithm 3.1: Kth-Best Algorithm**

---

[Begin]

**Step 1:** Put $i \leftarrow 1$. Solve the problem (3.8) with the simplex method to obtain an optimal solution $(x_{[1]}, y_{[1]})$. Let $W = \{(x_{[1]}, y_{[1]})\}$ and $T = \emptyset$.

**Step 2:** Solve the problem (3.10) with the simplex method, and denote its optimal solution by $\tilde{y}$. If $\tilde{y} = y_{[i]}$, stop; $(x_{[i]}, y_{[i]})$ is a globally optimal solution to the problem (3.1) with $K^* = i$. Otherwise, go to Step 3.

**Step 3:** Let $W_{[i]}$ be the set of adjacent extreme points of $(x_{[i]}, y_{[i]})$ such that $(x, y) \in W_{[i]}$ implies that

$$c_1 x + d_1 y \geq c_1 x_{[i]} + d_1 y_{[i]}.$$

Let $T = T \cup \{(x_{[i]}, y_{[i]})\}$ and $W = (W \cup W_{[i]}) \setminus T$.

**Step 4:** Set $i \leftarrow i + 1$ and choose $(x_{[i]}, y_{[i]})$ such that

$$c_1 x_{[i]} + d_1 y_{[i]} = \min\{c_1 x + d_1 y \mid (x, y) \in W\}.$$

Go to Step 2.

[End]

---

*Example 3.1* Let us consider the following linear bi-level program model:

$$\min_{x,y} \ F(x, y) = x - 4y$$

$$\text{s.t. } -x - y \leq -3,$$

$$-2x + y \leq 0,$$

$$\min_y f(x, y) = x + y$$

$$\text{s.t.} -3x + 2y \geq -4,$$

$$2x + y \leq 12,$$

$$x \geq 0, y \geq 0.$$

**Loop 1:** Let $i = 1$, and solve the following linear programming problem with the simplex method:

$$\min_{x,y} \ \ F(x, y) = x - 4y$$

$$\text{s.t.} \ \ -x - y \leq -3,$$

$$-2x + y \leq 0,$$

$$-3x + 2y \geq -4,$$

$$2x + y \leq 12,$$

$$x \geq 0, y \geq 0.$$

The optimal solution is $(x_{[i]}, y_{[i]}) = (3,6)$. Let $W = \{(3,6)\}$ and $T = \emptyset$. By the problem (3.10a–3.10d), we have

$$\min_{y} \quad f(x,y) = x + y$$

$$\text{s.t.} \quad 3x + 2y \geq -4,$$
$$2x + y \leq 12,$$
$$x = 3, y \geq 0.$$

Using the simplex method, we find that $\tilde{y} = 2.5$. Because of $\tilde{y} \neq y_{[i]}$, we go to Step 3.

It follows that $W_{[i]} = \{(4,4),(1,2)\}$, $T = \{(3,6)\}$ and $W = \{(4,4),(1,2)\}$, then go to Step 4.

Update $i = 2$, and choose $(x_{[i]}, y_{[i]}) = (4,4)$, then go to Step 2.

**Loop 2:** By the problem (3.10a–3.10d), we have

$$\min_{y} \quad f(x,y) = x + y$$

$$\text{s.t.} \quad -3x + 2y \geq -4,$$
$$2x + y \leq 12,$$
$$x = 4, y \geq 0.$$

Using the simplex method, we have $\tilde{y} = 4$. Because of $\tilde{y} = y_{[i]}$, we stop here. $(x_{[i]}, y_{[i]}) = (4,4)$ is a globally optimal solution.

By examining the above procedure, we find that the optimal solution for this example occurs at an extreme point $(x^*, y^*) = (4,4)$ with $F^* = -12$ and $f^* = 8$.

## 3.4 Kuhn-Tucker Approach for Linear Bi-level Programming

Proposition 3.1 means that the most direct approach for solving the problem (3.1a–3.1d) is to solve the non-linear programming problem (3.7a–3.7f). One advantage that it offers is that it allows for a more robust model to be solved without introducing new computational difficulties.

We use the Kuhn-Tucker approach to solve Example 3.1.

$$g_1(x,y) = -3x + 2y + 4 \geq 0, \tag{3.11a}$$

$$g_2(x,y) = -2x - y + 12 \geq 0, \tag{3.11b}$$

$$g_3(x, y) = y \geq 0. \tag{3.11c}$$

From the problem (3.7a–3.7f), we have

$$\min_{x,y,u_1,u_2,u_3} \quad x - 4y \tag{3.12a}$$

$$\text{s.t.} \quad -x - y \leq -3, \tag{3.12b}$$

$$-2x + y \leq 0, \tag{3.12c}$$

$$-3x + 2y \geq -4, \tag{3.12d}$$

$$2x + y \leq 12, \tag{3.12e}$$

$$-2u_1 + u_2 - u_3 = -1, \tag{3.12f}$$

$$u_1 g_1(x, y) + u_2 g_2(x, y) + u_3 g_3(x, y) = 0, \tag{3.12g}$$

$$x, y, u_1, u_2, u_3 \geq 0. \tag{3.12h}$$

From formulas (3.12f, 3.12h), we have the following two possibilities.

**Case 1** $\left(u_1^*, u_2^*, u_3^*\right) = (0, 0, 1)$.

From problem (3.12a–3.12h), we find that $g_3(x, y) = 0$, i.e., $y = 0$. Consequently, problem (3.12a–3.12h) can be rewritten as follows:

$$\min_{x,y} \quad x - 4y$$
$$\text{s.t.} \quad -x - y \leq -3,$$
$$-2x + y \leq 0,$$
$$-3x + 2y \geq -4,$$
$$2x + y \leq 12,$$
$$x \geq 0, y = 0.$$

It is easy to check that the above problem is infeasible.

**Case 2** $\left(u_1^*, u_2^*, u_3^*\right) = (0.5, 0, 0)$.

From the problem (3.12a–3.12h), it follows that $g_1(x, y) = 0$. Consequently, problem (3.12a–3.12h) can be rewritten as follows:

$$\min_{x,y} \quad x - 4y$$
$$\text{s.t.} \quad -x - y \leq -3,$$
$$-2x + y \leq 0,$$
$$-3x + 2y = -4,$$
$$2x + y \leq 12,$$
$$x \geq 0, y \geq 0.$$

Using the simplex method, we find that an optimal solution of the above linear programming problem occurs at the point $(x^*, y^*) = (4, 4)$ with $F^* = -12$ and $f^* = 8$.

By examining the above procedure, we find that the optimal solution for Example 3.1 occurs at the point $(x^*, y^*) = (4, 4)$ with $F^* = -12$ and $f^* = 8$, and only Case 2 is feasible.

## 3.5 Branch-and-Bound Algorithm for Linear Bi-level Programming

Before presenting the Branch-and-bound algorithm (Bard and Moore 1990), we first introduce some additional notations.

---

**Algorithm 3.2: Branch-and-bound Algorithm**

[Begin]

**Step 0:** (Initialization) Set $k = 0$, $S_k^+ = \emptyset$, $S_k^- = \emptyset$, $S_k^0 = \{1, ..., q + m\}$, and $\underline{F} = -\infty$.

**Step 1:** (Iteration $k$) Set $g_i = 0$ for $i \in S_k^-$ and $w_i = 0$ for $i \in S_k^+$. Attempt to solve the problem (3.7) without (3.7d). If the resultant problem is infeasible, go to Step 5; Otherwise, put $k \leftarrow k + 1$ and label the solution $(x^k, y^k, w^k)$.

**Step 2:** (Fathoming) If $F(x^k, y^k) \leq \underline{F}$, go to Step 5.

**Step 3:** (Branching) If $w_i^k g_i(x^k, y^k) = 0, i = 1, ..., q + m$, go to Step 4. Otherwise, select $i$ for which $w_i^k g_i(x^k, y^k) \neq 0$ is the largest and label it $i_1$. Put $S_k^+ \leftarrow S_k^+ \cup \{i_1\}$, $S_k^0 \leftarrow S_k^0 \setminus \{i_1\}$, $S_k^- \leftarrow S_k^-$, append $i_1$ to $P_k$, and go to Step 1.

**Step 4:** (Updating) $\underline{F} = F(x^k, y^k)$.

**Step 5:** (Backtracking) If no live node exists, go to Step 6. Otherwise, branch to the newest live vertex and update $S_k^+$, $S_k^-$, $S_k^0$ and $P_k$. Go to Step 1.

**Step 6:** (Termination) If $\underline{F} = -\infty$, there is no feasible solution to (3.1). Otherwise, declare that the feasible point $(x^k, y^k)$ associated with $\underline{F}$ is an optimal solution to (3.1).

[End]

---

Let $w = (u, v)$ and $W = \{1, \ldots, q + m\}$ be the index set for the terms in formula (3.7d), and let $\underline{F}$ be the incumbent lower bound on the leader's objective function. At the $k$th level of the search tree we define a subset of indices $W_k \subset W$, and a path $P_k$ corresponding to an assignment of either $w_i = 0$ or $g_i = 0$ for $i \in W_k$. Now let

$$
\begin{aligned}
S_k^+ &= \{i | i \in W_k, w_i = 0\}, \\
S_k^- &= \{i | i \in W_k, g_i = 0\}, \\
S_k^0 &= \{i | i \notin W_k\}.
\end{aligned}
$$

For $i \in S_k^0$, the variables $w_i$ or $g_i$ are free to assume any nonnegative value in the solution of problem (3.7) with (3.7d) omitted, so complementary slackness (3.7d) will not necessarily be satisfied.

We use the Branch-and-bound algorithm to solve Example 3.1.

$$g_1(x, y) = -3x + 2y + 4 \geq 0, \tag{3.13a}$$

$$g_2(x, y) = -2x - y + 12 \geq 0, \tag{3.13b}$$

$$g_3(x, y) = y \geq 0. \tag{3.13c}$$

From the problem (3.7a–3.7f), we have

$$\min_{x, y, u_1, u_2, u_3} \quad x - 4y \tag{3.14a}$$

$$\text{s.t.} \ -x - y \leq -3, \tag{3.14b}$$

$$-2x + y \leq 0, \tag{3.14c}$$

$$-3x + 2y \geq -4, \tag{3.14d}$$

$$2x + y \leq 12, \tag{3.14e}$$

$$-2u_1 + u_2 - u_3 = -1, \tag{3.14f}$$

$$u_1 g_1(x, y) + u_2 g_2(x, y) + u_3 g_3(x, y) = 0, \tag{3.14g}$$

$$x, y, u_1, u_2, u_3 \geq 0. \tag{3.14h}$$

Finally we obtain the following linear programming problem with (3.14g) omitted.

$$\min_{x, y, u_1, u_2, u_3} \quad x - 4y \tag{3.15a}$$

$$\text{s.t.} \quad -x - y \leq -3, \tag{3.15b}$$

$$-2x + y \leq 0, \tag{3.15c}$$

$$-3x + 2y \geq -4, \tag{3.15d}$$

$$2x + y \leq 12, \tag{3.15e}$$

$$-2u_1 + u_2 - u_3 = -1, \tag{3.15f}$$

$$x, y, u_1, u_2, u_3 \geq 0. \tag{3.15g}$$

At each iteration, the check is made to see if the following condition:

$$u_1 g_1(x, y) + u_2 g_2(x, y) + u_3 g_3(x, y) = 0 \tag{3.16}$$

is satisfied.

After initializing the data, the algorithm finds a feasible solution to the KKT representation with the complementary slackness conditions omitted and proceeds to Step 3.

The current point $x^1 = 3$, $y^1 = 6$, $u^1 = (0, 0, 1)$ with $F(x^1, y^1) = -21$ does not satisfy complementarity so a branching variable $u_3$ is selected and the index sets are updated, giving $S_1^+ = \{3\}, S_1^- = \emptyset, S_1^0 = \{1, 2\}$ and $P_1 = \{3\}$.

In the next iteration, the algorithm branches on $u_1$. Now, two levels down in the tree, the current sub-problem at Step 1 turns out to be infeasible so the algorithm goes to Step 5 and backtracks. The index sets are $S_2^+ = \{3\}, S_2^- = \{1\}, S_2^0 = \{2\}$ and $P_2 = \{3, 1\}$.

Go to Step 1, a feasible solution is found. At Step 2, however, the value of the leader's objective function is less than the incumbent lower bound, so the algorithm goes to Step 5 and backtracks, giving $S_3^+ = \emptyset, S_3^- = \{3\}, S_3^0 = \{1, 2\}$ and $P_3 = \{3\}$.

The current sub-problem at Step 1 turns out to be infeasible so the algorithm goes to Step 5 and backtracks. However, no live vertices exist. We have found an optimal solution, occurring at the point $(x^*, y^*) = (4, 4)$, $(u_1^*, u_2^*, u_3^*) = (0.5, 0, 0)$ with $F^* = -12$ and $f^* = 8$.

## 3.6 Penalty Function Method for Linear Bi-level Programming

For the linear bi-level programming problem (3.1a–3.1d) with $X = \{x | x \geq 0\}$ and $Y = \{y | y \geq 0\}$, ignoring the constant term $c_2 x$, the dual of the follower's problem is

$$\max_u \quad u(A_2 x - b_2) \tag{3.17a}$$

$$\text{s.t.} \quad u B_2 \geq -d_2, \tag{3.17b}$$

$$u \geq 0. \tag{3.17c}$$

Given $x$, $y$ solve problem (3.1c, 3.1d) if and only if there exists $u$ such that it satisfies (3.17b, 3.17c) and $\pi(x, y, u) = d_2 y - u(A_2 x - b_2) = 0$. Thus, it can be used to formulate a penalty function method for solving problem (3.1a–3.1d) (Campelo et al. 2000):

$$\min_{x,y,u} \quad c_1 x + d_1 y + K[d_2 y - u(A_2 x - b_2)] \tag{3.18a}$$

$$\text{s.t.} \quad A_1 x + B_1 y \leq b_1, \tag{3.18b}$$

$$u B_2 \geq -d_2, \tag{3.18c}$$

$$A_2 x + B_2 y \leq b_2, \tag{3.18d}$$

$$x, y, u \geq 0, \tag{3.18e}$$

where $K \in R_+$ is called the penalty parameter.

---

**Algorithm 3.3: Penalty Function Method**

[Begin]

**Step 0:** Choose $K$ (large) and $\eta > 1$.

**Step 1:** Obtain an optimal solution $(x^k, y^k, u^k)$ to problem (3.18) using a classical optimization method such as the trust region method or Newton-Raphson numerical approach.

**Step 2:** If $\pi(x^k, y^k, u^k) = 0$, then stop, and $(x^k, y^k)$ is an optimal solution to problem (3.1). Otherwise, $K = \eta K$, and go to Step 1.

[End]

---

The difficulty with this approach is that for each value of the penalty parameter, the nonconvex problem (3.18a–3.18e) has to be solved globally. Fortunately, Campelo et al. (2000) used a decomposition approach to solve this problem. Furthermore, under the assumptions, i.e., the following problem

$$\min_{(x,y) \in S} \quad F(x, y) = c_1 x + d_1 y$$

has an optimal solution, and $U = \{u | u B_2 \geq -d_2, u \geq 0\}$ is not empty, the following results are obtained.

**Theorem 3.4** (Campelo et al. 2000) There exists a finite value $K^* \in R_+$ of $K$ for which an optimal solution to the penalty function problem (3.18a–3.18e) yields an optimal solution to the problem (3.1a–3.1d) for all $K \geq K^*$.

## 3.7 Multi-level Programming Model

A *multi-level programming* (MLP) problem is identified as a kind of mathematical programming that solves decentralized planning problems with decision makers in multi-level or hierarchical systems. The execution of decisions is sequential, from top to middle levels and then to the bottom level. Each decision maker independently maximizes (or minimizes) their own objective but is affected by the actions of other decision makers at the different levels. Generally, the MLP problem can be formulated as follows:

$$
\min_{x_1 \in X_1} \quad f_1(x_1, x_2, \ldots, x_n)
$$
$$
\text{s.t.} \quad g_1(x_1, x_2, \ldots, x_n) \leq 0,
$$
$$
\min_{x_2 \in X_2} \quad f_2(x_1, x_2, \ldots, x_n)
$$
$$
\text{s.t.} \quad g_2(x_1, x_2, \ldots, x_n) \leq 0,
$$
$$
\vdots
$$
$$
\min_{x_n \in X_n} \quad f_n(x_1, x_2, \ldots, x_n)
$$
$$
\text{s.t.} \quad g_n(x_1, x_2, \ldots, x_n) \leq 0, \tag{3.19}
$$

where $x_i \in X_i \subset R^{m_i}$, $f_i : R^{m_1} \times R^{m_2} \times \cdots \times R^{m_n} \to R$ and $g_i : R^{m_1} \times R^{m_2} \times \cdots \times R^{m_n} \to R^{m_i}$ ($i = 1, 2, \ldots, n$) are continuous and twice differentiable functions.

Clearly, when $n = 2$, the MLP problem (3.19) becomes a bi-level programming problem. Based on the definitions of the bi-level programming given in Sect. 3.2, we give the definitions of the multi-level programming problem (3.19) as follows.

**Definition 3.4**

(a) Constraint region of the $i$th level programming problem:

$$
J_i(x_1, x_2, \ldots, x_{i-1}) = \left\{ (x_i, x_{i+1}, \ldots, x_n) \mid x_j \in X_j, g_j(x_1, x_2, \ldots, x_n) \leq 0, i \leq j \leq n \right\}.
$$

(b) Feasible set for the $i$th level programming problem:

$$
S_i(x_1, x_2, \ldots, x_{i-1}) = \{ (x_i, x_{i+1}, \ldots, x_n) \in J_i(x_1, x_2, \ldots, x_{i-1}) \mid (x_{i+1}, \ldots, x_n) \in \text{argmin}
$$
$$
\{ f_{i+1}(x_1, x_2, \ldots, x_n) \mid (x_{i+1}, \ldots, x_n) \in S_{i+1}(x_1, x_2, \ldots, x_i) \} \}
$$

Note that, the feasible set of the $n$th level programming problem, i.e., $S_n(x_1, x_2, \ldots, x_{n-1})$ is equivalent to the set $J_n(x_1, x_2, \ldots, x_{n-1})$.

(c)  Inducible region:

$$IR = \{(x_1, x_2, \ldots, x_n) | x_1 \in X_1, (x_2, x_3, \ldots, x_n) \in S_2(x_1), g_1(x_1, x_2, \ldots, x_n) \leq 0\}.$$

Finally, problem (3.19) is equivalent to the following problem:

$$\min_{(x_1, x_2, \ldots, x_n) \in IR} f_1(x_1, x_2, \ldots, x_n).$$

Some authors have discussed the optimality and geometric properties of the linear MLP problem. For details, see Bard (1985), Benson(1989), Ruan et al. (2004). The methods for solving problem (3.19) are mainly limited to the linear *tri-level programming* (TLP) problem because it reflects the main features of MLP. Note that the tri-level programming problem is a typical form of MLP. In a tri-level decision model, the decision entity at the top level is called the leader, and entities at the middle and bottom levels are its followers, but a decision entity at the middle level is also the leader if it has followers at the bottom level. The basic tri-level programming problem can be described as follows:

$$\min_{x_1 \in X_1} f_1(x_1, x_2, x_3)$$

$$\text{s.t. } g_1(x_1, x_2, x_3) \leq 0,$$

$$\min_{x_2 \in X_2} f_2(x_1, x_2, x_3)$$

$$\text{s.t. } g_2(x_1, x_2, x_3) \leq 0,$$

$$\min_{x_3 \in X_3} f_3(x_1, x_2, x_3)$$

$$\text{s.t. } g_3(x_1, x_2, x_3) \leq 0.$$

The details about tri-level programming and its solution methods are referred to Chap. 6 of this book.

## 3.8  Summary

This chapter introduces the basic concepts, models, solution definitions, and solution algorithms of bi-level programming and basic models of multi-level/tri-level programming. These knowledge and approaches, particularly the $K$th-Best algorithm, the Kuhn-Tucker approach, and the Branch-and-bound algorithm introduced in Sects. 3.3–3.5, will be used for the development of the approaches for dealing with fuzzy bi-level/multi-level programming, bi-level/multi-level multi-follower programming and other more complex decision situations in subsequent chapters of this book.

# Part II
# Multi-level Multi-follower Decision Making

# Chapter 4
# Bi-level Multi-follower Decision Making

A bi-level decision problem may involve multiple decision entities (decision units or decision makers) at the lower level, and these followers may have different reactions for a possible decision made by the leader. This is called a bi-level multi-follower (BLMF) decision problem. These followers may have different relationships, such as sharing decision variables, objectives and/or constraints for a possible decision made by the leader; the leader's decision will be affected, not only by the reactions of individual followers, but also by the relationships among these followers. We therefore need to establish different models to describe these situations in which followers have different relationships. We also need to develop solution methods to solve these decision situations.

In this chapter, we first identify the BLMF issue in Sect. 4.1 and then present a framework in Sect. 4.2 for the BLMF decision problem, which defines nine different kinds of relationship (Situations $S1$ to $S9$) amongst the followers. Under this BLMF framework, for each of the nine kinds of relationship, a corresponding BLMF decision model is developed in Sect. 4.3 using two BLMF modeling approaches: the decision entity-relationship diagrams (DERD) approach and the programming approach. For these BLMF decision models, we give related solution concepts and solution methods. Since some models are similar in solution methods, we select only three typical models: the uncooperative model $S1$, the semi-cooperative model $S5$ and the reference-uncooperative model $S9$ to provide more details. Related solution concepts, the existence of solutions to these models, a set of BLMF $K$th-Best algorithms and BLMF Kuhn-Tucker approaches for solving the three models are presented in Sects. 4.4, 4.5 and 4.6 respectively. In addition, some numerical examples are adopted to illustrate the feasibility of these models, algorithms and approaches. Discussions and further remarks are given in Sect. 4.7.

## 4.1 Problem Identification

Although much research on bi-level decision-making has been carried out, the existing techniques have been mainly focused on a specific situation comprising one leader and one follower. However, in a real-world bi-level decision problem,

© Springer-Verlag Berlin Heidelberg 2015
G. Zhang et al., *Multi-Level Decision Making*,
Intelligent Systems Reference Library 82, DOI 10.1007/978-3-662-46059-7_4

the lower level may involve multiple decision entities. The leader's decision is therefore affected by the objectives and strategies of the multiple followers. For each possible decision of the leader, those followers may have their own, different reactions. The relationships between these followers can vary. They may or may not share their decision variables. They may have individual objectives and constraints but interact with others cooperatively, or may have common objectives or common constraints. We can illustrate these situations through an example.

*Example 4.1* A university has five faculties. The university (as the leader) aims to improve its research quality through new research development strategies. The strategies made at university level directly affect the research strategies made in its faculties (as followers). In the meantime, the reactions at the faculty level may affect the research development strategies sought by the university. Each faculty wishes to optimize its individual research development objective in view of the partial control exercised at university level. Some faculties make research income the objective and some set the number of publications; some faculties have lab limitations (constraint) and some do not. The university's decision makers can control this effect by exercising pre-emptive partial control over the university through budget modifications or regulations, but are subject to possible reactions from the faculties.

This decision problem is a bi-level multi-follower decision problem, otherwise called a bi-level decentralized optimization problem.


## 4.2  Framework for Bi-level Multi-follower Decision Making

Different reactions by followers to each possible action conducted by the leader might be generated when multiple followers are involved in a bi-level decision-making problem. Moreover, different kinds of relationships between these followers could cause multiple different processes for deriving an optimal solution for the leader's decision-making. Therefore, the leader's decision will not only be affected by the reactions from followers, but also by the relationships between these followers.

Basically, there are four main kinds of relationship between followers which are determined by the form of sharing of decision variables (Lu et al. 2006).

1. *The uncooperative situation*: there is no sharing of decision variables between the followers. In such a situation, there are obviously neither shared objective functions nor shared constraint conditions among the followers.
2. *The cooperative situation*: the followers totally share the decision variables, objective functions and constraints.
3. *The semi-cooperative situation*: the followers share the decision variables but may have individual objective functions and constraints. There are several different sub-cases within this semi-cooperative situation which are determined

by the relationships between the objective functions and constraints of the followers. Each follower may have an individual objective, whatever constraints are shared. For example, the Computer Science Department has an objective of maximizing its income from international students' fees, and the Engineering Department's objective is to maximize total student number, to satisfy the university's policy on education development. The two followers have different objectives and may also have different constraints but share all decision variables, such as marketing investment.

4. *The reference-uncooperative situation*: the followers have individual decision variables but take other followers' variables as references when making their own decisions. Four sub-cases are involved within this case that involve individual objective functions respective or irrespective of constraints, and common objective function respective or irrespective of constraints.

Based on the four basic situations (cases) of relationships amongst followers, determined by decision variable (the first relationship factor as shown in Table 4.1), and their various sub-cases determined by two other relationship factors, objective functions and constraints, nine different kinds of situation in total are identified among the followers. These are named $S1, S2, \ldots, S9$. A framework is established to describe these cases in Table 4.1.

Each situation shown in the framework requires a specific BLMF decision model for description. We need to point out that these nine BLMF models are basic models, based on which we can establish hybrid BLMF models in which, for example, followers 1 and 2 have a cooperative relationship while followers 3 and 4 have an uncooperative relationship.

Based on the general linear bi-level decision-making model proposed in Chap. 3, the models of the nine specific BLMF decision situations shown in Table 4.1 are presented in the following Sect. 4.3.

**Table 4.1**  A framework for bi-level multi-follower decision-making

| Relationships among followers | Relationship factor | | | Situation ($Si$) |
|---|---|---|---|---|
| | Decision variables | Objectives | Constraints | |
| Uncooperative | Individual | Individual | Individual | $S1$ |
| Cooperative | Sharing | Sharing | Sharing | $S2$ |
| Semi-cooperative | Sharing | Sharing | Individual | $S3$ |
| | | Individual | Sharing | $S4$ |
| | | | Individual | $S5$ |
| Reference-uncooperative | Individual, but take other followers' variables as references | Sharing | Sharing | $S6$ |
| | | | Individual | $S7$ |
| | | Individual | Sharing | $S8$ |
| | | | Individual | $S9$ |

## 4.3  Bi-level Multi-follower Decision Models

This section introduces two modeling approaches for BLMF decision problems: the *decision entity-relationship diagrams* (DERD) and the programming approach. The DERD approach is a conceptual modeling approach.

### 4.3.1  BLMF Decision Entity-Relationship Diagram

In order to identify and classify BLMF decision problems, we first introduce the following concepts (Lu et al. 2012).

1. *Neighborhood entities* (*NES*): two decision entities, also called neighborhood followers, are at the same level and led by the same decision entity (leader).
2. *Cooperative entities*: two neighborhood entities share their decision variables and have the same objective function and constraints.
3. *Semi-cooperative entities*: two neighborhood entities share their decision variables but have individual objective functions and constraints.
4. *Uncooperative entities*: two neighborhood entities have individual decision variables, objective functions, and constraints.
5. *Reference-uncooperative entities*: two neighborhood entities have individual decision variables, objectives and constraints but take others' decision variables into consideration as references, that is, they include others' decision variables in their objective functions or constraints, but not as control variables.

The five concepts identified above fully reflect the features of a BLMF decision problem and any combinations of these features.

Concept modeling is a very important method in modeling a decision problem and diagrams are powerful tools for describing such models. We therefore develop a DERD approach and use it to model the BLMF decision problem. Figure 4.1 presents diagrammatic notations.

This DERD approach truly reflects real-world decision problems and is easy to use. In the following sections, we will show how a linear BLMF decision problem is described by DERD and also is described by bi-level programming.

### 4.3.2  Linear BLMF Decision Models

This section will describe the nine BLMF decision models identified in the framework of BLMF decision-making (Table 4.1) by using both DERD and programming approaches (Lu et al. 2006).

**Fig. 4.1** Notations for BLMF decision entity-relationship diagrams (DERD)

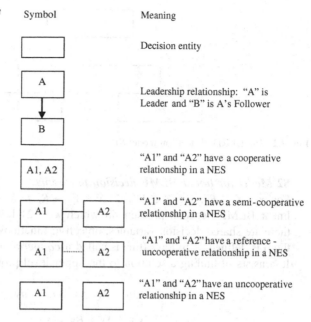

| Symbol | Meaning |
| --- | --- |
| | Decision entity |
| A → B | Leadership relationship: "A" is Leader and "B" is A's Follower |
| A1, A2 | "A1" and "A2" have a cooperative relationship in a NES |
| A1 — A2 | "A1" and "A2" have a semi-cooperative relationship in a NES |
| A1 ···· A2 | "A1" and "A2" have a reference - uncooperative relationship in a NES |
| A1   A2 | "A1" and "A2" have an uncooperative relationship in a NES |

## 1. *S1 Model for linear BLMF decision problems*

For $x \in X \subset R^n$, $y_i \in Y_i \subset R^{m_i}$, $F: X \times Y_1 \times \cdots \times Y_k \to R$, $f_i: X \times Y_i \to R$, $i = 1, \ldots, k$, a linear BLMF decision problem in which $k(\geq 2)$ followers are involved and there are no shared decision variables, objective functions or constraints between them is defined as follows (this is called an *uncooperative BLMF decision model*).

It consists of finding a solution to the upper level problem:

$$\min_{x \in X} \quad F(x, y_1, \ldots, y_k) = cx + \sum_{i=1}^{k} d_i y_i$$

$$\text{s.t.} \quad Ax + \sum_{i=1}^{k} B_i y_i \leq b,$$

where $y_i$ is the solution to the $i$th follower's problem:

$$\min_{y_i \in Y_i} \quad f_i(x, y_i) = c_i x + e_i y_i$$

$$\text{s.t.} \quad A_i x + C_i y_i \leq b_i,$$

where $c, c_i \in R^n, d_i, e_i \in R^{m_i}, A \in R^{p \times n}$, $B_i \in R^{p \times m_i}, b \in R^p, A_i \in R^{q_i \times n}$, $C_i \in R^{q \times m_i}, b_i \in R^q, i = 1, \ldots, k$.

The uncooperative BLMF decision model is described by DERD in Fig. 4.2.

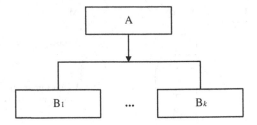

**Fig. 4.2** The DERD of decision model $S1$

2. **$S2$ *Model for linear BLMF decision problems***
   For $x \in X \subset R^n$, $y \in Y \subset R^m$, $F: X \times Y \to R$, $f_i: X \times Y \to R$, $i = 1, \ldots, k$, a linear BLMF decision problem in which $k(\geq 2)$ followers are involved and there are shared decision variables, objective functions and constraints between them is defined as follows (this is called a *cooperative BLMF decision model*). It consists of finding a solution to the upper level problem:

$$\min_{x \in X} \quad F(x, y) = cx + dy$$

$$\text{s.t.} \quad Ax + By \leq b,$$

where $y$ is the solution to the $i$th follower's problem

$$\min_{y \in Y} \quad f_i(x, y) = c'x + e'y$$

$$\text{s.t.} \quad A'x + C'y \leq b',$$

where $c, c' \in R^n, d, e' \in R^m, A \in R^{p \times n}, B \in R^{p \times m}, b \in R^p, A' \in R^{q \times n}, C' \in R^{q \times m}$, $b' \in R^q, i = 1, \ldots, k$.

The cooperative BLMF decision model is described by DERD in Fig. 4.3.

3. **$S3$ *Model for linear BLMF decision problems***
   For $x \in X \subset R^n$, $y \in Y_i \subset R^m$, $F: X \times Y_i \to R$, $f_i: X \times Y_i \to R$, $i = 1, \ldots, k$, a linear BLMF decision problem in which $k(\geq 2)$ followers are involved and

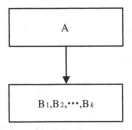

**Fig. 4.3** The DERD of decision model $S2$

there are shared decision variables (which means that followers control the same decision variable $y$) and objective functions but separate constraints (which implies that each follower has individual constraints) between them is defined as follows (this is called a *semi-cooperative BLMF decision model*).
It consists of finding a solution to the upper level problem:

$$\min_{x \in X} \quad F(x,y) = cx + dy$$

$$\text{s.t.} \quad Ax + By \le b,$$

where $y$ is the solution to the $i$th follower's problem

$$\min_{y \in Y_i} \quad f_i(x,y) = c'x + e'y$$

$$\text{s.t.} \quad A_ix + C_iy \le b_i,$$

where $c, c' \in R^n, d, e' \in R^m, A \in R^{p \times n}, \quad B \in R^{p \times m}, b \in R^p, A_i \in R^{q_i \times n}, \quad C_i \in R^{q_i \times m}, b_i \in R^{q_i}, i = 1, \ldots, k.$

The semi-cooperative BLMF decision model is described by DERD in Fig. 4.4.

4. **S4 *Model for linear BLMF decision problems***
For $x \in X \subset R^n$, $y \in Y \subset R^m$, $F: X \times Y \to R$, $f_i: X \times Y \to R$, $i = 1, \ldots, k$, a linear BLMF decision problem in which $k(\ge 2)$ followers are involved and there are shared decision variables (which means that followers control the same decision variable $y$) and constraints but separate objective functions between them is defined as follows (this is also called a *semi-cooperative BLMF decision model*).
It consists of finding a solution to the upper level problem:

$$\min_{x \in X} \quad F(x,y) = cx + dy$$

$$\text{s.t.} \quad Ax + By \le b,$$

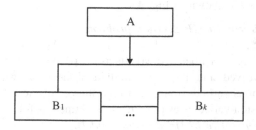

**Fig. 4.4** The DERD of decision model $S3$

where $y$ is the solution to the $i$th follower's problem

$$\min_{y \in Y_i} \quad f_i(x, y) = c_i x + e_i y$$

$$\text{s.t.} \quad A'x + C'y \le b',$$

where $c, c_i \in R^n, d, e_i \in R^m, A \in R^{p \times n}, \quad B \in R^{p \times m}, b \in R^p, A' \in R^{q \times n}, \quad C' \in R^{q \times m}, b' \in R^q, i = 1, \ldots, k.$

This semi-cooperative BLMF decision model $S4$ can be described by DERD in the same way as $S3$, shown in Fig. 4.4.

5. ***S5 Model for linear BLMF decision problems***
   For $x \in X \subset R^n$, $y \in Y_i \subset R^m$, $F: X \times Y_i \to R$, $f_i: X \times Y_i \to R$, $i = 1, \ldots, k$, a linear BLMF decision problem in which $k(\ge 2)$ followers are involved and there are shared decision variables (which means that followers control the same decision variable $y$) but separate objective and constraints (which means that an individual constraint exists for all followers) between them is defined as follows (this is also called a *semi-cooperative BLMF decision model*).
   It consists of finding a solution to the upper level problem:

$$\min_{x \in X} \quad F(x, y) = cx + dy$$

$$\text{s.t.} \quad Ax + By \le b,$$

where $y$ is the solution to the $i$th follower's problem

$$\min_{y \in Y_i} \quad f_i(x, y) = c_i x + e_i y$$

$$\text{s.t.} \quad A_i x + C_i y \le b_i,$$

where $c, c_i \in R^n, d, e_i \in R^m, A \in R^{p \times n}, \quad B \in R^{p \times m}, b \in R^p, A_i \in R^{q_i \times n}, \quad C_i \in R^{q_i \times m}, b_i \in R^{q_i}, i = 1, \ldots, k.$

This semi-cooperative BLMF decision model $S5$ can be described by DERD in the same way as $S3$, shown in Fig. 4.4.

6. ***S6 Model for linear BLMF decision problems***
   For $x \in X \subset R^n$, $y_i \in Y_i \subset R^{m_i}$, $F: X \times Y_1 \times \cdots \times Y_k \to R$, $f_i: X \times Y_1 \times \cdots \times Y_k \to R, i = 1, \ldots, k$, a linear BLMF decision problem in which $k(\ge 2)$ followers are involved and there are individual decision variables in shared objective functions and constraints between them, but the followers take other followers' decision variables as references, is defined as follows (this is called a *reference-uncooperative BLMF decision model*).

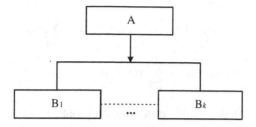

**Fig. 4.5** The DERD of decision model $S6$

It consists of finding a solution to the upper level problem:

$$\min_{x \in X} \quad F(x, y_1, \ldots, y_k) = cx + \sum_{i=1}^{k} d_i y_i$$

$$\text{s.t.} \quad Ax + \sum_{i=1}^{k} B_i y_i \leq b,$$

where $y_i$ is the solution to the $i$th follower's problem

$$\min_{y_i \in Y_i} \quad f_i(x, y_1, \ldots, y_k) = c'x + \sum_{i=1}^{k} e_i y_i$$

$$\text{s.t.} \quad A'x + \sum_{i=1}^{k} C_i y_i \leq b',$$

where $c, c' \in R^n, d_i, e_i \in R^{m_i}, A \in R^{p \times n}, \quad B_i \in R^{p \times m_i}, b \in R^p, A' \in R^{q \times n}, \quad C_i \in R^{q \times m_i}, b' \in R^q, i = 1, \ldots, k.$

Note that the shared objective functions for all the followers only imply that their mathematical expressions are the same as one another's because each follower has individual decision variables. The reference-uncooperative BLMF decision model is described by DERD in Fig. 4.5.

7. *S7 Model for linear BLMF decision problems*
For $x \in X \subset R^n$, $y_i \in Y_i \subset R^{m_i}$, $F: X \times Y_1 \times \cdots \times Y_k \to R$, $f_i: X \times Y_1 \times \cdots \times Y_k \to R, i = 1, \ldots, k$, a linear BLMF decision problem in which $k (\geq 2)$ followers are involved and there are individual decision variables in shared objective functions and separate constraints between them, but the followers take other followers' decision variables as references, is defined as follows (this is also called a *reference-uncooperative BLMF decision model*).
It consists of finding a solution to the upper level problem:

$$\min_{x \in X} \quad F(x, y_1, \ldots, y_k) = cx + \sum_{i=1}^{k} d_i y_i$$

$$\text{s.t.} \quad Ax + \sum_{i=1}^{k} B_i y_i \leq b,$$

where $y_i$ is the solution to the $i$th follower's problem

$$\min_{y_i \in Y_i} \quad f_i(x, y_1, \ldots, y_k) = c'x + \sum_{i=1}^{k} e_i y_i$$

$$\text{s.t.} \quad A_i x + \sum_{s=1}^{k} C_{is} y_s \leq b_i,$$

where $c, c' \in R^n, d_i, e_i \in R^{m_i}, A \in R^{p \times n}$, $B_i \in R^{p \times m_i}, b \in R^p, A_i \in R^{q_i \times n}$, $C_{is} \in R^{q_i \times m_s}$, $b_i \in R^{q_i}, i, s = 1, \ldots, k$.

This reference-uncooperative BLMF decision model can be described by DERD in the same way as $S6$, shown in Fig. 4.5.

8. **$S8$ Model for linear BLMF decision problems**
   For $x \in X \subset R^n$, $y_i \in Y_i \subset R^{m_i}$, $F: X \times Y_1 \times \cdots \times Y_k \to R$, $f_i: X \times Y_1 \times \cdots \times Y_k, \to R, i = 1, \ldots, k$, a linear BLMF decision problem in which $k (\geq 2)$ followers are involved and there are individual decision variables in separate objective functions and shared constraints between them, but the followers take other followers' decision variables as references, is defined as follows (this is also called a *reference-uncooperative BLMF decision model*).
   It consists of finding a solution to the upper level problem:

$$\min_{x \in X} \quad F(x, y_1, \ldots, y_k) = cx + \sum_{i=1}^{k} d_i y_i$$

$$\text{s.t.} \quad Ax + \sum_{i=1}^{k} B_i y_i \leq b,$$

where $y_i$ is the solution to the $i$th follower's problem

$$\min_{y_i \in Y_i} \quad f_i(x, y_1, \ldots, y_k) = c_i x + \sum_{s=1}^{k} e_{is} y_s$$

$$\text{s.t.} \quad A'x + \sum_{i=1}^{k} C_i y_i \leq b',$$

where $c, c_i \in R^n, d_i \in R^{m_i}, e_{is} \in R^{m_s}, A \in R^{p \times n}, \ B_i \in R^{p \times m_i}, b \in R^p, A' \in R^{q \times n}, C_i \in R^{q \times m_i}, b' \in R^q, i = 1, \ldots, k.$

This reference-uncooperative BLMF decision model can be described by DERD in the same way as $S6$, shown in Fig. 4.5.

9. **$S9$ Model for linear BLMF decision problems**

For $x \in X \subset R^n$, $y_i \in Y_i \subset R^{m_i}$, $F: X \times Y_1 \times \cdots \times Y_k \to R$, $f_i: X \times Y_1 \times \cdots \times Y_k \to R, i = 1, \ldots, k$, a linear BLMF decision problem in which $k(\geq 2)$ followers are involved and there are individual decision variables in separate objective functions and constraints between them, but the followers take other followers' decision variables as references, is defined as follows (this is also called a reference-uncooperative BLMF decision model).

It consists of finding a solution to the upper level problem:

$$\min_{x \in X} \quad F(x, y_1, \ldots, y_k) = cx + \sum_{i=1}^{k} d_i y_i$$

$$\text{s.t.} \quad Ax + \sum_{i=1}^{k} B_i y_i \leq b,$$

where $y_i$ is the solution to the $i$th follower's problem

$$\min_{y_i \in Y_i} \quad f_i(x, y_1, \ldots, y_k) = c_i x + \sum_{s=1}^{k} e_{is} y_s$$

$$\text{s.t.} \quad A_i x + \sum_{s=1}^{k} C_{is} y_s \leq b_i,$$

where $c, c_i \in R^n, d_i \in R^{m_i}, e_{is} \in R^{m_s}, A \in R^{p \times n}, \ B_i \in R^{p \times m_i}, b \in R^p, A_i \in R^{q_i \times n}, C_{is} \in R^{q_i \times m_s}, b_i \in R^{q_i}, i, s = 1, \ldots, k.$

This reference-uncooperative BLMF decision model can be described by DERD in the same way as $S6$, shown in Fig. 4.5.

The above BLMF decision models define the nine basic kinds of relationship between the followers described in the framework of the BLMF decision problem. These models clearly require individual definitions for optimal solutions, and related solution approaches to derive the optimal solutions. In the following sections, we will introduce different solution approaches to these BLMF decision models.

We need to point out that for any of the nine models, it is assumed that the leader has full knowledge of the objective functions and constraints of these followers and the relationships between these objective functions and constraints. The control for decision variables is partitioned between the leader and the followers. The leader must anticipate all possible responses of followers based on their relationships.

## 4.4 Uncooperative Bi-level Multi-follower Decision Making

The uncooperative situation model ($S1$) is the most basic form of all nine decision models for BLMF decision problems. This section focuses on this model by giving the definition for an optimal solution, related theorems and solution algorithms for solving the BLMF decision model S1 (Calvete and Galé 2007; Shi et al. 2005).

### 4.4.1 Solution Concepts

**Definition 4.1**

(a)  Constraint region of a linear BLMF decision problem:

$$S = \{(x, y_1, \ldots, y_k) \in X \times Y_1 \times \cdots \times Y_k | Ax$$

$$+ \sum_{i=1}^{k} B_i y_i \leq b, A_i x + C_i y_i \leq b_i, i = 1, \ldots, k \}.$$

The constraint region of the linear BLFM programming problem refers to all possible combinations of choices that the leader and followers may make.

(b)  Feasible set of the $i$th follower for $\forall x \in X$:

$$S_i(x) = \{y_i \in Y_i | C_i y_i \leq b_i - A_i x\}.$$

The feasible region for the follower is affected by the leader's choice and the allowable choices of each follower are the elements of $S_i(x)$.

(c)  Projection of $S$ onto the leader's decision space:

$$S(X) = \{x \in X | \exists (y_1, \ldots, y_k) \in Y_1 \times \cdots \times Y_k, (x, y_1, \ldots, y_k) \in S\}.$$

Unlike the rules in uncooperative game theory, in which each player must choose a strategy simultaneously, the definition of the BLMF model requires that the leader moves first by selecting an $x$ in attempt to minimize its objective subject to the constraints of the upper and each lower level.

(d)  The $i$th follower's rational reaction set for $x \in S(X)$:

$$P_i(x) = \{y_i \in Y_i | y_i \in \text{argmin}[f_i(x, z) | z \in S_i(x)]\},$$

where $\text{argmin}[f_i(x, z) | z \in S_i(x)] = \{y_i \in S_i(x) | f_i(x, y_i) \leq f_i(x, z), \forall z \in S_i(x)\}$. The followers observe the leader's action and simultaneously react by selecting from its feasible set to minimize its objective function.

(e)  Inducible region:

$$IR = \{(x, y_1, \ldots, y_k) | (x, y_1, \ldots, y_k) \in S, y_i \in P_i(x), i = 1, \ldots, k\}.$$

Thus in terms of the above notations, the linear uncooperative BLMF problem can be written as

$$\min\{F(x, y_1, \ldots, y_k) | (x, y_1, \ldots, y_k) \in IR\}. \qquad (4.1)$$

For the sake of assuring an optimal solution to the model $S1$, it is necessary that the solution to the problem (4.1) exists which implies that further assumptions about the uncooperative BLMF decision model should be satisfied.

**Assumption 4.1**

1. $S$ is non-empty and compact.
2. $IR$ is non-empty.
3. $P_i(x)$ is a point-to-point map with respect to $x$, where $i = 1, \ldots, k$.

**Theorem 4.1** *If the above assumptions are satisfied, there exists an optimal solution to the linear uncooperative BLMF decision model S1.*

*Proof* Since neither $S$ nor $IR$ is empty, there is at least one parameter value $x^* \in S(X)$ and $P_i(x^*) \neq \emptyset$. Consider a sequence $\{(x^t, y_1^t, \ldots, y_k^t)\}_{t=1}^{\infty} \subseteq IR$ converging to $(x^*, y_1^*, \ldots, y_k^*)$. Then, by the well-known results of linear parametric optimization, we have $y_1^* \in P_i(x^*)$. Hence, $(x^*, y_1^*, \ldots, y_k^*) \in IR$ shows that $IR$ is closed. By Assumption 4.1(1) and $IR \subseteq S$, $IR$ is also bounded, and $IR$ is non-empty, so the problem (4.1) consists of minimizing a continuous function over a compact non-empty set, which implies that the model $S1$ has an optimal solution. $\qquad\square$

## 4.4.2 Theoretical Properties

Several theoretical properties for the model $S1$ are presented to support the principle of algorithms for solving the uncooperative BLMF problem.

**Theorem 4.2** *The inducible region IR can be expressed equivalently as a piecewise linear equality constraint comprised of support hyperplanes of S.*

*Proof* First, we define

$$Q_i(x) = \min\{e_i \hat{y}_i | \hat{y}_i \in S_i(x)\}, \quad i = 1, \ldots, k.$$

Since $Q_i(x)$ can be seen as a linear programming problem with the parameter $x$, the dual problem of $Q_i(x)$ can be written as

$$\max\{(A_i x - b_i)u_i | C_i u_i \geq -e_i, u_i \geq 0\}, \quad i = 1, \ldots, k. \tag{4.2}$$

If both $Q_i(x)$ and problem (4.2) have feasible solutions, according to the dual theorem of linear programming, then both have optimal solutions and the same optimal objective function value. We know that a solution to problem (4.2) occurs at a vertex of its constraint region $U_i = \{u_i | C_i u_i \geq -e_i, u_i \geq 0\}, i = 1, \ldots, k$. Adopting $u_i^1, \ldots, u_i^{q_i}$ to express all the vertices of $U_i$, problem (4.2) can be rewritten as:

$$\max\{(A_i x - b_i)u_i^j | u_i^j \in \{u_i^1, \ldots, u_i^{q_i}\}\}, \quad i = 1, \ldots, k. \tag{4.3}$$

Evidently, $Q_i(x)$ is a piecewise linear function according to problem (4.3).

$$
\begin{aligned}
IR &= \{(x, y_1, \ldots, y_k) | (x, y_1, \ldots, y_k) \in S, y_i \in P_i(x), \quad i = 1, \ldots, k\} \\
&= \{(x, y_1, \ldots, y_k) \in S | e_i y_i = \min\{e_i \hat{y}_i | \hat{y}_i \in S_i(x)\}, \quad i = 1, \ldots, k\} \\
&= \{(x, y_1, \ldots, y_k) \in S | e_i y_i = Q_i(x), \quad i = 1, \ldots, k\}
\end{aligned}
\tag{4.4}
$$

and it can be seen that it is a piecewise linear equality constraint.                    □

**Corollary 4.1** *The uncooperative BLMF decision model S1 is equivalent to optimizing F over a feasible region comprised of a piecewise linear equality constraint.*

**Corollary 4.2** *An optimal solution to the uncooperative BLMF decision model S1 occurs at a vertex of IR.*

*Proof* According to the equivalent form (4.1) of the BLMF decision model, also, since $F(x, y_1, \ldots, y_k)$ is linear, an optimal solution to the problem must occur at a vertex of *IR* if it exists.                    □

**Theorem 4.3** *The optimal solution $(x^*, y_1^*, \ldots, y_k^*)$ to the uncooperative BLMF decision model S1 occurs at a vertex of S.*

*Proof* Let $(x^1, y_1^1, \ldots, y_k^1), \ldots, (x^t, y_1^t, \ldots, y_k^t)$ express the distinct vertices of $S$. Since any point in $S$ can be written as a convex combination of these vertices, we can get

$$\left(x^*, y_1^*, \ldots, y_k^*\right) = \sum_{r=1}^{\bar{t}} \delta_r \left(x^r, y_1^r, \ldots, y_k^r\right)$$

where $\sum_{r=1}^{\bar{t}} \delta_r = 1$, $\delta_r > 0$, $r = 1, \ldots, \bar{t}$ and $\bar{t} \leq t$.

Let us write the constraints of S1 in the piecewise linear form (4.4) discussed in Theorem 4.2:

$$0 = Q_i(x^*) - e_i y_i^* = Q_i\left(\sum_{r=1}^{\bar{t}} \delta_r x^r\right) - e_i \sum_{r=1}^{\bar{t}} \delta_r y_i^r, \quad i = 1, \ldots, k.$$

Because of the convexity of $Q_i(x)$, we have

$$0 \le \sum_{r=1}^{\bar{i}} \delta_r Q_i(x^r) - e_i \sum_{r=1}^{\bar{i}} \delta_r y_i^r = \sum_{r=1}^{\bar{i}} \delta_r \big( Q_i(x^r) - e_i y_i^r \big). \tag{4.5}$$

By the definition of $Q_i(x)$, we have:

$$Q_i(x^r) = \min e_i y_i \le e_i y_i^r, \quad r = 1, \ldots, \bar{i}, i = 1, \ldots, k.$$

Thus, $Q_i(x^r) - e_i y_i^r \le 0, r = 1, \ldots, \bar{i}, i = 1, \ldots, k.$

Because the preceding expression (4.5) must be held with $\delta_r > 0, r = 1, \ldots, \bar{i}$, it follows that $Q_i(x^r) - e_i y_i^r = 0, r = 1, \ldots, \bar{i}, i = 1, \ldots, k.$

These statements imply that $(x^r, y_1^r, \ldots, y_k^r) \in IR, \; r = 1, \ldots, \bar{i},$ and that $(x^*, y_1^*, \ldots, y_k^*)$ can be denoted as a convex combination of the points in the $IR$. Since $(x^*, y_1^*, \ldots, y_k^*)$ is a vertex of the $IR$ according to Corollary 4.2 and Assumption 4.1(3), there must exist $\bar{i} = 1$, which implies is a vertex of $S$.    $\square$

**Corollary 4.3** If $(x^*, y_1^*, \ldots, y_k^*)$ is a vertex of IR, it is also a vertex of S.


### 4.4.3  Uncooperative BLMF Kth-Best Algorithm

The well-known $K$th-Best algorithm has been successfully applied to a one-leader-and-one-follower linear bi-level decision problem, as shown in Chap. 3. This section extends this algorithm to deal with the abovementioned linear uncooperative BLMF decision problem, called uncooperative BLMF $K$th-Best algorithm.

Theorem 4.3 provides a theoretical foundation for searching an algorithm to solve the uncooperative BLMF decision problem $S1$. It means that we can efficiently find an optimal solution for a linear BLMF decision problem by searching vertices on the constraint region $S$. The basic idea of the algorithm is that we arrange all the vertices of $S$ in ascending order according to the objective function of the upper level, and select the first vertex to check whether it is on the inducible region $IR$. If it is, the current vertex is the optimal solution. Otherwise, the next one will be selected and checked.

Let $(x^1, y_1^1, \ldots, y_k^1), \ldots, (x^N, y_1^N, \ldots, y_k^N)$ denote the $N$ ordered vertices to the linear programming problem:

$$\min\{F(x, y_1, \ldots, y_k) | (x, y_1, \ldots, y_k) \in S\}. \tag{4.6}$$

such that:

$$F(x^j, y_1^j, \ldots, y_k^j) \le F(x^{j+1}, y_1^{j+1}, \ldots, y_k^{j+1}), \quad j = 1, \ldots, N-1.$$

Let $\tilde{y}_i, i = 1, \ldots, k$, denote the optimal solution to the following problem

$$\min_{y_i \in S_i(x^j)} f_i(x^j, y_i) = c_i x^j + e_i y_i. \tag{4.7}$$

We only need to find the smallest $j$ under which $y_i^j = \tilde{y}_i, i = 1, \ldots, k$.
From Definition 4.1(b), we rewrite (4.7) as follows:

$$\min_{y_i \in Y_i} \quad f_i(x, y_i) = c_i x + e_i y_i$$
$$\text{s.t.} \quad A_i x + C_i y_i \leq b_i, \tag{4.8}$$
$$x = x^j.$$

where $i = 1, \ldots, k$.

Solving problem (4.1) is equivalent to selecting one ordered vertex $\left(x^j, y_1^j, \ldots, y_k^j\right)$ and then solving (4.8) to obtain the optimal solution $\tilde{y}_i$. If $y_i^j = y_i$ for all $i$ then $\left(x^j, y_1^j, \ldots, y_k^j\right)$ is the globally optimal solution to $S1$. Otherwise, check the next vertex.

Based on the above results, a multi-follower $K$th-Best algorithm (Shi et al. 2005) which can solve an uncooperative BLMF decision problem is described as follows.

---

**Algorithm 4.1: Uncooperative BLMF $K$th-Best Algorithm**

[Begin]

**Step 1:** Put $j \leftarrow 1$. Solve (4.6) with the simplex method to obtain an optimal solution $(x^1, y_1^1, \ldots, y_k^1)$. Let $W = \{(x^1, y_1^1, \ldots, y_k^1)\}$ and $T = \emptyset$. Go to Step 2.

**Step 2:** Solve (4.8) with the simplex method. Let $\tilde{y}_i$ denote the optimal solution to (4.8). If $y_i^j = \tilde{y}_i$ for all $i, i = 1, \ldots, k$, $\left(x^j, y_1^j, \ldots, y_k^j\right)$ is the globally optimal solution to $S1$. Otherwise, go to Step 3.

**Step 3:** Let $W_{[j]}$ denote the set of adjacent vertices of $\left(x^j, y_1^j, \ldots, y_k^j\right)$. Let $T = T \cup \left\{\left(x^j, y_1^j, \ldots, y_k^j\right)\right\}$ and $W = W \cup W_{[j]}/T$. Go to Step 4.

**Step 4:** Set $j \leftarrow j + 1$ and choose $\left(x^j, y_1^j, \ldots, y_k^j\right)$ so that

$$F\left(x^j, y_1^j, \ldots, y_k^j\right) = \min\{F(x, y_1, \ldots, y_k) | (x, y_1, \ldots, y_k) \in W\}.$$

Go back to Step 2.

[End]

---

The uncooperative BLMF $K$th-Best algorithm is easy to use to solve a linear uncooperative BLMF decision problem.

Let us give the following example to show how the uncooperative BLMF $K$th-Best algorithm works.

*Example 4.2* Consider the following linear BLMF problem with $x \in R$, $y \in R$, $z \in R$ and $X = \{x|x \geq 0\}$, $Y = \{y|y \geq 0\}$, $Z = \{z|z \geq 0\}$.

$$\min_{x \in X} \quad F(x,y,z) = x - 2y - 4z$$

$$\text{s.t.} \quad x + y + z \leq 4,$$
$$-x + 3y \leq 4,$$
$$-x + z \leq 1,$$
$$\min_{y \in Y} \quad f_1(x,y,z) = x + y$$
$$\text{s.t.} \quad x - y \leq 0,$$
$$\min_{z \in Z} \quad f_2(x,y,z) = -2x + z$$
$$\text{s.t.} \quad 2x - 5z \leq 1,$$
$$2x + z \geq 1.$$

This model can be rewritten in the formulation of (4.6) as follows:

$$\min_{x,y,z} \quad F(x,y,z) = x - 2y - 4z$$

$$\text{s.t.} \quad x + y + z \leq 4,$$
$$-x + 3y \leq 4,$$
$$-x + z \leq 1,$$
$$x - y \leq 0,$$
$$2x - 5z \leq 1,$$
$$2x + z \geq 1.$$

Now we go through this uncooperative BLMF $K$th-Best algorithm from Step 1 to Step 4.

In Step 1, set $j = 1$, and solve the above problem with the simplex method to obtain an optimal solution $\left(x_{[1]}, y_{[1]}, z_{[1]}\right) = (0.71, 1.57, 1.71)$. Let $W = \{(0.71, 1.57, 1.71)\}$ and $T = \emptyset$. Go to Step 2.

In Loop 1:

Setting $i \leftarrow 1$ and by the formulation (4.8), we have

$$\min_{y \in Y} \quad f_1(x,y,z) = x + y$$

$$\text{s.t.} \quad x - y \leq 0,$$
$$x = 0.71.$$

Using the simplex method, we have $\tilde{y}_j = 0.71$. Because of $\tilde{y}_j \neq y_{[j]}$, we go to Step 3 and then have $W_{[j]} = \{(0, 1.33, 1), (1, 1, 2), (1.65, 1.88, 0.46)\}$ and $T = \{(0.71, 1.57, 1.71)\}$ and $W = \{(0, 1.33, 1), (1, 1, 2), (1.65, 1.88, 0.46)\}$.

We then go to Step 4. Update $j = 2$, and choose $(x_{[j]}, y_{[j]}, z_{[j]}) = (1, 1, 2)$, go back to Step 2.

In Loop 2:

Setting $i \leftarrow 1$ and the formulation (4.8), we have

$$\min_{y \in Y} \quad f_1(x, y, z) = x + y$$

$$\text{s.t.} \quad x - y \leq 0,$$

$$x = 1.$$

Using the simplex method, we obtain $\tilde{y}_j = 1$ and $\tilde{y}_j = y_{[j]}$. This is a different situation from the last loop. We thus set $i \leftarrow 2$ and have a new expression of $f_2$ by the formulation (4.8):

$$\min_{z \in Z} \quad f_2(x, y, z) = -2x + z$$

$$\text{s.t.} \quad 2x - 5z \leq 1,$$

$$2x + z \geq 1,$$

$$x = 1.$$

The same as before, by using the simplex method again, we have $\tilde{z}_j = 0.2$. Because $\tilde{z}_j \neq z_{[j]}$, we go to Step 3, and have $W_{[j]} = \{(0, 0, 1), (1.75, 1.75, 0.5)\}$, $T = \{(0.71, 1.57, 1.71), (1, 1, 2)\}$ and $W = \{(0, 1.33, 1), (1.65, 1.88, 0.46), (0, 0, 1), (1.75, 1.75, 0.5)\}$.

We then go to Step 4. Update $j = 3$ and choose $(x_{[j]}, y_{[j]}, z_{[j]}) = (0, 1.33, 1)$, then we go back to Step 2.

In Loop 3:

Setting $i \leftarrow 1$ and the formulation (4.8), we have

$$\min_{y \in Y} \quad f_1(x, y, z) = x + y$$

$$\text{s.t.} \quad x - y \leq 0,$$

$$x = 0.$$

Using the simplex method, we obtain $\tilde{y}_j = 0$. Since $\tilde{y}_j \neq y_{[j]}$, go to Step 3, and we have $W_{[j]} = \{(0, 0, 1), (0.71, 1.57, 1.71)\}$, $T = \{(0.71, 1.57, 1.71), (1, 1, 2), (0, 1.33, 1)\}$, $W = \{(1.65, 1.88, 0.46), (0, 0, 1), (1.75, 1.75, 0.5)\}$. We then go to Step 4. Update $j = 4$ and we get $(x_{[j]}, y_{[j]}, z_{[j]}) = (0, 0, 1)$.

In Loop 4:

Setting $i \leftarrow 1$ and the formulation (4.8), we have

$$\min_{y \in Y} \quad f_1(x, y, z) = x + y$$

$$\text{s.t.} \quad x - y \leq 0,$$

$$x = 0.$$

Using the simplex method, we obtain $\tilde{y}_j = 0, \tilde{y}_j \neq y_{[j]}$. We set $i = 2$, and have

$$\min_{z \in Z} \quad f_2(x, y, z) = -2x + z$$

$$\text{s.t.} \quad 2x - 5z \leq 1,$$
$$2x + z \geq 1,$$
$$x = 0.$$

We have $\tilde{z}_j = 1$ and $\tilde{z}_j = z_{[j]}$ The optimal solution of the uncooperative BLMF problem occurs at the point $(0, 0, 1)$ with the leader's objective function value $F^* = -4$, and two followers' objective function values $f_1^* = 0$, and $f_2^* = 1$ respectively.

### 4.4.4 Uncooperative BLMF Kuhn-Tucker Approach

A natural idea for dealing with uncooperative BLMF decision problems is to replace each follower's problem with its Kuhn-Tucker (or KKT) conditions and append the resultant system to the leader's problem. Omitting or relaxing the complementary constraints leaves a standard linear programming problem that can be solved by using the simplex method. The Kuhn-Tucker approach is the most popular method for solving one-leader one-follower bi-level decision problems. Based on the definition of an optimal solution, a multi-follower Kuhn-Tucker approach for the uncooperative BLMF decision problem is proposed and described as follows (Lu et al. 2006).

Let $v_i \in R^{q_i}$ and $w_i \in R^{m_i} (i = 1, \ldots, k)$ be the dual variables associated with constraints, $A_i x + C_i y_i \leq b_i$, $y_i \geq 0$, $i = 1, \ldots, k$, respectively. We have the following theorem.

**Theorem 4.4** *A necessary condition that* $(x^*, y_1^*, \ldots, y_k^*)$ *solves the linear BLMF problem (4.1) is that there exist (row) vectors* $(v_1^*, v_2^*, \ldots, v_k^*)$ *and* $(w_1^*, w_2^*, \ldots, w_k^*)$ *such that* $(x^*, y_1^*, \ldots, y_k^*, v_1^*, \ldots, v_k^*, w_1^*, \ldots, w_k^*)$ *solve the following problem:*

$$\min_{x, y_1, \ldots, y_k, v_1, \ldots, v_k, w_1, \ldots, w_k} \quad F(x, y_1, \ldots, y_k) = cx + \sum_{i=1}^{k} d_i y_i$$

$$\text{s.t.} \quad Ax + \sum_{t=1}^{k} B_i y_i \leq b,$$

$$A_i x + C_i y_i \leq b_i, \tag{4.9}$$
$$v_i C_i - w_i = -e_i,$$
$$v_i (b_i - A_i x - C_i y_i) + w_i y_i = 0,$$
$$x \geq 0, y_i \geq 0, v_i \geq 0, w_i \geq 0, \quad i = 1, 2, \ldots, k$$

Theorem 4.4 indicates that the most direct approach for solving (4.1) is to solve the equivalent problem (4.9). One of its advantages is that it allows a more robust model to be solved without introducing new computational difficulties.

Now, we use Example 4.2 to illustrate the feasibility of the uncooperative BLMF Kuhn-Tucker approach for solving the uncooperative BLMF decision problem.

Using the Kuhn-Tucker condition of all followers' problems, we can transform Example 4.2 into the following problem:

$$
\begin{aligned}
\min_{x,y,z,u,v} \quad & x - 2y - 4z \\
\text{s.t.} \quad & 1 - u_1 - u_2 = 0, \\
& u_1(x - y) = 0, \\
& u_2 y = 0, \\
& 1 - 5v_1 - v_2 - v_3 = 0, \\
& v_1(2x - 5z - 1) = 0, \\
& v_2(-2x - z + 1) = 0, \\
& v_3 z = 0, \\
& x + y + z \le 4, \\
& -x + 3y \le 4, \\
& -x + z \le 1, \\
& x - y \le 0, \\
& 2x - 5z \le 1, \\
& 2x + z \ge 1, \\
& x, y, z, u = (u_1, u_2), v = (v_1, v_2, v_3) \ge 0.
\end{aligned}
$$

It follows the formulation (4.9) that we have the following six possibilities.

Case 1: $(u, v) = (0, 1, 0, 0, 1)$,
Case 2: $(u, v) = (0, 1, 0, 1, 0)$,
Case 3: $(u, v) = (0, 1, 0.2, 0, 0)$,
Case 4: $(u, v) = (1, 0, 0, 0, 1)$,
Case 5: $(u, v) = (1, 0, 0, 1, 0)$,
Case 6: $(u, v) = (1, 0, 0.2, 0, 0)$.

**Table 4.2** Procedures of uncooperative BLMF Kuhn-Tucker approach

| Case | $u$ | $v$ | $x$ | $y$ | $z$ | $F$ |
|------|------|-----------|------|------|-----|------------|
| 1 | (0, 1) | (0, 0, 1) | | | | Infeasible |
| 2 | (0, 1) | (0, 1, 0) | 0 | 0 | 1 | -4 |
| 3 | (0, 1) | (0.2, 0, 0) | | | | Infeasible |
| 4 | (1, 0) | (0, 0, 1) | 0.5 | 0.5 | 0 | -0.5 |
| 5 | (1, 0) | (0, 1, 0) | 0 | 0 | 1 | -4 |
| 6 | (1, 0) | (0.2, 0, 0) | 1.75 | 1.75 | 0.5 | -3.75 |

For Case 2, the above problem can be rewritten as follows:

$$\min_{x,y,z} \quad x - 2y - 4z$$

$$\text{s.t.} \quad y = 0$$
$$-2x - z + 1 = 0,$$
$$x + y + z \leq 4,$$
$$-x + 3y \leq 4,$$
$$-x + z \leq 1,$$
$$x - y \leq 0,$$
$$2x - 5z \leq 1,$$
$$2x + z \geq 1,$$
$$x, y, z \geq 0$$

Using the simplex method, we find that a solution occurs at the point $(x, y, z) = (0, 0, 1)$. By using the same approach as that used in Case 2, we obtain a solution for each case as shown in Table 4.2.

From Table 4.2, we find that the optimal solution for this example occurs at the point (0, 0, 1) with the optimal value −4. Therefore, we achieve the same solution as the Result attained by the uncooperative BLMF $K$th-Best algorithm.

## 4.5 Semi-cooperative Bi-level Multi-follower Decision Making

Within the framework of BFML decision-making, the semi-cooperative relationship is also a common situation in which followers control the same decision variables but have individual objective functions and/or constraint conditions. The linear models ($S3$–$S5$) for this kind of BLMF decision-making are presented in Sect. 4.3.2. Since model $S5$ is the most representative and typical in the set of models, this section mainly presents solution concepts, related theoretical properties and algorithms for model $S5$.

## 4.5.1 Solution Concepts

**Definition 4.2**

(a)  Constraint region of a linear BLMF decision problem:

$$S = \{(x,y) \in X \times Y | Ax + By \leq b, A_i x + C_i y \leq b_i, i = 1, \ldots, k\}.$$

where $Y = Y_1 \cap \cdots \cap Y_k$.

(b)  Feasible set of the $i$th follower for $\forall x \in X$:

$$S_i(x) = \{y \in Y_i | C_i y \leq b_i - A_i x\}, \quad i = 1, 2, \ldots, k.$$

(c)  Projection of $S$ onto the leader's decision space:

$$S(X) = \{x \in X | \exists y \in Y, (x,y) \in S\}.$$

(d)  The $i$th follower's rational reaction set for $x \in S(X)$:

$$P_i(x) = \{y \in Y_i | y \in \text{argmin } [f_i(x,z) | z \in S_i(x)]\},$$

where $\text{argmin}[f_i(x,z) | z \in S_i(x)] = \{y \in S_i(x) | f_i(x,y) \leq f_i(x,z), \forall z \in S_i(x)\}$.

(e)  Inducible region:

$$IR = \{(x,y) | (x,y) \in S, y \in P_i(x), i = 1, \ldots, k\}.$$

Thus, in terms of the above notations, the linear BLMF decision problem can be written as

$$\min\{F(x,y) | (x,y) \in IR\}. \tag{4.10}$$

For the sake of assuring an optimal solution to the model S5, it requires that the solution to the formulation (4.10) exists which implies that further assumptions should be satisfied for the semi-cooperative BLMF decision model in order to give the solution concepts.

**Assumption 4.2**

1. $S$ is non-empty and compact.
2. $IR$ is non-empty.
3. $P_i(x)$ is a point-to-point map with respect to $x$, where $i = 1, \ldots, k$.

**Theorem 4.5** *If the above assumptions are satisfied, there exists an optimal solution to the linear semi-cooperative BLMF decision model S5.*

*Proof* Since neither $S$ nor $IR$ is empty, there exists a point $x^* \in S(X)$ such that $P(x^*) = \{y | y \in P_i(x^*), i = 1, \ldots, k\} \neq \emptyset$. Consider a sequence $\{(x^t, y^t)\}_{t=1}^{\infty} \subseteq IR$ converging to $(x^*, y^*)$. Then, by the well-known results of linear parametric optimization, $y^* \in P(x^*)$. Hence, $(x^*, y^*) \in IR$ that shows $IR$ is closed. By Assumption 4.2(1) and $IR \subseteq S$, thus $IR$ is also bounded, and $IR$ is non-empty. Therefore, the problem (4.10) consists of minimizing a continuous function over a compact non-empty set, which implies that model $S5$ has an optimal solution.                 $\square$

### 4.5.2 Theoretical Properties

Theoretical properties and related proofs for the semi-cooperative BLMF decision model $S5$ are similar to those of the uncooperative BLMF decision model $S1$. In this section, we provide only the related theorems and corollaries for the semi-cooperative BLMF decision model $S5$. Readers can demonstrate these theorems by taking the proofs in Sect. 4.4.2 as references.

**Theorem 4.6** *The inducible region IR of the semi-cooperative BLMF decision model S5 can be expressed equivalently as a piecewise linear equality constraint comprised of support hyperplanes of S.*

**Corollary 4.4** *The semi-cooperative BLMF decision model S5 is equivalent to optimizing F over a feasible region comprised of a piecewise linear equality constraint.*

**Corollary 4.5** *An optimal solution to the semi-cooperative BLMF decision model S5 occurs at a vertex of IR.*

**Theorem 4.7** *The optimal solution $(x^*, y^*)$ to the semi-cooperative BLMF decision model S5 occurs at a vertex of S.*

**Corollary 4.6** *If $(x^*, y^*)$ is a vertex of the IR, it is also a vertex of S.*

### 4.5.3 Semi-cooperative BLMF Kth-Best Algorithm

The solution method for model $S5$ is similar to that for model $S1$. The multi-follower $K$th-Best algorithm proposed in Sect. 4.4.3 can also be used to solve the model $S5$. This section will adopt a simple example to illustrate the multi-follower $K$th-Best algorithm for solving a semi-cooperative BLMF decision problem.

*Example 4.3* Consider the following linear BLMF problem with $x \in R$, $y \in R$, and $X = \{x | x \geq 0\}$, $Y_1 = \{y | y \geq 0\}$, $Y_2 = \{y | y \geq 0\}$:

$$\min_{x \in X} \quad F(x,y) = x + 2y$$

$$\text{s.t.} \quad x \le 3,$$

$$x \ge 1,$$

$$\min_{y \in Y_1} \quad f_1(x,y) = x - 2y$$

$$\text{s.t.} \quad x + y \le 4,$$

$$y \le 2,$$

$$\min_{y \in Y_2} \quad f_2(x,y) = x + y$$

$$\text{s.t.} \quad x + y \ge 3,$$

$$y \ge 1.$$

According to the multi-follower $K$th-Best algorithm, we consider the following problem:

$$\min_{x \in X} \quad F(x,y) = x + 2y$$

$$\text{s.t.} \quad x \le 3,$$

$$x \ge 1,$$

$$x + y \le 4,$$

$$y \le 2,$$

$$x + y \ge 3,$$

$$y \ge 1.$$

Now we go through the multi-follower $K$th-Best algorithm from Step 1 to Step 4.

In Step 1, set $j = 1$, and solve the above problem with the simplex method to obtain an optimal solution $(x_{[1]}, y_{[1]}) = (2,1)$. Let $W = \{(2,1)\}$ and $T = \emptyset$. Go to Step 2.

In Loop 1:

Setting $i \leftarrow 1$ and the formulation (4.8), we have

$$\min_{y \in Y_1} \quad f_1(x,y) = x - 2y$$

$$\text{s.t.} \quad x + y \le 4,$$

$$y \le 2,$$

$$x = 2.$$

Using the simplex method, we have $\tilde{y} = 2$. Because of $\tilde{y} \ne y_{[j]}$, we go to Step 3 and then have $W_{[j]} = \{(1,2),(3,1)\}$, $T = \{(2,1)\}$, $W = \{(1,2),(3,1)\}$. We then go to Step 4. Update $j = 2$, and choose $(x_{[j]}, y_{[j]}) = (1,2)$, go back to Step 2.

In Loop 2:

Setting $i \leftarrow 1$ and the formulation (4.8), we have

$$\min_{y \in Y_1} \quad f_1(x, y) = x - 2y$$

$$\text{s.t.} \quad x + y \leq 4,$$
$$y \leq 2,$$
$$x = 1.$$

Through using the simplex method, we obtain $\tilde{y} = 2$ and $\tilde{y} = y_{[j]}$. We set $i \leftarrow 2$ and have

$$\min_{y \in Y_2} \quad f_2(x, y) = x + y$$

$$\text{s.t.} \quad x + y \geq 3,$$
$$y \geq 1,$$
$$x = 1.$$

We have $\tilde{y} = 2$ and $\tilde{y} = y_{[j]}$.

It has been found that the optimal solution of the semi-cooperative BLMF problem occurs at the point $(x_{[2]}, y_{[2]}) = (1, 2)$ with the leader's objective value $F^* = 5$, and two followers' objective values $f_1^* = -3$ and $f_2^* = 3$ respectively.

### 4.5.4 Semi-cooperative BLMF Kuhn-Tucker Approach

Similar to Theorem 4.4 for the uncooperative linear BLMF decision problem, we have the following result for the linear semi-cooperative BLMF decision problem.

**Theorem 4.8** *A necessary condition that* $(x^*, y^*)$ *solves the linear BLMF problem S5 is that there exist (row) vectors* $(v_1^*, v_2^*, \ldots, v_k^*)$ *and* $(w_1^*, w_2^*, \ldots, w_k^*)$ *such that* $(x^*, y^*, v_1^*, \ldots, v_k^*, w_1^*, \ldots, w_k^*)$ *solves*

$$\min_{x, y, v_1, \ldots, v_k, w_1, \ldots, w_k} \quad F(x, y) = cx + dy$$

$$\text{s.t.} \quad Ax + By \leq b,$$
$$A_i x + C_i y \leq b_i,$$
$$v_i C_i - w_i = -e_i,$$
$$v_i(b_i - A_i x - C_i y) + w_i y = 0,$$
$$x \geq 0, y \geq 0, v_i \geq 0, w_i \geq 0, \quad i = 1, 2, \ldots, k$$

Theorem 4.8 indicates that the most direct approach for solving $S5$ is to solve the above equivalent mathematical programming problem.

Now, we use Example 4.3 proposed in Sect. 4.5.3 to illustrate the feasibility of the multi-follower *Kuhn-Tucker approach* for solving the linear semi-cooperative BLMF decision problem.

Using the Kuhn-Tucker conditions of all followers' problems, we can transform Example 4.3 into the following problem.

$$\min_{x,y,v_1,\ldots,v_4,w} \quad F(x,y) = x + 2y$$

$$\text{s.t.} \quad 1 \le x \le 3,$$
$$-2 + v_1 + v_2 - w = 0,$$
$$v_1(x + y - 4) = 0,$$
$$v_2(y - 2) = 0,$$
$$wy = 0,$$
$$1 - v_3 - v_4 = 0,$$
$$v_3(x + y - 3) = 0,$$
$$v_4(y - 1) = 0,$$
$$x + y \le 4,$$
$$y \le 2,$$
$$x + y \ge 3,$$
$$y \ge 1,$$
$$x, y, v_1, \ldots, v_4, w \ge 0.$$

It follows from Theorem 4.8 that we have four possibilities.

$$\text{Case 1: } (v, w) = (2, 0, 1, 0, 0),$$
$$\text{Case 2: } (v, w) = (2, 0, 0, 1, 0),$$
$$\text{Case 3: } (v, w) = (0, 2, 1, 0, 0),$$
$$\text{Case 4: } (v, w) = (0, 2, 0, 1, 0),$$

For Case 2, the above problem can be rewritten as follows:

$$\min_{x,y} \quad F(x,y) = x + 2y$$

$$\text{s.t.} \quad 1 \le x \le 3,$$
$$2(x + y - 4) = 0,$$
$$y - 1 = 0,$$
$$x + y \le 4,$$
$$y \le 2,$$
$$x + y \ge 3,$$
$$y \ge 1,$$
$$x, y \ge 0.$$

**Table 4.3** Procedures of the semi-cooperative BLMF Kuhn-Tucker approach

| Case | $v$ | $w$ | $x$ | $y$ | $F$ |
|------|------|-----|-----|-----|-----|
| 1 | (2, 0, 1, 0) | 0 | | | Infeasible |
| 2 | (2, 0, 0, 1) | 0 | 3 | 1 | 5 |
| 3 | (0, 2, 1, 0) | 0 | 1 | 2 | 5 |
| 4 | (0, 2, 0, 1) | 0 | | | Infeasible |

Using the simplex method, we find that an optimal solution of the above linear programming problem occurs at the point $(x, y) = (3, 1)$. By using the same approach as that applied in other cases, we obtain the results for each case as shown in Table 4.3.

By examining the procedure shown in Table 4.3, we find that the optimal solution for this example occurs at points (3, 1) and (1, 2) with the optimal value 5.

## 4.6 Reference-Uncooperative Bi-level Multi-follower Decision Making

The reference-uncooperative relationship is another common situation in which the followers uncooperatively make decisions while cross-referencing the decision information between them. Of the BLMF decision-making models $S6$–$S9$, model $S9$ is the most representative. This section will provide more details about solution concepts, related theoretical properties and algorithms for model $S9$.

### 4.6.1 Solution Concepts

**Definition 4.3**

(a) Constraint region of a linear BLMF decision problem:

$$S = \left\{ (x, y_1, \ldots, y_k) \in X \times Y_1 \times \cdots \times Y_k | Ax \right.$$

$$\left. + \sum_{i=1}^{k} B_i y_i \leq b, A_i x + \sum_{s=1}^{k} C_{is} y_s \leq b_i, i = 1, \ldots, k \right\}$$

The problem's constraint region refers to all possible choice combinations that the leader and followers might make.

(b)  Feasible set for the $i$th follower:

$$S_i(x, y_1, \ldots, y_{i-1}, y_{i+1}, \ldots, y_k)$$
$$= \left\{ y_i \in Y_i | A_i x + \sum_{s=1}^{k} C_{is} y_s \le b_i \right\}, \quad i = 1, \ldots, k.$$

The feasible region for each follower is affected by the leader's choice of $x$, and also uses other followers' decisions for reference.

(c)  Projection of $S$ onto the leader's decision space:

$$S(X) = \{ x \in X | \exists (y_1, \ldots, y_k) \in Y_1 \times \cdots \times Y_k, (x, y_1, \ldots, y_k) \in S \}.$$

(d)  The $i$th follower's rational reaction set:

$$P_i(x, y_1, \ldots, y_{i-1}, y_{i+1}, \ldots, y_k)$$
$$= \{ y_i | y_i \in \text{argmin}[f_i(x, y_1, \ldots, y_k) | y_i \in S_i(x, y_1, \ldots, y_{i-1}, y_{i+1}, \ldots, y_k)] \},$$

where

$$\text{argmin}[f_i(x, y_1, \ldots, y_k) | y_i \in S_i(x, y_1, \ldots, y_{i-1}, y_{i+1}, \ldots, y_k)]$$
$$= \{ y_i \in S_i(x, y_1, \ldots, y_{i-1}, y_{i+1}, \ldots, y_k) | f_i(x, y_1, \ldots, y_k)$$
$$\le f_i(x, y_1, \ldots, y_{i-1}, \hat{y}_i, y_{i+1}, \ldots, y_k), \forall \hat{y}_i \in S_i(x, y_1, \ldots, y_{i-1}, y_{i+1}, \ldots, y_k) \}.$$

(e)  Inducible region:

$$IR = \{ (x, y_1, \ldots y_k) | (x, y_1, \ldots, y_k) \in S,$$
$$y_i \in P_i(x, y_1, \ldots, y_{i-1}, y_{i+1}, \ldots, y_k), i = 1, \ldots, k \}.$$

Thus in terms of the above notations, the linear BLMF decision problem can be written as:

$$\min\{ F(x, y_1, \ldots, y_k) | (x, y_1, \ldots, y_k) \in IR \}. \tag{4.11}$$

For the sake of assuring an optimal solution to model $S9$, it is a requirement that the optimal solution to the formulation (4.11) exists which implies that further assumptions should be satisfied for the reference-uncooperative BLMF decision model to give the solution concepts.

**Assumption 4.3**

1.  $S$ is non-empty and compact.
2.  $IR$ is non-empty.

3. $P_i(x, y_1, \ldots, y_{i-1}, y_{i+1}, \ldots, y_k)$ is a point-to-point map with respect to $(x, y_1, \ldots, y_{i-1}, y_{i+1}, \ldots, y_k)$, where $i = 1, \ldots, k$.

**Theorem 4.9** *If the above assumptions are satisfied, there exists an optimal solution to the linear reference-uncooperative BLMF decision model S9.*

*Proof* Let

$$P(x) = \{(y_1, \ldots, y_k) | y_i \in P_i(x, y_1, \ldots, y_{i-1}, y_{i+1}, \ldots, y_k), i = 1, \ldots, k\}.$$

Since neither $S$ nor $IR$ is empty, there exists at least one point $x^* \in S(X)$ and $P(x^*) \neq \emptyset$ for all $i = 1, \ldots, k$. Consider a sequence $\{(x^t, y_1^t, \ldots, y_k^t)\}_{t=1}^{\infty} \subseteq IR$ converging to $(x^*, y_1^*, \ldots, y_k^*)$. Then, by the well-known results of linear parametric optimization, $(y_1^*, \ldots, y_k^*) \in P(x^*)$. Hence, $(x^*, y_1^*, \ldots, y_k^*) \in IR$ shows that $IR$ is closed. Thus by Assumption 4.3(1) and $IR \subseteq S$, $IR$ is also bounded, and $IR$ is non-empty, therefore the problem (4.11) consists of minimizing a continuous function over a compact non-empty set, which implies that model $S9$ has a solution. $\square$

### 4.6.2 Theoretical Properties

A number of theoretical properties for model $S9$ are presented to support the principle of algorithms for solving the reference-uncooperative BLMF problem.

**Theorem 4.10** *The inducible region IR can be expressed equivalently as a piecewise linear equality constraint comprised of support hyperplanes of S.*

*Proof* First, $i = 1, \ldots, k$, define

$$Q_i(x, y_1, \ldots, y_{i-1}, y_{i+1}, \ldots, y_k) = \min\{e_{ii}\hat{y}_i | \hat{y}_i \in S_i(x, y_1, \ldots, y_{i-1}, y_{i+1}, \ldots, y_k)\}.$$

Since $Q_i(x, y_1, \ldots, y_{i-1}, y_{i+1}, \ldots, y_k)$ can be seen as a linear programming problem with parameters $x, y_1, \ldots, y_{i-1}, y_{i+1}, \ldots, y_k$, its dual problem can be written as

$$\max\left\{(A_ix + \sum_{s=1, s\neq i}^{k} C_{is}y_s - b_i)u_i | C_{ii}u_i \geq -e_{ii}, u_i \geq 0\right\}, \quad i = 1, \ldots, k. \quad (4.12)$$

If both $Q_i(x, y_1, \ldots, y_{i-1}, y_{i+1}, \ldots, y_k)$ and problem (4.12) have feasible solutions, according to the dual theorem of linear programming, both of them have optimal solutions and the same optimal objective function value. We know that a solution to problem (4.12) occurs at a vertex of its constraint region $U_i = \{u_i | C_{ii}u_i \geq -e_{ii}, u_i \geq 0\}, i = 1, \ldots, k$. Adopting $u_i^1, \ldots, u_i^{q_i}$ to express all the vertices of $U_i$, problem (4.12) can be written as:

$$\max\left\{ (A_i x + \sum_{s=1, s \neq i}^{k} C_{is} y_s - b_i) u_i | u_i \in \{u_i^1, \ldots, u_i^{q_i}\} \right\}, \quad i = 1, \ldots, k. \quad (4.13)$$

Clearly, $Q_i(x, y_1, \ldots, y_{i-1}, y_{i+1}, \ldots, y_k)$ is a piecewise linear function according to problem (4.13).

$$\begin{aligned}
IR &= \{(x, y_1, \ldots, y_k) \in S | y_i \in P_i(x, y_1, \ldots, y_{i-1}, y_{i+1}, \ldots, y_k), i = 1, \ldots, k\} \\
&= \{(x, y_1, \ldots, y_k) \in S | e_{ii} y_i = Q_i(x, y_1, \ldots, y_{i-1}, y_{i+1}, \ldots, y_k), i = 1, \ldots, k\}.
\end{aligned}$$
$$(4.14)$$

and it can be seen as a piecewise linear equality constraint.                    □

**Corollary 4.7** *The reference-uncooperative BLMF decision model S9 is equivalent to optimizing F over a feasible region comprised of a piecewise linear equality constraint.*

**Corollary 4.8** *An optimal solution to the reference-uncooperative BLMF decision model S9 occurs at a vertex of IR.*

*Proof* According to the equivalent form (4.11) of the BLMF decision model, also, since $F(x, y_1, \ldots, y_k)$ is linear, an optimal solution to the problem must occur at a vertex of *IR* if it exists.                    □

**Theorem 4.11** *The optimal solution $(x^*, y_1^*, \ldots, y_k^*)$ to the reference-uncooperative BLMF decision model occurs at a vertex of S.*

*Proof* Let $(x^1, y_1^1, \ldots, y_k^1), \ldots, (x^t, y_1^t, \ldots, y_k^t)$ express the distinct vertices of $S$. Since any point in $S$ can be written as a convex combination of these vertices, we can get

$$\left( x^*, y_1^*, \ldots, y_k^* \right) = \sum_{r=1}^{\bar{t}} \delta_r \left( x^r, y_1^r, \ldots, y_k^r \right)$$

where $\sum_{r=1}^{\bar{t}} \delta_r = 1, \delta_r > 0, r = 1, \ldots, \bar{t}$ and $\bar{t} \leq t$.

Let us write the constraints of $S9$ in the piecewise linear form (4.14) discussed in Theorem 4.10:

$$\begin{aligned}
0 &= Q_i \left( x^*, y_1^*, \ldots, y_{i-1}^*, y_{i+1}^*, \ldots, y_k^* \right) - e_{ii} y_i^* \\
&= Q_i \left( \sum_{r=1}^{\bar{t}} \delta_r x^r, \sum_{r=1}^{\bar{t}} \delta_r y_1^r, \ldots, \sum_{r=1}^{\bar{t}} \delta_r y_{i-1}^r, \sum_{r=1}^{\bar{t}} \delta_r y_{i+1}^r, \ldots, \sum_{r=1}^{\bar{t}} \delta_r y_k^r \right) \\
&\quad - e_{ii} \sum_{r=1}^{\bar{t}} \delta_r y_i^r, \quad i = 1, \ldots, k.
\end{aligned}$$

Because of the convexity of $Q_i(x, y_1, \ldots, y_{i-1}, y_{i+1}, \ldots, y_k)$, we have

$$0 \leq \sum_{r=1}^{\bar{\imath}} \delta_r Q_i\left(x^r, y_1^r, \ldots, y_{i-1}^r, y_{i+1}^r, \ldots, y_k^r\right) - e_{ii} \sum_{r=1}^{\bar{\imath}} \delta_r y_i^r \tag{4.15}$$
$$= \sum_{r=1}^{\bar{\imath}} \delta_r (Q_i\left(x^r, y_1^r, \ldots, y_{i-1}^r, y_{i+1}^r, \ldots, y_k^r\right) - e_{ii} y_i^r).$$

By the definition of $Q_i(x, y_1, \ldots, y_{i-1}, y_{i+1}, \ldots, y_k)$, we have

$$Q_i\left(x^r, y_1^r, \ldots, y_{i-1}^r, y_{i+1}^r, \ldots, y_k^r\right)$$
$$= \min e_{ii} y_i \leq e_{ii} y_i^r, \quad r = 1, \ldots, \bar{\imath}, i = 1, \ldots, k.$$

Thus, $Q_i\left(x^r, y_1^r, \ldots, y_{i-1}^r, y_{i+1}^r, \ldots, y_k^r\right) - e_{ii} y_i^r \leq 0, r = 1, \ldots, \bar{\imath}, i = 1, \ldots, k.$
Because the preceding expression (4.15) must be held with $\delta_r > 0, r = 1, \ldots, \bar{\imath}$, we have $Q_i\left(x^r, y_1^r, \ldots, y_{i-1}^r, y_{i+1}^r, \ldots, y_k^r\right) - e_{ii} y_i^r = 0, r = 1, \ldots, \bar{\imath}, i = 1, \ldots, k.$
These statements imply that $(x^r, y_1^r, \ldots, y_k^r) \in IR, r = 1, \ldots, \bar{\imath}$, and that $(x^*, y_1^*, \ldots, y_k^*)$ can be denoted as a convex combination of the points in the $IR$. Since $(x^*, y_1^*, \ldots, y_k^*)$ is a vertex of the $IR$ and according to Corollary 4.8 and Assumption 4.3(3), there must exist $\bar{\imath} = 1$, which implies $(x^*, y_1^*, \ldots, y_k^*)$ is a vertex of $S$.    $\square$

**Corollary 4.9** *If* $(x^*, y_1^*, \ldots, y_k^*)$ *is a vertex of the IR, it is also a vertex of S.*

### 4.6.3 Reference-Uncooperative BLMF Kth-Best Algorithm

In Sect. 4.4, the well-known $K$th-Best algorithm has been successfully applied to solve model $S1$. This section will extend this algorithm to deal with the above linear reference-uncooperative BLMF decision model $S9$.

Theorem 4.11 provides a theoretical foundation for searching an algorithm to solve the reference-uncooperative BLMF decision problem. It also means that we can efficiently find an optimal solution for a linear BLMF decision problem by searching vertices on the constraint region $S$. The basic idea of the algorithm is that we arrange all the vertices of $S$ in ascending order according to the objective function value of the upper level, and select the first vertex to check whether it is on the inducible region $IR$. If it is, the current vertex is the optimal solution. Otherwise, the next one will be selected and checked.

More specifically, let $(x^1, y_1^1, \ldots, y_k^1), \ldots, (x^N, y_1^N, \ldots, y_k^N)$ denote the $N$ ordered vertices to the linear programming problem

$$\min\{F(x, y_1, \ldots, y_k) | (x, y_1, \ldots, y_k) \in S\}, \tag{4.16}$$

such that:

$$F\left(x^j, y_1^j, \ldots, y_k^j\right) \leq F\left(x^{j+1}, y_1^{j+1}, \ldots, y_k^{j+1}\right), \quad j = 1, \ldots, N - 1.$$

Let $\tilde{y}_i$, $i = 1, \ldots, k$, denote the optimal solution to the following problem:

$$\min_{y_i \in S_i(x^j, y_1, \ldots, y_{i-1}, y_{i+1}, \ldots, y_k)} f_i\left(x^j, y_1, \ldots, y_k\right) = c_i x^j + \sum_{s=1}^{k} e_{is} y_s. \qquad (4.17)$$

We only need to find the smallest $j$ under which $y_i^j = \tilde{y}_i, i = 1, \ldots, k$.
From Definition 4.3(b), we rewrite (4.17) as follows:

$$\min_{y_i} \quad f_i(x, y_1, \ldots, y_k) = c_i x + \sum_{s=1}^{k} e_{is} y_s$$

$$\text{s.t.} \quad A_i x + \sum_{s=1}^{k} C_{is} y_s \leq b_i, \qquad (4.18)$$

$$x = x^j,$$

$$y_l = y_l^j, l \neq i,$$

where $i = 1, \ldots, k$.

Solve problem (4.18) and obtain the optimal solution $\tilde{y}_i$. If $y_i^j = \tilde{y}_i$ for all $i$ then $\left(x^j, y_1^j, \ldots, y_k^j\right)$ is the globally optimal solution to $S9$. Otherwise, check the next vertex.

Based on the above results, a multi-follower $K$th-Best algorithm (Zhang et al. 2008) which can solve the reference-uncooperative BLMF decision problems is described as follows.

It is easy to use the reference-uncooperative BLMF $K$th-Best algorithm to solve the linear reference-uncooperative BLMF decision problem.

Let us present a logistics planning problem which is modeled as a reference-uncooperative BLMF decision model to show how the proposed reference-uncooperative BLMF $K$th-Best algorithm works.

*Example 4.4* A logistics chain often involves a series of units such as supplier and distributor. All the units involved in the chain are interrelated in such a way that a decision made at one unit affects the performance of the next unit. In the meantime, when one unit tries to optimize its objective, it may need to consider the objective of the next unit, and its decision will be affected by the next unit's reaction. Both supplier and distributor, two important units in a logistics chain, have their own objectives such as maximizing their benefits and minimizing their costs; constraints such as time, locations and facilities; and variables such as prices. For each possible decision made by the supplier, the distributor finds a way to optimize its objective value. The optimal solution of the distributor allows the supplier to compute its objective function value. As the main purpose of making a logistics plan is to optimize the supplier's objective function value, the supplier is the leader, and the distributor is the follower in this case.

---

**Algorithm 4.2: Reference-uncooperative BLMF Kth-Best Algorithm**

[Begin]

**Step 1:** Put $j \leftarrow 1$. Solve (4.16) with the simplex method to obtain an optimal solution $(x^1, y_1^1, \dots, y_k^1)$. Let $W = \{(x^1, y_1^1, \dots, y_k^1)\}$ and $T = \emptyset$. Go to Step 2.

**Step 2:** Solve (4.18) with the simplex method. Let $\tilde{y}_i$ denote the optimal solution to (4.18). If $y_i^j = \tilde{y}_i$ for all $i = 1, \dots, k$, $(x^j, y_1^j, \dots, y_k^j)$ is a globally optimal solution to $S9$. Otherwise, go to Step 3.

**Step 3:** Let $W_{[j]}$ denote the set of adjacent vertices of $(x^j, y_1^j, \dots, y_k^j)$ such that $(x, y_1, \dots, y_k) \in W_{[j]}$ implies $F(x^j, y_1^j, \dots, y_k^j) \leq F(x, y_1, \dots, y_k)$. Let $T = T \cup \{(x^j, y_1^j, \dots, y_k^j)\}$ and $W = W \cup W_{[j]}/T$. Go to Step 4.

**Step 4:** Set $j \leftarrow j + 1$ and choose $(x^j, y_1^j, \dots, y_k^j)$ such that
$$F(x^j, y_1^j, \dots, y_k^j) = \min\{F(x, y_1, \dots, y_k) | (x, y_1, \dots, y_k) \in W\}.$$
Go back to Step 2.

[End]

---

We assume that there are two kinds of distributors, A and B, in this case. They have their own decision variables, objectives and constraints, but they cross-reference information by considering other followers' decision results in each of their own decision objectives and constraints. For example, distributor A considers distributor B's transportation price.

Before establishing a reference-cooperative BLMF model for the above problem, we first give the following notations.

$x$: the supplier's (leader's) decision variable;
$y$: the distributor A's (follower A's) decision variable;
$z$: the distributor B's (follower A's) decision variable;
$F(x, y, z)$: the supplier's objective function;
$f_1(x, y, z)$: the distributor A's objective function;
$f_2(x, y, z)$: the distributor B's objective function.

Let $X = \{x | x \geq 0\}$, $Y = \{y | y \geq 0\}$, $Z = \{z | z \geq 0\}$ with $x \in R$, $y \in R$, $z \in R$.

The supplier's objective is to minimize, over the set $X$, the total transportation cost of the system described by $\min F(x, y, z)$. Distributor A seeks to minimize its transportation time delay described by $\min f_1(x, y, z)$ over the set $Y$, and distributor B seeks to do the same by $\min f_2(x, y, z)$ over the set $Z$. Although the two distributors have different decision variables, decision objectives and constraints, each of them takes the other's decision variable into their objective and constraints as a reference.

This is a typical reference-uncooperative BLMF decision problem. The problem's model is presented as follows:

$$\min_{x \in X} \quad F(x, y, z) = -x + 2y + 3z$$

$$\text{s.t.} \quad x \geq 1,$$

$$\min_{y \in Y} \quad f_1(x, y, z) = x - y + z$$

$$\text{s.t.} \quad x + y + z \geq 1,$$

$$y \leq 1,$$

$$\min_{z \in Z} \quad f_2(x, y, z) = x + y - z$$

$$\text{s.t.} \quad x + y + z \leq 8,$$

$$x \leq 2,$$

$$z \leq 1.$$

According to the reference-uncooperative BLMF $K$th-Best algorithm, we consider the following problem:

$$\min_{x, y, z} \quad F(x, y, z) = -x + 2y + 3z$$

$$\text{s.t.} \quad x \geq 1,$$

$$x + y + z \geq 1,$$

$$y \leq 1,$$

$$x + y + z \leq 8,$$

$$x \leq 2,$$

$$z \leq 1.$$

Now we go through the reference-uncooperative BLMF $K$th-Best algorithm from Step 1 to Step 4.

In Loop 1:

In Step 1, set $j = 1$, and solve the above problem with the simplex method to obtain an optimal solution $\left(x_{[1]}, y_{[1]}, z_{[1]}\right) = (2, 0, 0)$. Let $W = \{(2, 0, 0)\}$ and $T = \emptyset$. Go to Step 2.

Setting $i \leftarrow 1$ and the formulation (4.18), we have

$$\min_{y \in Y} \quad f_1(x, y, z) = x - y + z$$

$$\text{s.t.} \quad x + y + z \geq 1,$$

$$y \leq 1,$$

$$x = 2,$$

$$z = 0.$$

Using the simplex method, we have $\tilde{y}_j = 1$. Because of $\tilde{y}_j \neq y_{[j]}$, we go to Step 3 and then have $W_{[j]} = \{(1,0,0),(2,1,0),(2,0,1)\}, T = \{(2,0,0)\}$ and $W = \{(1,0,0),(2,1,0),(2,0,1)\}$. We then go to Step 4. Update $j = 2$, and choose $(x_{[j]}, y_{[j]}, z_{[j]}) = (1,0,0)$, go back to Step 2.

In Loop 2:

Setting $i \leftarrow 1$ and the formulation (4.18), we have

$$\min_{y \in Y} \quad f_1(x,y,z) = x - y + z$$

$$\text{s.t.} \quad x + y + z \geq 1,$$

$$y \leq 1,$$

$$x = 1,$$

$$z = 0.$$

The same as Loop 1, by using the simplex method, we have $\tilde{y}_j = 1$. Because of $\tilde{y}_j \neq y_{[j]}$, we go to Step 3, and obtain $W_{[j]} = \{(2,0,0),(1,1,0),(1,0,1)\}, T = \{(2,0,0),(1,0,0)\}$, $W = \{(2,1,0),(2,0,1),(1,1,0),(1,0,1)\}$.

Then go to Step 4. Update $j = 3$, and choose $(x_{[j]}, y_{[j]}, z_{[j]}) = (2,1,0)$, then go to Step 2 again.

In Loop 3:

Setting $i \leftarrow 1$ and the formulation (4.18), we have

$$\min_{y \in Y} \quad f_1(x,y,z) = x - y + z$$

$$\text{s.t.} \quad x + y + z \geq 1,$$

$$y \leq 1,$$

$$x = 2,$$

$$z = 0.$$

Using the simplex method, we obtain $\tilde{y}_j = 1$ and $\tilde{y}_j = y_{[j]}$. This is a different situation from the last loop. We thus set $i \leftarrow 2$ and have a new expression of the distributor's function $f_2$ by the formulation (4.18):

$$\min_{z \in Z} \quad f_2(x,y,z) = x + y - z$$

$$\text{s.t.} \quad x + y + z \leq 8,$$

$$x \leq 2,$$

$$z \leq 1,$$

$$x = 2,$$

$$y = 1.$$

By using the simplex method as before, we have $\tilde{z}_j = 1$. Because $\tilde{z}_j \neq z_{[j]}$, we go to Step 3, and have $W_{[j]} = \{(2,0,0),(1,1,0),(2,1,1)\}$, $T = \{(2,0,0),(1,0,0),(2,1,0)\}$, $W = \{(2,0,1),(1,1,0),(1,0,1),(2,1,1)\}$.

We then go to Step 4. Update $j = 4$ and choose $\left(x_{[j]},y_{[j]},z_{[j]}\right) = (2,0,1)$, then we go back to Step 2.

In Loop 4:

Setting $i \leftarrow 1$ and by the formulation (4.18), we have

$$\min_{y \in Y} \quad f_1(x,y,z) = x - y + z$$

$$\text{s.t.} \quad x + y + z \geq 1,$$

$$y \leq 1,$$

$$x = 2,$$

$$z = 0.$$

Using the simplex method, we obtain $\tilde{y}_j = 1$ and $\tilde{y}_j \neq y_{[j]}$. We go to Step 3, and have $\quad W_{[j]} = \{(2,0,0),(1,0,1),(2,1,1)\}$, $\quad T = \{(2,0,0),(1,0,0),(2,1,0),(2,0,1)\}$, $W = \{(1,1,0),(1,0,1),(2,1,1)\}$.

Go to Step 4. Update $j = 5$ and we get $\left(x_{[j]},y_{[j]},z_{[j]}\right) = (1,1,0)$

In Loop 5:

Setting $i \leftarrow 1$ and by the formulation (4.18), we have:

$$\min_{y \in Y} \quad f_1(x,y,z) = x - y + z$$

$$\text{s.t.} \quad x + y + z \geq 1,$$

$$y \leq 1,$$

$$x = 1,$$

$$z = 0.$$

Using the simplex method, we obtain $\tilde{y}_j = 1$ and $\tilde{y}_j = y_{[j]}$. Set $i \leftarrow 2$, we have

$$\min_{z \in Z} \quad f_2(x,y,z) = x + y - z$$

$$\text{s.t.} \quad x + y + z \leq 8,$$

$$x \leq 2,$$

$$z \leq 1,$$

$$x = 1,$$

$$y = 1.$$

We have $\tilde{z}_j = 1$, $\tilde{z}_j \neq z_{[j]}$, go to Step 3. We obtain $W_{[j]} = \{(1,0,0),(1,1,1),$
$(2,1,0)\}$, $T = \{(2,0,0),(1,0,0),(2,1,0),(2,0,1),(1,1,0)\}$, $W = \{(1,1,1),(1,$
$0,1),(2,1,1)\}$.

We then go to Step 4. Update $j = 6$ and we get $\left(x_{[j]}, y_{[j]}, z_{[j]}\right) = (1,0,1)$.

In Loop 6:

Setting $i \leftarrow 1$ and by the formulation (4.18), we have

$$\min_{y \in Y} \; f_1(x,y,z) = x - y + z$$

$$\text{s.t.} \quad x + y + z \geq 1,$$
$$y \leq 1,$$
$$x = 1,$$
$$z = 1.$$

Using the simplex method, we obtain $\tilde{y}_j = 1$ and $\tilde{y}_j \neq y_{[j]}$. We go to Step 3, and have
$W_{[j]} = \{(1,0,0),(1,1,1),(2,0,1)\}$,          $T = \{(2,0,0),(1,0,0),(2,1,0),(2,0,1),$
$(1,1,0),(1,0,1)$, $W = \{(1,1,1),(2,1,1)\}$.

We then go to Step 4. Update $j = 7$ and we get $\left(x_{[j]}, y_{[j]}, z_{[j]}\right) = (2,1,1)$.

In Loop 7:

Setting $i \leftarrow 1$ and by the formulation (4.18), we have

$$\min_{y \in Y} \; f_1(x,y,z) = x - y + z$$

$$\text{s.t.} \quad x + y + z \geq 1,$$
$$y \leq 1,$$
$$x = 2,$$
$$z = 1.$$

Through using the bounded simplex method, we obtain $\tilde{y}_j = 1$ and $\tilde{y}_j = y_{[j]}$. Set
$i \leftarrow 2$, we have

$$\min_{z \in Z} \; f_2(x,y,z) = x + y - z$$

$$\text{s.t.} \quad x + y + z \leq 8,$$
$$x \leq 2,$$
$$z \leq 1,$$
$$x = 2,$$
$$y = 1.$$

We have $\tilde{z}_j = 1$, $\tilde{z}_j = z_{[j]}$.

It has been found that from Loop 7, the optimal solution of the reference-uncooperative BLMF problem occurs at the point (2, 1, 1) with the leader's objective function value $F^* = 3$, and two followers' objective function values $f_1^* = 2$ and $f_2^* = 2$ respectively.

### 4.6.4  Reference-Uncooperative BLMF Kuhn-Tucker Approach

Similar to Theorem 4.4 for the uncooperative linear BLMF decision problem, we have the following result for the BLMF in a reference-uncooperative situation (Lu et al. 2007).

**Theorem 4.12** *A necessary condition that* $(x^*, y_1^*, \ldots, y_k^*)$ *solves the linear BLMF problem* (4.11) *is that there exist (row) vectors* $(v_1^*, v_2^*, \ldots, v_k^*)$ *and* $(w_1^*, w_2^*, \ldots, w_k^*)$ *such that* $(x^*, y_1^*, \ldots, y_k^*, v_1^*, \ldots, v_k^*, w_1^*, \ldots, w_k^*)$ *solve:*

$$\min_{x, y_1, \ldots, y_k, v_1, \ldots, v_k, w_1, \ldots, w_k} F(x, y_1, \ldots, y_k) = cx + \sum_{i=1}^{k} d_i y_i$$

$$\text{s.t.} \quad Ax + \sum_{i=1}^{k} B_i y_i \leq b,$$

$$A_i x + \sum_{s=1}^{k} C_{is} y_s b_i, \qquad (4.19)$$

$$v_i C_{ii} - w_i = -e_{ii},$$

$$v_i \left( b_i - A_i x - \sum_{s=1}^{k} C_{is} y_s \right) + w_i y_i = 0,$$

$$x \geq 0, y_i \geq 0, v_i \geq 0, w_i \geq 0, i = 1, 2, \ldots, k.$$

Theorem 4.12 indicates that the most direct approach for solving (4.11) is to solve the equivalent problem (4.19). One of its advantages is that it allows a more robust model to be solved without introducing new computational difficulties.

Now, we use Example 4.4 proposed in Sect. 4.6.3 to illustrate the feasibility of the reference-uncooperative BLMF Kuhn-Tucker approach for solving the linear BLMF decision problem in a reference-uncooperative situation.

Using the Kuhn-Tucker conditions of all followers' problems, we can transform Example 4.4 into the following problem:

$$\min_{x,y,z,u,v} \quad F(x,y,z) = -x + 2y + 3z$$

$$\text{s.t.} \quad -1 - u_1 + u_2 - u_3 = 0,$$
$$u_1(x+y+z-1) = 0,$$
$$u_2(y-1) = 0,$$
$$u_3 y = 0,$$
$$-1 + v_1 + v_2 - v_3 = 0,$$
$$v_1(x+y+z-8) = 0,$$
$$v_2(z-1) = 0,$$
$$v_3 z = 0,$$
$$x \geq 1,$$
$$x+y+z \geq 1,$$
$$y \leq 1,$$
$$x+y+z \leq 8,$$
$$x \leq 2,$$
$$z \leq 1,$$
$$x, y, z, u = (u_1, u_2, u_3), v = (v_1, v_2, v_3) \geq 0.$$

Clearly, we have the following two possibilities:

$$\text{Case 1: } (u, v) = (0, 1, 0, 0, 1, 0),$$
$$\text{Case 2: } (u, v) = (0, 1, 0, 1, 0, 0).$$

For Case 1, the above problem can be rewritten as follows:

$$\min_{x,y,z} \quad F(x,y,z) = -x + 2y + 3z$$

$$\text{s.t.} \quad y - 1 = 0,$$
$$z - 1 = 0,$$
$$x \geq 1,$$
$$x+y+z \geq 1,$$
$$y \leq 1,$$
$$x+y+z \leq 8,$$
$$x \leq 2,$$
$$z \leq 1,$$
$$x, y, z \geq 0.$$

**Table 4.4** Procedures of
reference-uncooperative
BLMF Kuhn-Tucker
approach

| Case | $u$ | $v$ | $x$ | $y$ | $z$ | $F$ |
|---|---|---|---|---|---|---|
| 1 | (0, 1, 0) | (0, 1, 0) | 2 | 1 | 1 | 3 |
| 2 | (0, 1, 0) | (1, 0, 0) | | | | Infeasible |

Using the simplex method, we find that a solution occurs at the point $(x, y, z) = (2, 1, 1)$. By using the same approach as that used in Case 1, we obtain a solution for each case as shown in Table 4.4.

As shown in Table 4.4, the optimal solution occurs at the point (2, 1, 1) with the optimal value 3. We obtain the same solution through the reference-uncooperative BLMF Kuhn-Tucker approach and the reference-uncooperative BLMF $K$th-Best algorithm for Model $S9$.

## 4.7 Summary

Bi-level multi-follower decision-making is a common issue in organizational management. This chapter establishes a framework for the BLMF decision problem which identifies nine kinds of relationship between the followers. For each of the nine relationships, corresponding DERD and BLMF programming decision models are proposed. In particular, this chapter proposes related theories that focus on the uncooperative, semi-cooperative and reference-uncooperative BLMF decision models $S1$, $S5$ and $S9$. To solve these BLMF decision models, a set of multi-follower $K$th-Best algorithms and Kuhn-Tucker approaches are presented. Some examples are adopted to illustrate how the proposed algorithms work. A decision support system implements the proposed techniques and will be discussed in Chap. 11.

# Chapter 5
# Bi-level Multi-leader Decision Making

In real-world applications, a bi-level decision problem may involve multiple decision entities on the upper level, that is, the bi-level decision problem has multiple leaders. The leaders may have their individual decision variables, objective functions and/or constraint conditions. This kind of bi-level decision problem is called a *bi-level multi-leader* (BLML) decision problem.

In this chapter, we first give a case-based example in Sect. 5.1 to illustrate what a BLML decision problem is. We then introduce a framework for the BLML decision problem in Sect. 5.2. Section 5.3 presents nine BLML decision models, *ML-S*1, *ML-S*2, ..., *ML-S*9, to describe different situations of a BLML decision process. Section 5.4 proposes related concepts and definitions of a typical BLML model, the reference-uncooperative BLML decision model *ML-S*9. A generalized Nash equilibrium solution concept for a BLML decision problem is developed in Sect. 5.5. Based on the solution concept, a *bi-level multi-leader particle swarm optimization* (BLML-PSO) algorithm is given in Sect. 5.6. A numerical example is ultimately adopted to illustrate the BLML decision model and algorithm in Sect. 5.7.

## 5.1 Problem Identification

In a bi-level decision problem, the leader at the upper level might consist of multiple decision entities.

*Example 5.1* There are several suppliers (also called leaders) in a supply chain market, but one consumer (also called follower). These suppliers have their individual variables, objective functions and/or constraint conditions. They decide on factors such as pricing, contracting, scheduling, inventory, and transportation strategies within space constraints, stack ability constraints, load and unloading rules, warehouse efficiency, load stability, and other constraint limitations, and their objective is to maximize individual profit. The consumer makes a decision which maximizes its own objective of 'paying less, but achieving higher quality commodity and service'.

© Springer-Verlag Berlin Heidelberg 2015
G. Zhang et al., *Multi-Level Decision Making*,
Intelligent Systems Reference Library 82, DOI 10.1007/978-3-662-46059-7_5

This is a BLML decision problem. The suppliers (leaders) and the consumer (follower) make sequential and independent decisions, but in making those decisions, they will be affected by the choices of the other parties involved.

When a bi-level decision problem involves multiple leaders and multiple followers, it becomes a *bi-level multi-leader and multi-follower* (BLMLMF) decision problem. In mathematical programming, it is described as an *equilibrium problem with equilibrium constraints* (EPEC). Related research can be found in Sherali (1984), Pang and Fukushima (2005), Hu and Ralph (2007), De et al. (2009) and will be not discussed in this book.

## 5.2 Framework for Bi-level Multi-leader Decision Making

The various relationships between leaders can result in different processes for deriving an optimal solution for the leaders' decision-making. The leaders' decisions will therefore not only be affected by the reactions of the follower, but also by the relationships between these leaders. Similar to the bi-level multi-follower decision, we can consider four main kinds of relationship between the leaders in a BLMF decision problem:

1. The *uncooperative situation*: there is no sharing of decision variables between the leaders.
2. The *cooperative situation*: the leaders share their decision variables, objectives and constraints, similar to group decision-making at the upper level.
3. The *semi-cooperative situation*: the leaders share the decision variables but may have individual objective functions and constraints. There are three sub-cases within this situation, determined by the relationships between the objective functions and constraints.
4. The *reference-uncooperative situation*: the leaders have individual decision variables but take other leaders' variables as references when making their own decisions. Four sub-cases are included in this case that involve individual objectives irrespective of constraints and common objective functions irrespective of constraints.

Based on the four basic cases and their sub-cases of relationships between followers, a framework for BLML decision problems is established to describe these nine situations in Table 5.1.

Each situation shown in the framework requires a specific BLML decision model for description and a specific approach for deriving an optimal solution for the decision model.

**Table 5.1** A framework for BLML (one follower) decision-making

| Relationships between leaders | Relationship factor | | | Multi-leader Situation (ML-Si) |
| --- | --- | --- | --- | --- |
| | Decision variables | Objectives | Constraints | |
| Uncooperative | Individual | Individual | Individual | ML-S1 |
| Cooperative | Sharing | Sharing | Sharing | ML-S2 |
| Semi-cooperative | Sharing | Sharing | Individual | ML-S3 |
| | | Individual | Sharing | ML-S4 |
| | | | Individual | ML-S5 |
| Reference-uncooperative | Individual, but take other leaders' variables as a reference | Sharing | Sharing | ML-S6 |
| | | | Individual | ML-S7 |
| | | Individual | Sharing | ML-S8 |
| | | | Individual | ML-S9 |

# 5.3   Linear Bi-level Multi-leader Decision Models

This section will describe all nine BLML decision models identified in the framework shown in Table 5.1. We can use the BLMF concepts (Neighborhood entities, Cooperative entities, Semi-cooperative entities, Uncooperative entities, and Reference-uncooperative entities) described in Sect. 4.3 and the DERD approach (see Fig. 4.1) to describe the leader relationships in a BLML decision problem.

Let $L \geq 2$ be the number of leaders, $x_i$, $i = 1, \ldots, L$, be the $i$th leader decision variable, and all the leaders' decision variables are abbreviated by $x = (x_1, \ldots, x_L)$. Otherwise, if the variables are shared, the leader decision variables are identical to each other and can be denoted by $x$. We use $y$ to denote the follower's decision variable. We will list the nine BLML decision models in programming form and also in DERD, as follows:

## 1. *ML-S1 Model for linear BLML decision problems*
For $x_i \in X_i \subset R^{m_i}, y \in Y \subset R^n, F_i : R^{m_i} \times R^n \to R, f : R^m \times R^n \to R, i = 1, \ldots, L$, a linear BLML decision problem in which $L$ leaders are involved and no decision variables, objective functions or constraints are shared between them is called an uncooperative BLML decision model. Since the leaders' decision involves a Nash Equilibrium, this model solved by the $i$th $(i = 1, \ldots, L)$ leader gives rise to the following problem:

$$\min_{x_i \in X_i} F_i(x_i, y) = c_i x_i + d_i y$$

$$\text{s.t.} \, A_i x_i + B_i y \leq b_i,$$

$$\min_{y \in Y} f(x, y) = cx + ey$$

$$\text{s.t.} \, Ax + By \leq b,$$

where $c_i \in R^{m_i}, d_i, e \in R^n, A_i \in R^{p_i \times m_i}, B_i \in R^{p_i \times n}, b_i \in R^{p_i}, c \in R^m, A \in R^{q \times m}, B \in R^{q \times n}, b \in R^q$.

**Fig. 5.1**  The DERD of
decision model *ML-S*1

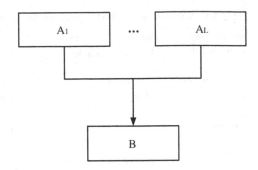

**Fig. 5.1**  The DERD of
decision model *ML-S*1

The uncooperative BLML decision model *ML-S*1 is described by DERD in Fig. 5.1, where $A_1$, $A_2$, …, $A_L$ are leaders and B is the follower.

## 2. *ML-S2 Model for linear BLML decision problems*

For $x \in X \subset R^m, y \in Y \subset R^n, F : R^m \times R^n \to R, f : R^m \times R^n \to R$, a linear BLML decision problem in which $L$ leaders are involved and the decision variables, objective functions and constraints are shared between them is called a *cooperative BLML decision model* and is defined as follows.

$$\min_{x \in X}\ F(x,y) = ax + dy$$

$$\text{s.t. } Cx + Dy \le h,$$

$$\min_{y \in Y}\ f(x,y) = cx + ey$$

$$\text{s.t. } Ax + By \le b,$$

where $ca, c \in R^m, d, e \in R^n, C \in R^{p \times m}, D \in R^{p \times n}, h \in R^p, A \in R^{q \times m}, B \in R^{q \times n}, b \in R^q$.

In fact, the cooperative BLML decision model reduces to a general (one leader and one follower) bi-level programming problem.

The cooperative BLML decision model *ML-S*2 is described by DERD in Fig. 5.2.

**Fig. 5.2**  The DERD of
decision model *ML-S*2

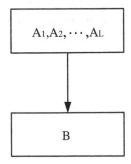

### 3. *ML-S3 Model for linear BLML decision problems*

For $x \in X \subset R^m, y \in Y \subset R^n, F : R^m \times R^n \to R, f : R^m \times R^n \to R$, a linear BLML decision problem in which $L$ leaders are involved and there are shared decision variables and objective functions but separate constraints between them is defined as follows (this is called a *semi-cooperative BLML decision model*).

$$\min_{x \in X} \ F(x, y) = ax + dy$$

$$\text{s.t. } A_i x + B_i y \le b_i,$$

$$\min_{y \in Y} f(x, y) = cx + ey$$

$$\text{s.t. } Ax + By \le b,$$

where $a, c \in R^m, d, e \in R^n, A_i \in R^{p_i \times m}, B_i \in R^{p_i \times n}, b_i \in R^{p_i}, A \in R^{q \times m}, B \in R^{q \times n}, b \in R^q$.

The semi-cooperative BLML decision model *ML-S3* is described by DERD in Fig. 5.3.

### 4. *ML-S4 Model for linear BLML decision problems*

For $x \in X \subset R^m, y \in Y \subset R^n, F_i : R^m \times R^n \to R, f : R^m \times R^n \to R, i = 1, \ldots, L$, a linear BLML decision problem in which $L$ leaders are involved and there are shared decision variables and constraints between them but separate objective functions is defined as follows (this is also called a *semi-cooperative BLML decision model*).

$$\min_{x \in X} \ F(x, y) = (a_1 x + d_1 y, \ldots, a_L x + d_L y)$$

$$\text{s.t. } Cx + Dy \le h,$$

$$\min_{y \in Y} f(x, y) = cx + ey$$

$$\text{s.t. } Ax + By \le b,$$

where $a_i, c \in R^m, d_i, e \in R^n, C \in R^{p \times m}, D \in R^{p \times n}, h \in R^p, A \in R^{q \times m}, B \in R^{q \times n}, b \in R^q$.

The *ML-S4* model is, in fact, a bi-level programming problem in which the upper level is a multi-objective optimization problem. It can be described by DERD in the same way as the *ML-S3* model in Fig. 5.3.

**Fig. 5.3** The DERD of decision model *ML-S3*

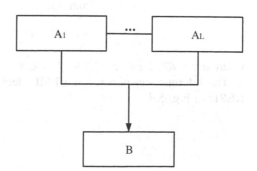

### 5. *ML-S5 Model for linear BLML decision problems*

For $x \in X \subset R^m$, $y \in Y \subset R^n$, $F_i : R^m \times R^n \to R$, $f : R^m \times R^n \to R$, $i = 1, \ldots, L$, a linear BLML decision problem in which $L$ leaders are involved and there are shared decision variables but separate objective functions and constraints among them is defined as follows (this is also called a *semi-cooperative BLMF decision model*).

$$\min_{x \in X} F(x, y) = (a_1 x + d_1 y, \ldots, a_L x + d_L y)$$

$$\text{s.t. } A_i x + B_i y \leq b_i,$$

$$\min_{y \in Y} f(x, y) = cx + ey$$

$$\text{s.t. } Ax + By \leq b,$$

where $a_i, c \in R^m, d_i, e \in R^n, A_i \in R^{p_i \times m}, B_i \in R^{p_i \times n}, b_i \in R^{p_i}, A \in R^{q \times m}, B \in R^{q \times n}$, $b \in R^q$.

The upper level of the semi-cooperative BLML decision model *ML-S5* is also a multi-objective programming; the difference between *ML-S5* and *ML-S4* models is that *ML-S5* has $L$ individual constraints while *ML-S4* does not. The *ML-S5* model can be described by DERD in the same way as *ML-S3* in Fig. 5.3.

### 6. *ML-S6 Model for linear BLML decision problems*

For $x_i \in X_i \subset R^{m_i}, y \in Y \subset R^n$, $F : R^m \times R^n \to R$, $f : R^m \times R^n \to R$, $i = 1, \ldots, L$, consider a linear BLML decision problem in which $L$ leaders are involved and individual decision variables are present in shared objective functions and constraints among them, but all leaders take others' decision variables as references (this is called a *reference-uncooperative BLML decision model*). In this case, the model solved by the $i$th leader gives rise to the following problem where the $i$th leader has to consider other leaders' variables and shares the objective function $F(x, y)$ and constraints $g(x, y)$, but only controls its own variable $x_i$:

$$\min_{x_i \in X_i} F(x, y) = ax + dy$$

$$\text{s. t. } g(x, y) = Cx + Dy - h \leq 0,$$

$$\min_{y \in Y} f(x, y) = cx + ey$$

$$\text{s.t. } Ax + By \leq b,$$

where $a, c \in R^m, d, e \in R^n, C \in R^{p \times m}, D \in R^{p \times n}, h \in R^p, A \in R^{q \times m}, B \in R^{q \times n}, b \in R^q$.

The reference-uncooperative BLML decision model *ML-S6* is described by DERD in Fig. 5.4.

**Fig. 5.4** The DERD of
decision model *ML-S6*

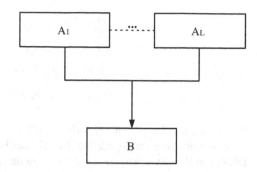

### 7. *ML-S7 Model for linear BLML decision problems*

For $x_i \in X_i \subset R^{m_i}, y \in Y \subset R^n, F : R^m \times R^n \to R, f : R^m \times R^n \to R, i = 1, \ldots, L$, consider a linear BLML decision problem in which $L$ leaders are involved and individual decision variables are present in shared objective functions and separate constraints, but the leaders take other leaders' decision variables as references (this is also called a *reference-uncooperative BLML decision model*). In this case, the model solved by the $i$th leader gives rise to the following problem where the $i$th leader has to consider other leaders' variables to optimize the shared objective function $F(x, y)$ but it only controls its own variable $x_i$ and subjects only to its constraints $g_i(x, y)$.

$$\min_{x_i \in X_i} F(x, y) = ax + dy$$

$$\text{s.t. } g_i(x, y) = A_i x + B_i y - b_i \leq 0,$$

$$\min_{y \in Y} f(x, y) = cx + ey$$

$$\text{s.t. } Ax + By \leq b,$$

where $a, c \in R^m, d, e \in R^n, A_i \in R^{p_i \times m}, B_i \in R^{p_i \times n}, b_i \in R^{p_i}, A \in R^{q \times m}, B \in R^{q \times n}, b \in R^q$ .

The reference-uncooperative BLML decision model *ML-S7* can be described by DERD in the same way as the *ML-S6* model in Fig. 5.4.

### 8. *ML-S8 Model for BLML decision problems*

For  $x_i \in X_i \subset R^{m_i}, y \in Y \subset R^n, F_i : R^m \times R^n \to R, f : R^m \times R^n \to R, i = 1, \ldots, L$, we consider a linear BLML decision problem in which $L$ leaders are involved and individual decision variables exist in separate objective functions and shared constraints between them, but each leader takes others' decision variables as references. This is also called a *reference-uncooperative BLML decision model*. In this case, the model solved by the $i$th leader gives rise to the following problem where the $i$th leader has to consider other leaders' variables, but shares the constraints $g(x, y)$ with other leaders, and independently optimizes its own objective function $F_i(x, y)$.

$$\min_{x_i \in X_i} F_i(x, y) = a_i x + d_i y$$

$$\text{s.t. } g(x, y) = Cx + Dy - h \leq 0,$$

$$\min_{y \in Y} f(x, y) = cx + ey$$

$$\text{s.t. } Ax + By \leq b,$$

where $a_i, c \in R^m, d_i, e \in R^n, C \in R^{p \times m}, D \in R^{p \times n}, h \in R^p, A \in R^{q \times m}, B \in R^{q \times n}, b \in R^q$.

The reference-uncooperative BLML decision model *ML-S8* is described by DERD in the same way as the *ML-S6* model in Fig. 5.4.

### 9. *ML-S9 Model for linear BLML decision problems*

For $x_i \in X_i \subset R^{m_i}, y \in Y \subset R^n, F_i : R^m \times R^n \to R, f : R^m \times R^n \to R, i = 1, \ldots, L,$ consider a linear BLML decision problem in which $L$ leaders are involved and individual decision variables exist in separate objective functions and constraints between them, but the leaders take other leaders' decision variables as references (this is also called a reference-uncooperative BLML decision model).In this case, this model solved by the $i$th leader gives rise to the following problem where the $i$th leader has to consider other leaders' variables $x$ but has its own objective function $F_i(x, y)$ and constraints $g_i(x, y)$.

$$\min_{x_i \in X_i} F_i(x, y) = a_i x + d_i y$$

$$\text{s.t. } g_i(x, y) = A_i x + B_i y - b_i \leq 0,$$

$$\min_{y \in Y} f(x, y) = cx + ey$$

$$\text{s.t. } Ax + By \leq b,$$

where $a_i, c \in R^m, d_i, e \in R^n, A_i \in R^{p_i \times m}, B_i \in R^{p_i \times n}, b_i \in R^{p_i}, A \in R^{q \times m}, B \in R^{q \times n}, b \in R^q$.

The reference-uncooperative BLML decision model *ML-S9* is described by DERD in the same way as *ML-S6* model in Fig. 5.4.

## 5.4  Concepts and Definitions

In this section, we only describe the definitions for the reference-uncooperative BLML decision problem, that is the *ML-S9* model, since it is the most general of the nine models (*ML-S1* to *ML-S9*) defined in Sect. 5.3. Interested readers can figure similar definitions for other BLML decision problems.

We reformulate the *ML-S9* model into the following problem:

$$\min_{x_i \in X_i} F_i(x_1, \ldots, x_L, y)$$

$$\text{s.t. } g_i(x_1, \ldots, x_L, y) \leq 0,$$

$$\min_{y \in Y} f(x, y) = cx + ey \tag{5.1}$$

$$\text{s.t. } Ax + By \leq b,$$

For problem (5.1), the multiple leaders and the follower have individual control variables, objective functions and constraints. Below, we provide some basic terms and symbols:

(a) Constraint region of problem (5.1):

$$S = \{(x_1, \ldots, x_L, y) | x_i \in X_i, g_i(x, y) \leq 0, Ax + By \leq b, y \in Y, i = 1, \ldots, L\}.$$

(b) Feasible set for the follower for each x:

$$S(x) = \{y | Ax + By \leq b, y \in Y\}.$$

(c) Projection of $S$ onto the $i$th leader's decision space:

$$S(X) = \{x_i | \exists y \in Y, (x_1, \ldots, x_L, y) \in S\}.$$

(d) The follower's rational reaction set:

$$P(x) = \{y \in Y | y \in \text{argmin}[f(x, \hat{y}) : \hat{y} \in S(x)]\}$$

where $\text{argmin}[f(z) : z \in S(x)] = \{z^* \in S(x) | f(z^*) \leq f(z), z \in S(x)\}$.

(e) The $i$th leader's feasible region:

$$Q(x_{-i}) = \{(x_i, y) | x_i \in X_i, g_i(x, y) \leq 0, y \in P(x), Ax + By \leq b\}$$

where $x_{-i} = (x_1, \ldots, x_{i-1}, x_{i+1}, \ldots, x_L)$.

(f) Inducible region or feasible region:

$$IR = \{(x_1, \cdots, x_L, y) : (x_1, \cdots, x_L, y) \in S, (x_i, y) \in Q(x_{-i}), \quad i = 1, \cdots, L\}$$

In terms of the above notions, for $i = 1, \ldots, L$, the BLML decision problem (5.1) can be written as:

$$\min_{(x_i, y) \in Q(x_{-i})} F_i(x_1, \ldots, x_L, y). \tag{5.2}$$

## 5.5  Generalized Nash Equilibrium Solution

In a BLML decision problem, the leaders have their individual variables, objective functions and constraints. However, each leader's decision will inevitably be made by guessing other leaders' strategies. This means that the upper level problem is a kind of game problem, that is, a *BLML game* problem. Given the objective of searching for equilibrium solutions, this BLML game problem is different to common single objective, multiple objectives or bi-level decision problems. A BLML game problem is also different from a conventionally generalized Nash equilibrium problem, which has no hierarchical structure. Here, it can be referred to as an extended generalized Nash equilibrium problem. Before addressing a BLML game problem, we first need to provide the definition of the solution, and then develop an algorithm by this definition.

Below, we give the solution concept for the BLML decision problem (5.1).

**Definition 5.1** A tuple $\left(x_1^*, \ldots, x_L^*, y^*\right)$ is called a generalized Nash equilibrium optimal solution to the BLML game problem (5.1), if it satisfies the following conditions: $\forall (x_i, y) \in Q\left(x_{-i}^*\right)$, $i = 1, \ldots, L$,

$$F_i\left(x_1^*, \ldots, x_L^*, y^*\right) \le F_i\left(x_1^*, \ldots, x_{i-1}^*, x_i, x_{i+1}^*, \ldots, x_L^*, y^*\right),$$

where $x^* = \left(x_1^*, \ldots, x_L^*, y^*\right), x_{-i}^* = \left(x_1^*, \ldots, x_{i-1}^*, x_i, x_{i+1}^*, \ldots, x_L^*, y^*\right)$.

To obtain a Nash equilibrium optimal solution for the BLML game problem (5.1), we define the optimal reaction from a leader as follows:

If the $i$th leader knows the other leaders' strategies $x_{-i}$, then let the optimal reaction of the $i$th leader be represented as follows $(x_i, y) \in \gamma_i(x_{-i})$ which solves the $i$th leader's problem:

$$\min_{x_i \in X_i} F_i(x_1, \ldots, x_L, y)$$

$$\text{s.t. } g_i(x_1, \ldots, x_L, y) \le 0,$$

$$\min_{y \in Y} f(x, y) = cx + ey$$

$$\text{s.t. } Ax + By \le b,$$

Our aim is to make a choice from each leader that is as close to the rational reaction as possible. A feasible solution is supposed to be a Nash equilibrium optimal solution when we achieve the situation where the choices of all the leaders are close enough to their corresponding rational reactions. Based on this strategy, we re-define a BLML decision problem as a BLML game problem:

$$\min_{x_1,\ldots,x_L,y} H(x_1,\ldots,x_L,y) = \sum_{i=1}^{L} \|(x_i,y) - \gamma_i(x_{-i})\|$$

$$\text{s.t. } x_i \in X_i$$

$$g_i(x_1,\ldots,x_L,y) \leq 0, i = 1,\ldots,L, \tag{5.3}$$

$$\min_{y \in Y} f(x,y) = cx + ey$$

$$\text{s.t. } Ax + By \leq b,$$

We obtain the following result, which shows the relationship between problems (5.1) and (5.3).

**Theorem 5.1** *Suppose that* $(x_1^*,\ldots,x_L^*,y^*)$ *is an optimal solution to problem* (5.3); *If* $H(x_1^*,\ldots,x_L^*,y^*) = 0$, *then* $(x_1^*,\ldots,x_L^*,y^*)$ *is also an optimal solution to problem* (5.1).

*Proof* Since $(x_1^*,\ldots,x_L^*,y^*)$ is an optimal solution to problem (5.3), we have $(x_1^*,\ldots,x_L^*,y^*) \in IR$. It follows from $H(x_1^*,\ldots,x_L^*,y^*) = 0$ that

$$(x_i^*,y^*) = \gamma_i(x_{-i}^*) \quad \text{for all } i \in \{1,\ldots,L\},$$

which implies that $(x_1^*,\ldots,x_L^*,y^*)$ solves problem (5.1). This completes the proof. □

The above theorem provides us with a new way to obtain a Nash equilibrium optimal solution to problem (5.1).

## 5.6 BLML Particle Swarm Optimization Algorithm

In this section, based on the definitions and the solution concept for a BLML decision problem (5.1), we use the strategy adopted in the *particle swarm optimization* (PSO) method (Biswal et al. 2008) to develop a *BLML particle swarm optimization* (BLML-PSO) algorithm to reach a Nash equilibrium solution for the BLML decision problem. The algorithm can be used to solve both linear and non-linear BLML decision models.

Suppose that the search space for the PSO is $n$-dimensional, and the $i$th particle of a swarm can be represented by a vector $x_i = (x_{i1},\ldots,x_{in})$. The velocity (position change) of this particle can be represented by another $n$-dimensional vector $v_i = (v_{i1},\ldots,v_{in})$. The best previously visited position of the $i$th particle is denoted as $p_i = (p_{i1},\ldots,p_{in})$. By defining $g$ as the index of the best particle in the swarm (i.e., the $g$th particle is the best), and letting the superscripts denote the iteration number, the swarm is manipulated according to the following two equations (Gao et al. 2008):

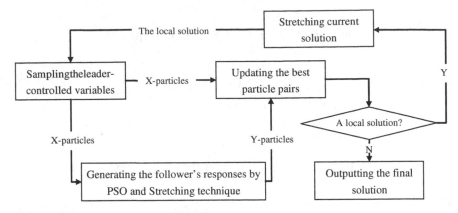

**Fig. 5.5** The outline of the BLML-PSO algorithm

$$
\begin{aligned}
v_{id}^{k+1} &= wv_{id}^{k} + cr_1^k\left(p_{id} - x_{id}^{k}\right) + cr_2^k\left(p_{gd}^k - x_{id}^k\right), \\
x_{id}^{k+1} &= x_{id}^k + v_{id}^{k+1},
\end{aligned}
\tag{5.4}
$$

where $d = 1, \ldots, n$, $i = 1, \ldots, N$ denotes the $i$th particle, $N$ is the size of the swarm, $w$ is the "inertia weight", $c$ is a positive constant, called the "acceleration constant", $r_1, r_2$ are random numbers, distributed uniformly in [0, 1], and $k$ determines the iteration number.

The BLML-PSO algorithm is outlined in Fig. 5.5 and the notations used in the subsequent paragraphs are detailed in Table 5.2.

**Table 5.2** Explanation of notation

| Variable | Description |
|---|---|
| $N_L$ | The number of candidate solutions (particles) for leaders |
| $N_f$ | The number of candidate solutions (particles) for the follower |
| $x_{ij}$ | The $j$th candidate solutions for the controlling variables from the $i$th leader |
| $p_{x_{ij}}$ | The best previously visited position of $x_{ij}$ |
| $x_{ij}^*$ | current best position for particle $x_{ij}$ |
| $v_{x_{ij}}$ | The velocity of $x_{ij}$ |
| $k_l$ | The current iteration number for the upper level problem |
| $y_i$ | The $i$th candidate solution for the controlling variables from the follower |
| $p_{y_i}$ | the best previously visited position of $y_i$ |
| $y^*$ | The current best position for particle $y$ |
| $v_{y_i}$ | The velocity of $y_i$ |
| $k_f$ | The current iteration number for the lower level problem |
| $MaxK_l$ | The predefined maximum iteration number for $k_l$ |
| $MaxK_f$ | The predefined maximum iteration number for $k_f$ |
| $w_l, w_f$ | The inertia weights for a leader and its follower respectively (coefficient for PSO) |
| $c_l, c_f$ | The acceleration constants for a leader and its follower respectively (coefficient for PSO) |
| $r_{1l}, r_{2l}, r_{1f}, r_{2f}$ | Random numbers uniformly distributed in [0, 1] for a leader and its follower respectively (coefficient for PSO) |

In the BLML-PSO algorithm, we first initiate a swarm comprised of the leader-controlled variables (X-particles) for each leader. For every X-particle from a leader, we construct the values of the X-particles from other leaders as constants and generate the optimal response from the follower by solving the follower's problem to obtain the leader's optimal strategy. To do this, we also need to generate a population (Y-particles) for the follower, each of which has a velocity. We then select the previously visited best positions for each Y-particle and the best position among the Y-particles. Having the current best positions, we adjust the velocities which are redirected towards these best positions. Every Y-particle will then be moved by its corresponding velocity in the manner illustrated in (5.4). Having obtained the optimal objective function value from the follower, the leader's objective function values can be calculated. Again with the help of the PSO optimization strategy, we obtain the leader's optimal strategy and find the difference between the optimal strategy and the samples (X-particles) for this leader. Once the sum of the differences between the X-particles and optimal strategies for all leaders is smaller than a pre-defined value, a Nash equilibrium optimal solution for the whole BLML decision problem is achieved.

The detailed BLML-PSO algorithm has two parts:

Algorithm 5.1, which generates the response from a follower, and

Algorithm 5.2, which generates optimal strategies for all leaders.

These two algorithms are specified as follows.

---

**Algorithm 5.1: Response Generation from a Follower**

[Begin]

**Step 1:** Input the values of $x_{lj}$ from $L$ leaders;

**Step 2:** Sample $N_f$ candidates $y_i$ and the corresponding velocities $v_{y_i}$, $i = 1, \ldots, N_f$;

**Step 3:** Initiate the follower's loop counter $k_f = 0$;

**Step 4:** Record the best particles $p_{y_i}$ and $y^*$ from $p_{y_i}$, $i = 1, \ldots, N_f$;

**Step 5:** Update velocities and positions using

$$v_{y_i}^{k+1} = w_f v_{y_i}^k + c_f r_{1i}^k \left( p_{y_i} - y_i^k \right) + c_f r_{2l}^k \left( y^{*k} - y_i^k \right)$$
$$y_i^{k+1} = y_i^k + v_{y_i}^{k+1}$$

**Step 6:** $k_f = k_f + 1$;

**Step 7:** If $k_f \geq MaxK_f$ or the solution changes for several consecutive generations are small enough, then we use Stretching technology to obtain the global solution and go to Step 8. Otherwise go to Step 5;

**Step 8:** Output $y^*$ as the response from the follower.

[End]

| **Algorithm 5.2: Optimal Strategy Generation for All Leaders** |
| --- |

[Begin]

**Step 1:** Sample $N_l$ particles of $x_{ij}$, and the corresponding velocities $v_{x_{ij}}$;

**Step 2:** Initiate the leaders' loop counter $k_l = 0$;

**Step 3:** For the $k$th particle, $k = 1,..., N_l$, calculate the optimal response for each leader;

    **Step 3.1:** Sample $N_l$ particles $x_{ij}$ within the constraints of $x_{ij}$;

    **Step 3.2:** By using Algorithm 5.1, we calculate the rational response from the follower;

    **Step 3.3:** Using the PSO technique, we obtain the optimal response for each leader;

**Step 4:** Calculate the function value of every particle by $F_i$;

**Step 5:** Record $p_{x_{ij}}$, $x_{ij}^*$, $j = 1,...,N$ for each $x_{ij}$, $j = 1,...,N$;

**Step 6:** Update velocities and positions using

$$v_{x_{ij}}^{k+1} = w_l v_{x_{ij}}^k + c_l r_{1l}^k \left(p_{x_{ij}} - x_{ij}^k\right) + c_l r_{2i}^k \left(x_{ij}^{*k} - x_{ij}^k\right),$$
$$x_{ij}^{k+1} = x_{ij}^k + v_{x_{ij}}^{k+1}.$$

**Step 7:** $k_l = k_l + 1$;

**Step 8:** If the sum of the differences between the samples and optimal strategies is smaller than $\varepsilon$ or $k_l \geq MaxK_l$, we use Stretching technology for the current leaders' solutions to obtain the global solution. Otherwise, goto Step 3.

[End]

This algorithm will be used in the next section to solve a numerical BLML decision problem.

## 5.7 A Numerical Example

We employ a numerical example to illustrate a BLML decision problem and use the BLML-PSO algorithm to reach solutions.

*Example 5.2* Suppose that a BLML decision problem has two leaders and one follower. Each leader has one objective and one constraint, and the follower has one objective and one constraint. The first leader controls the decision variable $x_1$, trying to minimize the objective $F_1$, and the second leader controls the decision variable $x_2$, trying to minimize the objective $F_2$. The follower controls the decision variable $y$ to minimize the objective function $f$. This BLML decision problem is specified as follows:

$$\min_{x_1} F_1(x_1, x_2, y) = 3x_1 + 4x_2 + 5y$$

s.t. $0 \le x_1 \le 3,$

$$\min_{x_2} F_2(x_1, x_2, y) = x_1 \cdot x_2 + y$$

s.t. $0 \le x_2 \le 5,$

$$\min_{y} f(x_1, x_2, y) = x_1 \cdot y + x_2$$

s.t. $0 \le y \le 10.$

Using the solution concept of the generalized Nash equilibrium, we have $(x_1, y) = \gamma_1(x_{-1})$ which solves the sub-problem:

$$\min_{x_1} F_1(x_1, x_2, y) = 3x_1 + 4x_2 + 5y$$

s.t. $0 \le x_1 \le 3,$

$$\min_{y} f(x_1, x_2, y) = x_1 \cdot y + x_2$$

s.t. $0 \le y \le 10.$

and $(x_2, y) = \gamma_2(x_{-2})$ which solves the sub-problem:

$$\min_{x_2} F_2(x_1, x_2, y) = x_1 \cdot x_2 + y$$

s.t. $0 \le x_2 \le 5,$

$$\min_{y} f(x_1, x_2, y) = x_1 \cdot y + x_2$$

s.t. $0 \le y \le 10.$

From the above problems, we have the following BLML game problem:

$$\min_{x_1, x_2, y} H(x_1, x_2, y) = \sum_{i=1}^{2} \|(x_i, y) - \gamma_i(x_{-i})\|$$

s.t. $0 \le x_1 \le 3,$

$0 \le x_2 \le 5,$

$$\min_{y} f(x_1, x_2, y) = x_1 \cdot y + x_2$$

s.t. $0 \le y \le 10.$

Using the BLML-PSO algorithm developed in this chapter, we reach a Nash equilibrium optimal solution for this BLML game problem as $(x_1^*, x_2^*, y) = (0, 0, 5.5432)$.

Under this solution, the objective values for the leaders and the follower are: $F_1 = 0, F_2 = 5.5432, f = 0.$

This solution means that when the first leader and the second leader choose their decision variables as 0, the follower will select a decision variable value of 5.5432. In this case, a Nash equilibrium status is achieved, and any movement by any leader will destroy this equilibrium.

## 5.8 Summary

This chapter presents bi-level decision problems with multiple leaders. After giving a mathematical definition of a BLML decision problem based on the Nash equilibrium concept, a BLML-PSO algorithm is developed. This algorithm is also used to solve a strategic bidding problem in electricity markets, which will be discussed in Chap. 12.

# Chapter 6
# Tri-level Multi-follower Decision Making

In a tri-level hierarchical decision problem, each decision entity at one level has its objective, constraints and decision variables affected in part by the decision entities at the other two levels. The choice of values for its variables may allow it to influence the decisions made at other levels, and thereby improve its own objective. We called this a tri-level decision problem. When multiple decision entities are involved at the middle and bottom levels, the top-level entity's decision will be affected not only by these followers' individual reactions but also by the relationships among the followers. We call this problem a *tri-level multi-follower* (TLMF) decision.

In this chapter, we first identify tri-level decision problems from real world cases in Sect. 6.1. We then introduce basic tri-level decision-making models in Sect. 6.2. Section 6.3 presents a framework for the TLMF decision through analyzing various kinds of relationships between decision entities in a tri-level decision problem. The TLMF decision framework contains 64 standard TLMF decision-making situations. To model these TLMF decision situations, we extend the bi-level *decision entity-relationship diagram* (DERD) approach introduced in Chap. 4 to describe tri-level decision problems. Furthermore, we establish a set of standard and hybrid TLMF decision models using a mathematical programming approach in Sect. 6.4. A set of case studies illustrates the development of TLMF decision models by DERD, as well as programming approaches, in Sect. 6.5. Section 6.6 gives solution concepts for a linear tri-level decision problem. It also presents a set of tri-level programming algorithms including a tri-level $K$th-Best algorithm. Section 6.7 focuses on solution methods for the proposed 64 kinds of TLMF decision model. To discuss this in detail, we take the TLMF decision model $S12$ in its linear version as a representative to illustrate solution concepts and theoretical properties, and to describe a TLMF $K$th-Best algorithm for TLMF decision-making. Finally, Sect. 6.8 summarizes this chapter.

© Springer-Verlag Berlin Heidelberg 2015
G. Zhang et al., *Multi-Level Decision Making*,
Intelligent Systems Reference Library 82, DOI 10.1007/978-3-662-46059-7_6

## 6.1 Problem Identification

Some decision problems require making a compromise between the objectives of several interacting *decision entities* (DE) allocated in a three-level hierarchy. The execution of decisions is sequential, from top to middle and then to bottom levels. Each decision entity independently optimizes (maximizes or minimizes) its own objective but is affected by the actions of other decision entities at the other two levels. Such a hierarchical decision process appears naturally in many organizations and business systems.

We use a university example here to explain the nature of the problem.

*Example 6.1* A university is organized with three faculties (Information, Business, Science) and each faculty has 2–4 departments. The university aims to improve its research quality through creating new research development strategies in 2013. The strategies made at university level directly affect the research strategy-making in its faculties. This process continues within a hierarchy of decision entities, including its departments and research centers. In the meantime, the actions at the faculty level may affect the research development strategies sought by the university and the actions at department level may affect those of its faculty. Each related decision entity in this university wishes to optimize its individual research development objective in view of the partial control exercised at other levels. The university's decision makers can control this effect by exercising preemptive-partial control over the university through budget modifications or regulations, but subject to possible reactions from its faculties and also departments. This kind of decision problem is called a multi-level decision problem or multi-level optimization problem.

The complexity of decision problems increases significantly when the number of levels ($n$) is greater than two (Blair 1992). The tri-level decision is the most typical form of multi-level decision ($n > 2$). In a tri-level decision, the decision entity at the top level is called the *leader,* while entities at the middle and bottom levels are the *followers.* However, a decision entity at the middle level is also the leader for associate entities at the bottom level. As a tri-level decision reflects the main features of multi-level decision problems, the models and methods developed for tri-level decisions can be easily extended to other multi-level decision problems.

The tri-level decision problem has been studied by researchers such as White (1997), Bard and Falk (1982a), Lai (1996) and Shih et al. (1996). The existing research results are mostly limited to the *one-level one-entity* situation. In real world tri-level decision applications, decisions are often made in situations where several decision entities are at the middle and bottom levels and interact with one another in some way. Consider Example 6.1. As these three faculties may have different objectives and different reactions to each possible decision made by the university, they should be treated as multiple entities at the middle level. These faculties may also have various relationships between each other, such as sharing their decision variables or not, and sharing their constraints or not, which may create different decision situations. As a result, the university's decision will be affected not only by its faculties' individual optimal reactions but also by the relationships between faculties and related departments. Some research, such as Shih et al. (1996), considered tri-level decision problems with multiple followers. However, very few studies classify the possible relationships among these followers and discuss different models to handle different situations.

Another issue related to tri-level decision-making is the relationship between the top-level decision entity and the bottom-level entities. In general, in a tri-level decision problem, the top-level decision entity's solution will be directly affected by the middle-level decision entities but indirectly affected by the bottom-level decision entities. However, in some cases, the solution of the top-level decision entity can be directly affected by the bottom-level entities' reactions as well. Considering Example 6.1, this university leader may also take a department's feedback in strategy making and in such a situation its decisions will be directly affected by its departments' reactions.

A more complex situation occurs when different entities at the same level have different decision situations. Considering Example 6.1, some faculties' departments make decisions (reactions) cooperatively while others do not. For example, all the departments in the Business Faculty react cooperatively to the decisions of the faculty, whereas the departments in the Information Faculty react uncooperatively to decisions made by the Faculty.

In summary, tri-level decisions involve a variety of situations caused by various possible relationships among multiple decision entities at two lower levels. The following sections will first provide basic tri-level decision models and will then model TLMF decision problems in various situations.

## 6.2  Basic Tri-level Decision Models

Basic tri-level decision focuses on a one-level one-entity situation and therefore has only three decision entities: DE1, DE2, and DE3. It can be described as follows (Bard and Falk 1982a):

$$\min_{x \in X} \quad f_1(x, y, z) \text{(DE1)}$$

$$\text{s.t.} \quad g_1(x, y, z) \leq 0,$$

where $y, z$ solve:

$$\min_{y \in Y} \quad f_2(x, y, z) \text{(DE2)}$$

$$\text{s.t.} \quad g_2(x, y, z) \leq 0,$$

where $z$ solves:

$$\min_{z \in Z} \quad f_3(x, y, z) \text{(DE3)}$$
$$\text{s.t.} \quad g_3(x, y, z) \leq 0,$$
(6.1)

where $x \in X \subset R^n$, $y \in Y \subset R^m$, $z \in Z \subset R^p$, $f_i : X \times Y \times Z \rightarrow R$, $i = 1, 2, 3$, variables $x$, $y$, $z$ are called the top-level, middle-level and bottom-level variables, and $f_1(x, y, z)$, $f_2(x, y, z)$, $f_3(x, y, z)$ are the top-level, middle-level and bottom-level objective functions respectively.

From the tri-level decision model (6.1), we can see that this decision problem has three optimization sub-problems (objective functions). Each level has individual control variables within its optimization sub-problem, but also considers other levels' variables in its optimization sub-problem. This decision process is sequential: decision entity DE1, at the top-level, selects an action within its specified constraint set, then DE2, at the middle-level, responds within its constraint set, and lastly DE3 responds.

To solve the tri-level decision problem, Bard and Falk (1982a) first developed a cutting plane algorithm and White (1997) developed a penalty function approach. In the meantime, Lai (1996) and Shih et al. (1996) extended the tri-level decision research in two aspects. One is that they developed a fuzzy approach to solve multi-level programming problems. The other is that a TLMF decision model is proposed in which multiple followers are at both middle and bottom levels. Below is a TLMF model. It assumes three sub-problems as centre $f_1 \rightarrow$ division $f_{2i} \rightarrow$ subdivision, $f_{3t}$, $t = 1, 2, \ldots, t_i$, $i = 1, 2, \ldots, s$ (Shih et al. 1996):

$$\min_{x_1} \quad f_1(x) = \sum_j c_{1j} x_j \text{ (top level)}$$

where $x_{2i}, x_{3i1}, \ldots, x_{3it_i}$ solve

$$\min_{x_{2i}} \quad f_{2i}(x) = \sum_j c_{2ij} x_j \text{ (middle level)}$$

where $x_{3i1}, \ldots, x_{3it_i}$ solve

$$\min_{x_{3i1}} \quad f_{3i1}(x) = \sum_j c_{3i1j} x_j \text{ (bottom level)}$$

$$\vdots$$

$$\min_{x_{3it_i}} \quad f_{3it_i}(x) = \sum_j c_{3it_ij} x_j$$
(6.2)

$$\text{s.t.} \quad A_1 x_1 + A_{2i} x_{2i} + A_{3i1} x_{3i1} + \cdots + A_{3it_i} x_{3it_i} \leq b,$$
$$x_j \geq 0, j = 1, 2, \ldots, n,$$

In this model, there is one decision entity at the top level, $s$ decision entities at the middle level and $t = \sum_i t_i$ decision entities at the bottom level. This is a general TLMF decision model with uncooperative relationships which adopts the decisions of other decision entities as references.

In the following section, we will provide more discussion on the TLMF decision models and solution methods.

## 6.3 Tri-level Multi-follower Decision Framework

This section first identifies seven issues which are related to the TLMF decision classification, and then presents a TLMF decision framework and a DERD modeling approach for TLMF decision situations.

### 6.3.1 TLMF Decision Concepts

When a tri-level decision problem has multiple followers at the middle level and/or the bottom level, we call it a TLMF decision problem. The model given in (6.1) describes a basic situation of tri-level decision, that is, each level has one decision entity only. Problem (6.2) presents the model for a general TLMF decision problem. In order to identify and classify TLMF decision situations, we first introduce the following concepts:

1. *Neighborhood entity*: two decision entities are at the same level, led by the same decision entity. All neighborhood entities under the same leader are called a neighborhood entity set (NES).
2. *Cooperative entity*: two neighborhood entities share their decision variables and have the same objective and constraint functions. In such a case, we consider the two entities as one.
3. *Semi-cooperative entity*: two neighborhood entities share their decision variables but have distinct objectives and constraint functions.
4. *Uncooperative entity*: two neighborhood entities have distinct decision variables, objectives, and constraints.
5. *Reference-uncooperative entity*: two neighborhood entities have distinct decision variables, objectives and constraints but take account of others' variables as references; that is, they include others' variables in their objective/constraint functions, but not as control variables.
6. *Direct* and *secondary follower*: all decision entities at the middle level are direct followers of the top-level decision entity (similarly, each bottom-level entity is a direct follower of an entity at the middle level); and all entities at the bottom level are secondary followers of the top-level decision entity.

7. *Direct leader and secondary leader*: a decision entity at the top level is the direct leader of all decision entities at the middle level (similarly, each bottom-level entity has a direct leader at the middle level) and is the secondary leader of all decision entities at the bottom level.

## 6.3.2 TLMF Decision Problem Classification

In a TLMF decision problem, a middle-level decision entity has two roles in decision-making process, that is, it reacts to each possible strategy made by the top-level entity and is influenced by the decisions of the followers at the bottom level. Different relationships between the decision entities at the middle level and bottom level could result in different processes for deriving an optimal solution for the decision entity at the top level. The top level's decision will also sometimes be affected by the reactions of its secondary followers as well as those of its direct followers. We therefore list the following relationships between decision entities for TLMF decision problems:

1. *Leader-follower relationship*: if an entity is a direct follower of another entity (leader), we say there is a leader-follower relationship or leadership relationship between the two entities.
2. *Secondary leadership relationship*: if the top-level decision entity directly considers the reactions of an entity at the bottom level, that is, includes a control variable of this bottom-level entity in its objective and/or constraints, we say that this top-level entity and the bottom-level entity's NES have a secondary leadership relationship.
3. *Uncooperative relationship*: if there are uncooperative entities but no reference-uncooperative entities in a NES, we say there is an uncooperative relationship in this NES.
4. *Reference-uncooperative relationship*: if there are reference-uncooperative entities in a NES and the rest are uncooperative, we say there is a reference-uncooperative relationship in this NES.
5. *Cooperative relationship at the middle level*: if all entities in a NES are cooperative, we say there is a cooperative relationship in this NES.
6. *Semi-cooperative relationship at the middle level*: if there are semi-cooperative entities in a NES and the rest, if any, are cooperative entities, we say there is a *semi-cooperative relationship* in this NES.
7. *Secondary followership relationship*: if a bottom-level decision entity includes the control variables of the top-level decision entity in its objective and/or constraints, we call the relationship between this bottom-level entity's NES and the top-level entity a secondary followership relationship.

### 6.3.3 TLMF Decision Framework

Based on the above seven relationships defined, a TLMF decision framework is established as shown in Table 6.1. The framework also presents a classification for TLMF decision problems. Under the eight features (SL, ML-V, ML-O, ML-R, SF, BL-V, BL-O, and BL-R) given in Table 6.1, "$Y$" means "yes", "$N$" means "no", and blank means 'not applicable'. A total of 64 standard situations of TLMF decision problems are identified, named $S1$, $S2$, ..., and $S64$ (note that some combinations of these features are not applicable). Each situation is described by using these seven relationships. We can describe any complex TLMF decision problem by combining two or more of these standard situations. For example, in a TLMF decision problem, a set of bottom-level entities are in the $S1$ situation and another set of bottom-level entities match the features of $S2$. We describe this problem of the combination of $S1$ and $S2$ as a hybrid situation.

The abbreviations used in Table 6.1 for the features are explained as follows:

1. SL: secondary leadership relationship;
2. ML-V: middle-level entities have the same variables;
3. ML-O: middle-level entities have the same objectives and constraints;
4. ML-R: middle-level entities include others' variables as references;
5. SF: secondary followership relationship;
6. BL-V: bottom-level entities have the same variables;
7. BL-O: bottom-level entities have the same objectives and constraints;
8. BL-R: bottom-level entities include others' variables as references.

### 6.3.4 TLMF Decision Entity-Relationship Diagrams

We have identified seven decision-entity relationships: a normal leader-follower relationship and six implicit relationships. These seven relationships are capable of fully reflecting the features of the TLMF decision problems identified in Table 6.1 and any of their combinations. Based on this, we introduce a TLMF *Decision Entity-Relationship Diagrams* (TLMF-DERD) approach and use it in TLMF modeling. Figure 6.1 presents diagrammatic notations of the TLMF-DERD approach.

This TLMF-DERD approach is a concept modeling of TLMF decision problems. In the following sections, we will show how a TLMF decision problem is first described by the DERD approach and then presented in a tri-level programming model.

**Table 6.1** TLMF decision framework with 64 standard situations

| Feature situation | SL | ML-V | ML-O | ML-R | SF | BL-V | BL-O | BL-R | Decision situation description |
|---|---|---|---|---|---|---|---|---|---|
| S1 | Y | Y | Y | | Y | Y | Y | | Both middle and bottom levels cooperative; both secondary leadership and followership |
| S2 | Y | Y | Y | | Y | Y | N | | Middle-level cooperative, bottom-level semi-cooperative; both secondary leadership and followership |
| S3 | Y | Y | Y | | Y | N | | Y | Middle-level cooperative, bottom-level reference-uncooperative; both secondary leadership and followership |
| S4 | Y | Y | Y | | Y | N | | N | Middle-level cooperative, bottom-level uncooperative; both secondary leadership and followership |
| S5 | Y | Y | Y | | N | Y | Y | | Both middle and bottom levels cooperative; secondary leadership only |
| S6 | Y | Y | Y | | N | Y | N | | Middle-level cooperative, bottom-level semi-cooperative; secondary leadership only |
| S7 | Y | Y | Y | | N | N | | Y | Middle-level cooperative, bottom-level reference-uncooperative; secondary leadership only |
| S8 | Y | Y | Y | | N | N | | N | Middle-level cooperative, bottom-level uncooperative; secondary leadership only |
| S9 | Y | Y | N | | Y | Y | Y | | Middle-level semi-cooperative; bottom-level cooperative; both secondary leadership and followership |
| S10 | Y | Y | N | | Y | Y | N | | Both middle and bottom levels semi-cooperative; both secondary leadership and followership |
| S11 | Y | Y | N | | Y | N | | Y | Middle-level semi-cooperative, bottom-level reference uncooperative; both secondary leadership and followership |
| S12 | Y | Y | N | | Y | N | | N | Middle-level semi-cooperative, bottom-level uncooperative; both secondary leadership and followership |

(continued)

**Table 6.1** (continued)

| Feature situation | SL | ML-V | ML-O | ML-R | SF | BL-V | BL-O | BL-R | Decision situation description |
|---|---|---|---|---|---|---|---|---|---|
| S13 | Y | Y | N |   | N | Y | Y |   | Middle-level semi-cooperative; bottom-level cooperative; secondary leadership only |
| S14 | Y | Y | N |   | N | Y | N |   | Both middle and bottom levels semi-cooperative; secondary leadership only |
| S15 | Y | Y | N |   | N | N |   | Y | Middle-level semi-cooperative, bottom-level reference-uncooperative; secondary leadership only |
| S16 | Y | Y | N |   | N | N |   | N | Middle-level semi-cooperative, bottom-level uncooperative; secondary leadership only |
| S17 | Y | N |   | Y | Y | Y | Y |   | Middle-level reference-uncooperative, bottom-level cooperative; both secondary leadership and followership |
| S18 | Y | N |   | Y | Y | Y | N |   | Middle-level reference-uncooperative, bottom-level semi-cooperative; both secondary leadership and followership |
| S19 | Y | N |   | Y | Y | N |   | Y | Both middle and bottom levels reference-uncooperative; both secondary leadership and followership |
| S20 | Y | N |   | Y | Y | N |   | N | Middle-level reference-uncooperative, bottom-level uncooperative; both secondary leadership and followership |
| S21 | Y | N |   | Y | N | Y | Y |   | Middle-level reference-uncooperative, bottom-level cooperative; secondary leadership only |
| S22 | Y | N |   | Y | N | Y | N |   | Middle-level reference-uncooperative, bottom-level semi-cooperative; secondary leadership only |
| S23 | Y | N |   | Y | N | N |   | Y | Both middle and bottom levels reference-uncooperative; secondary leadership only |
| S24 | Y | N |   | Y | N | N |   | N | Middle-level reference-uncooperative, bottom-level uncooperative; secondary leadership only |

(continued)

**Table 6.1** (continued)

| Feature situation | SL | ML-V | ML-O | ML-R | SF | BL-V | BL-O | BL-R | Decision situation description |
|---|---|---|---|---|---|---|---|---|---|
| S25 | Y | N |   | N | Y | Y | Y |   | Middle-level uncooperative; bottom-level cooperative; both secondary leadership and followership |
| S26 | Y | N |   | N | Y | Y | N |   | Middle-level uncooperative; bottom-level semi-cooperative; both secondary leadership and followership |
| S27 | Y | N |   | N | Y | N |   | Y | Middle-level uncooperative, bottom-level reference-uncooperative; both secondary leadership and followership |
| S28 | Y | N |   | N | Y | N |   | N | Both middle and bottom levels uncooperative; both secondary leadership and followership |
| S29 | Y | N |   | N | N | Y | Y |   | Middle-level uncooperative; bottom-level cooperative; secondary leadership only |
| S30 | Y | N |   | N | N | Y | N |   | Middle-level uncooperative, bottom-level semi-cooperative, secondary leadership only |
| S31 | Y | N |   | N | N | N |   | Y | Middle-level uncooperative, bottom-level reference-uncooperative; secondary leadership only |
| S32 | Y | N |   | N | N | N |   | N | Both middle and bottom levels uncooperative; secondary leadership only |
| S33 | N | Y | Y |   | Y | Y | Y |   | Both middle and bottom levels cooperative; secondary followership only |
| S34 | N | Y | Y |   | Y | Y | N |   | Middle-level cooperative, bottom-level semi-cooperative; secondary followership only |
| S35 | N | Y | Y |   | Y | N |   | Y | Middle-level cooperative, bottom-level reference-uncooperative; secondary followership only |
| S36 | N | Y | Y |   | Y | N |   | N | Middle-level cooperative, bottom-level uncooperative; secondary followership only |

(continued)

**Table 6.1** (continued)

| Feature situation | SL | ML-V | ML-O | ML-R | SF | BL-V | BL-O | BL-R | Decision situation description |
|---|---|---|---|---|---|---|---|---|---|
| S37 | N | Y | Y |  | N | Y | Y |  | Both middle and bottom levels cooperative; no secondary relationships |
| S38 | N | Y | Y |  | N | Y | N |  | Middle-level cooperative, bottom-level semi-cooperative; no secondary relationships |
| S39 | N | Y | Y |  | N | N |  | Y | Middle-level cooperative, bottom-level reference-uncooperative; no secondary relationships |
| S40 | N | Y | Y |  | N | N |  | N | Middle-level cooperative, bottom-level uncooperative; no secondary relationships |
| S41 | N | Y | N |  | Y | Y | Y |  | Middle-level semi-cooperative; bottom-level cooperative; secondary followership only |
| S42 | N | Y | N |  | Y | Y | N |  | Both middle and bottom levels semi-cooperative; secondary followership only |
| S43 | N | Y | N |  | Y | N |  | Y | Middle-level semi-cooperative, bottom-level reference-uncooperative; secondary followership only |
| S44 | N | Y | N |  | Y | N |  | N | Middle-level semi-cooperative, bottom-level uncooperative; secondary followership only |
| S45 | N | Y | N |  | N | Y | Y |  | Middle-level semi-cooperative; bottom-level cooperative; no secondary relationships |
| S46 | N | Y | N |  | N | Y | N |  | Both middle and bottom levels semi-cooperative, no secondary relationships |
| S47 | N | Y | N |  | N | N |  | Y | Middle-level semi-cooperative, bottom-level reference-uncooperative; no secondary relationships |
| S48 | N | Y | N |  | N | N |  | N | Middle-level semi-cooperative, bottom-level uncooperative; no secondary relationships |
| S49 | N | N |  | Y | Y | Y | Y |  | Middle-level reference-uncooperative, bottom-level cooperative; secondary followership only |
| S50 | N | N |  | Y | Y | Y | N |  | Middle-level reference-uncooperative, bottom-level semi-cooperative; secondary followership only |

(continued)

**Table 6.1** (continued)

| Feature situation | SL | ML-V | ML-O | ML-R | SF | BL-V | BL-O | BL-R | Decision situation description |
|---|---|---|---|---|---|---|---|---|---|
| S51 | N | N | | Y | Y | N | | Y | Both middle and bottom levels reference-uncooperative; secondary followership only |
| S52 | N | N | | Y | Y | N | | N | Middle-level reference-uncooperative, bottom-level uncooperative; secondary followership only |
| S53 | N | N | | Y | N | Y | Y | | Middle-level reference-uncooperative, bottom-level cooperative; no secondary relationships |
| S54 | N | N | | Y | N | Y | N | | Middle-level reference-uncooperative, bottom-level semi-cooperative; no secondary relationships |
| S55 | N | N | | Y | N | N | | Y | Both middle and bottom levels reference-uncooperative; no secondary relationships |
| S56 | N | N | | Y | N | N | | N | Middle-level reference-uncooperative, bottom-level uncooperative; no secondary relationships |
| S57 | N | N | | N | Y | Y | Y | | Middle-level uncooperative; bottom-level cooperative; secondary followership only |
| S58 | N | N | | N | Y | Y | N | | Middle-level uncooperative; bottom-level semi-cooperative; secondary followership only |
| S59 | N | N | | N | Y | N | | Y | Middle-level uncooperative, bottom-level reference-uncooperative; secondary followership only |
| S60 | N | N | | N | Y | N | | N | Both middle and bottom levels uncooperative; secondary followership only |
| S61 | N | N | | N | N | Y | Y | | Middle-level uncooperative; bottom-level cooperative; no secondary relationships |
| S62 | N | N | | N | N | Y | N | | Middle-level uncooperative, bottom-level semi-cooperative, no secondary relationships |
| S63 | N | N | | N | N | N | | Y | Middle-level uncooperative, bottom-level reference-uncooperative; no secondary relationships |
| S64 | N | N | | N | N | N | | N | Both middle and bottom levels uncooperative; no secondary relationships |

**Fig. 6.1** Notations for TLMF decision entity-relationship diagrams

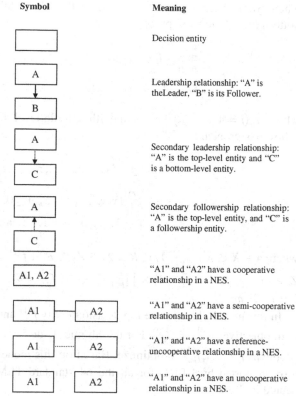

| Symbol | Meaning |
|---|---|
| | Decision entity |
| | Leadership relationship: "A" is theLeader, "B" is its Follower. |
| | Secondary leadership relationship: "A" is the top-level entity and "C" is a bottom-level entity. |
| | Secondary followership relationship: "A" is the top-level entity, and "C" is a followership entity. |
| | "A1" and "A2" have a cooperative relationship in a NES. |
| | "A1" and "A2" have a semi-cooperative relationship in a NES. |
| | "A1" and "A2" have a reference-uncooperative relationship in a NES. |
| | "A1" and "A2" have an uncooperative relationship in a NES. |

## 6.4 Tri-level Multi-follower Decision Models

This section first describes a general TLMF decision model using multi-level programming. It then presents a set of specific models for some standard TLMF decision problems including $S9$, $S12$, $S15$, $S18$, $S20$, $S25$ and $S32$ selected from Table 6.1. We also give a hybrid TLMF decision model for a decision situation which is the combination of $S63$ and $S64$.

### 6.4.1 General Model for TLMF Decision

A general TLMF decision model, which covers all the 64 TLMF decision situations, is given as follows:

$$\min_{x \in X} \quad f^{(1)}(x, y_1, \ldots, y_n, z_{11}, \ldots, z_{1m_1}, \ldots, z_{n1}, \ldots, z_{nm_n})$$
$$\text{s.t.} \quad g^{(1)}(x, y_1, \ldots, y_n, z_{11}, \ldots, z_{1m_1}, \ldots, z_{n1}, \ldots, z_{nm_n}) \leq 0,$$

where $y_i, z_{i1}, \ldots, z_{im_i} (i = 1, \ldots, n)$, solve the $i$th middle-level follower's and its bottom-level followers' problems:

$$\min_{y_i \in Y_i} \quad f_i^{(2)}(x, y_1, \ldots, y_i, \ldots, y_n, z_{i1}, \ldots, z_{im_i})$$

$$\text{s.t.} \quad g_i^{(2)}(x, y_1, \ldots, y_i, \ldots y_n, z_{i1}, \ldots, z_{im_i}) \leq 0,$$

where $z_{ij} (j = 1, \ldots, m_i)$ solves the $i$th middle-level follower's $j$th bottom-level follower's problem:

$$\min_{z_{ij} \in Z_{ij}} \quad f_{ij}^{(3)}(x, y_i, z_{i1}, \ldots, z_{ij}, \ldots, z_{im_i})$$

$$\text{s.t.} \quad g_{ij}^{(3)}(x, y_i, z_{i1}, \ldots, z_{im_i}) \leq 0, \tag{6.3}$$

$$i = 1, \ldots, n, \; j = 1, \ldots, m_i,$$

where $x \in X \subset R^{l_1}$, $y_i \in Y_i \subset R^{l_{2i}}$, $z_{ij} \in Z_{ij} \subset R^{l_{3ij}}$, $f^{(1)} : X \times \prod_{i=1}^{n} Y_i \times \prod_{i=1}^{n} \prod_{j=1}^{m_i} Z_{ij} \to R$, $f_i^{(2)} : X \times \prod_{i=1}^{n} Y_i \times \prod_{j=1}^{m_i} Z_{ij} \to R$, $f_{ij}^{(3)} : X \times Y_i \times \prod_{j=1}^{m_i} Z_{ij}$, $i = 1, \ldots, n$, $j = 1, \ldots, m_i$.

In this model, there is one top decision entity $f^{(1)}$ and $n$ middle decision entities with objectives $f_1^{(2)}, \ldots, f_n^{(2)}$. For the $i$th middle decision problem, there are $m_i$ sub-problems $f_{i1}^{(3)}, \ldots, f_{im_i}^{(3)}$ to optimize. Based on this model, we can establish models, also supported by DERD for all the 64 standard TLMF decision situations presented in Table 6.1.

## 6.4.2 Typical Standard Models for TLMF Decision

This section will present seven typical TLMF decision models from the 64 models proposed in Sect. 6.3.3 by using both DERD and tri-level programming approaches.

### 1. S9 Model

This model presents a TLMF decision problem which has the following features and is described by DERD in Fig. 6.2:

1. The top level entity takes the control variables of the decision entities at both middle and bottom levels into consideration in its objectives, that is, there is a secondary leadership relationship;
2. The middle-level decision entities have the same variables;
3. The middle-level decision entities have individual objective functions and constraints, that is, they have a semi-cooperative relationship;
4. The bottom-level decision entities include the control variables of the top-level entity, that is, there is a secondary followership relationship;
5. The bottom-level decision entities have the same variables;

**Fig. 6.2** The DERD of
TLMF decision situation $S9$

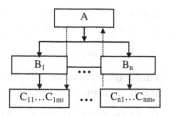

6. The bottom-level decision entities have the same objective functions and constraints, that is, they have a cooperative relationship.

We describe the $S9$ model by the tri-level programming approach as follows:

$$\min_{x \in X} \ f^{(1)}(x, y, z_1, \ldots, z_n)$$

$$\text{s.t.} \quad g^{(1)}(x, y, z_1, \ldots, z_n) \leq 0,$$

where $y, z_i (i = 1, \ldots, n)$ solve the $i$th middle-level follower's problem and its bottom-level followers' problems:

$$\min_{y \in Y_i} \ f_i^{(2)}(x, y, z_i)$$

$$\text{s.t.} \quad g_i^{(2)}(x, y, z_i) \leq 0,$$

where $z_i (i = 1, \ldots, n)$ solves the $i$th middle-level follower's bottom-level follower's problem:

$$\min_{z_i \in Z_i} \ f_i^{(3)}(x, y, z_i)$$

$$\text{s.t.} \quad g_i^{(3)}(x, y, z_i) \leq 0, \tag{6.4}$$

where $x \in X \subset R^{l_1}$, $y \in Y_i \subset R^{l_2}$, $z_i \in Z_i \subset R^{l_{3i}}$, $Y = Y_1 \cap \cdots \cap Y_n$, $f^{(1)} : X \times Y \times \prod_{i=1}^{n} Z_i \to R$, $f_i^{(2)} : X \times Y_i \times Z_i \to R$, $f_i^{(3)} : X \times Y_i \times Z_i \to R$, $i = 1, \ldots, n$.

In this model, there is one top-level decision entity $f^{(1)}$ and $n$ middle-level decision entities with objectives $f_1^{(2)}, \ldots, f_n^{(2)}$ respectively. Since these middle-level entities have a semi-cooperative relationship, we describe all middle-level followers as sharing a decision variable $y \in Y_i$ and having individual objective functions $f_i^{(2)}$ and the individual constraints $g_i^{(2)} \leq 0$. For any middle-level decision problem $f_i^{(2)}$, there are $m_i$ sub-problems $f_{i1}^{(3)}, \ldots, f_{im_i}^{(3)}$ at the bottom level. As all bottom-level neighborhood decision entities attached to the $i$th middle-level follower share variables, objective functions and constraints, that is, they are in a cooperative relationship. We describe this feature as the shared variable $z_i \in Z_i$ and $f_{i1}^{(3)} = \cdots = f_{im_i}^{(3)} = f_i^{(3)}$, $g_{i1}^{(3)} = \cdots = g_{im_i}^{(3)} = g_i^{(3)}$. To describe the secondary leadership relationship, we have $z_1, \ldots, z_n$ in the objective functions and constraints of the top-level decision entity.

### 2. *S12* Model

This model presents a TLMF decision problem which has the following features and is described by DERD in Fig. 6.3:

1. There is a secondary leadership relationship;
2. The decision entities at the middle level have the same variables;
3. The middle-level decision entities have a semi-cooperative relationship;
4. There is a secondary followership relationship;
5. The bottom-level decision entities have individual variables;
6. The bottom-level decision entities have an uncooperative relationship.

We describe the *S12* model by the tri-level programming approach as follows:

$$\min_{x \in X} \quad f^{(1)}\left(x, y, z_{11}, \ldots, z_{1m_1}, \ldots, z_{n1}, \ldots, z_{nm_n}\right)$$

$$\text{s.t.} \quad g^{(1)}\left(x, y, z_{11}, \ldots, z_{1m_1}, \ldots, z_{n1}, \ldots, z_{nm_n}\right) \leq 0,$$

where $y, z_{i1}, \ldots, z_{im_i} (i = 1, \ldots, n)$ solve the $i$th middle-level follower's problem and its bottom-level followers' problems:

$$\min_{y \in Y_i} \quad f_i^{(2)}\left(x, y, z_{i1}, \ldots, z_{im_i}\right)$$

$$\text{s.t.} \quad g_i^{(2)}\left(x, y, z_{i1}, \ldots, z_{im_i}\right) \leq 0,$$

where $z_{ij} (j = 1, \ldots, m_i)$ solves the $i$th middle-level follower's $j$th bottom-level follower's problem:

$$\min_{z_{ij} \in Z_{ij}} \quad f_{ij}^{(3)}\left(x, y, z_{ij}\right)$$

$$\text{s.t.} \quad g_{ij}^{(3)}\left(x, y, z_{ij}\right) \leq 0,$$

(6.5)

where $x \in X \subset R^{l_1}$, $y \in Y_i \subset R^{l_2}$, $z_{ij} \in Z_{ij} \subset R^{l_{3ij}}$, $Y = Y_1 \cap \cdots \cap Y_n$, $f^{(1)} : X \times Y \times \prod_{i=1}^{n} \prod_{j=1}^{m_i} Z_{ij} \to R$, $f_i^{(2)} : X \times Y_i \times \prod_{j=1}^{m_i} Z_{ij} \to R$, $f_{ij}^{(3)} : X \times Y_i \times Z_{ij} \to R$, $i = 1, \ldots, n, j = 1, \ldots, m_i$.

**Fig. 6.3** The DERD of TLMF decision situation S12

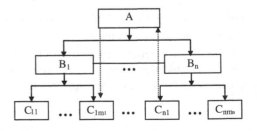

In this model, for the $i$th middle-level decision problem, there are $m_i$ sub-problems $f_{i1}^{(3)}, \ldots, f_{im_i}^{(3)}$ at the bottom level. As the bottom-level decision entities are uncooperative, that is, they have the individual decision variables $z_{ij} \in Z_{ij}$, objective $f_{ij}^{(3)}$ and constraint $g_{ij}^{(3)}$ for $i = 1, \ldots, n$, $j = 1, \ldots, m_i$.

### 3. S15 Model

This model presents a TLMF decision problem which has the following features and is described by DERD in Fig. 6.4:

1. There is a secondary leadership relationship;
2. The middle-level decision entities have the same variables;
3. The middle-level decision entities have a semi-cooperative relationship;
4. There is no secondary followership relationship;
5. The bottom-level decision entities have individual variables;
6. The bottom-level decision entities are reference-uncooperative.

We describe the S15 model by the tri-level programming approach as follows:

$$\min_{x \in X} \quad f^{(1)}(x, y, z_{11}, \ldots, z_{1m_1}, \ldots, z_{n1}, \ldots, z_{nm_n})$$
$$\text{s.t.} \quad g^{(1)}(x, y, z_{11}, \ldots, z_{1m_1}, \ldots, z_{n1}, \ldots, z_{nm_n}) \le 0,$$

where $y, z_{i1}, \ldots, z_{im_i}$ $(i = 1, \ldots, n)$ solve the $i$th middle-level follower's problem and its bottom-level followers' problems:

$$\min_{y \in Y_i} \quad f_i^{(2)}(x, y, z_{i1}, \ldots, z_{im_i})$$
$$\text{s.t.} \quad g_i^{(2)}(x, y, z_{i1}, \ldots, z_{im_i}) \le 0,$$

where $z_{ij}(j = 1, \ldots, m_i)$ solves the $i$th middle-level follower's $j$th bottom-level follower's problem:

$$\min_{z_{ij} \in Z_{ij}} \quad f_{ij}^{(3)}(y, z_{i1}, \ldots, z_{im_i})$$
$$\text{s.t.} \quad g_{ij}^{(3)}(y, z_{i1}, \ldots, z_{im_i}) \le 0, \tag{6.6}$$

**Fig. 6.4** The DERD of TLMF decision situation S15

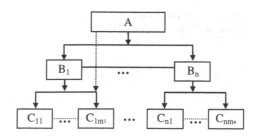

where $x \in X \subset R^{l_1}$, $y \in Y_i \subset R^{l_2}$, $z_{ij} \in Z_{ij} \subset R^{l_{3ij}}$, $Y = Y_1 \cap \cdots \cap Y_n$, $f^{(1)} : X \times Y \times \prod_{i=1}^{n} \prod_{j=1}^{m_i} Z_{ij} \to R$, $f_i^{(2)} : X \times Y_i \times \prod_{j=1}^{m_i} Z_{ij} \to R$, $f_{ij}^{(3)} : Y_i \times Z_{ij} \to R$, $i = 1, \ldots, n$, $j = 1, \ldots, m_i$.

In this model, the bottom level has no secondary followership relationship to the top level entity; there are only $y$, $z_{ij}$ as variables in the objectives $f_{ij}^{(3)}$ and constraints $g_{ij}^{(3)}$ of the bottom level. As the bottom-level decision entities attached to the $i$th middle-level follower are reference-uncooperative, we have $z_{i1}, \ldots, z_{im_i}$ in all objective functions $f_{i1}^{(3)}, \ldots, f_{im_i}^{(3)}$ and constraints $g_{i1}^{(3)}, \ldots, g_{im_i}^{(3)}$ of the bottom level for $i = 1, \ldots, n$.

### 4. *S*18 Model

This model will present a TLMF decision problem which has the following features and is described by DERD in Fig. 6.5:

1. There is a secondary leadership relationship;
2. The middle-level decision entities have individual variables;
3. The middle-level entities have a reference-uncooperative relationship;
4. There is a secondary followership relationship;
5. The bottom-level decision entities have the same variables;
6. The bottom-level entities have a semi-cooperative relationship.

We describe the *S*18 model by the tri-level programming approach as follows:

$$\min_{x \in X} \quad f^{(1)}(x, y_1, \ldots, y_n, z_1, \ldots, z_n)$$
$$\text{s.t.} \quad g^{(1)}(x, y_1, \ldots, y_n, z_1, \ldots, z_n) \le 0,$$

where $y_i, z_i (i = 1, \ldots, n)$ solve the $i$th middle-level follower's problem and its bottom-level followers' problems:

$$\min_{y_i \in Y_i} \quad f_i^{(2)}(x, y_1, \ldots, y_n, z_i)$$
$$\text{s.t.} \quad g_i^{(2)}(x, y_1, \ldots, y_n, z_i) \le 0,$$

where $z_i$ solves the $i$th middle-level follower's $j$th $(j = 1, 2, \ldots, m_i)$ bottom-level follower's problem:

**Fig. 6.5** The DERD of TLMF decision situation S18

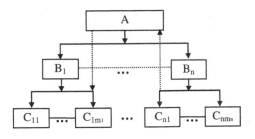

$$\min_{z_i \in Z_{ij}} \quad f_{ij}^{(3)}(x, y_i, z_i)$$

$$\text{s.t.} \quad g_{ij}^{(3)}(x, y_i, z_i) \le 0,$$

(6.7)

where $x \in X \subset R^{l_1}$, $y_i \in Y_i \subset R^{l_{2i}}$, $z_i \in Z_{ij} \subset R^{l_{3i}}$, $Z_i = Z_{i1} \cap \cdots \cap Z_{im_i}$, $f^{(1)} : X \times \prod_{i=1}^{n} Y_i \times \prod_{i=1}^{n} Z_i \to R$, $f_i^{(2)} : X \times Y_i \times Z_i \to R$, $f_{ij}^{(3)} : X \times Y_i \times Z_{ij} \to R$, $i = 1,$ $\ldots, n, j = 1, \ldots, m_i$.

In this model, the middle-level decision entities have a reference-uncooperative relationship so we have $y_1, \ldots, y_n$ in all objective functions $f_i^{(2)}$ and constraints $g_i^{(2)}$ of the middle level. As all decision entities at the bottom level have a semi-cooperative relationship, we have the shared variable $z_i \in Z_i$ for the $i$th middle-level follower's NES, $j = 1, \ldots, m_i, i = 1, \ldots, n$.

## 5. *S20* Model

This model will present a TLMF decision problem which has the following features and is described by DERD in Fig. 6.6:

1. There is a secondary leadership relationship;
2. The decision entities at the middle level have individual variables;
3. The middle-level entities have a reference-uncooperative relationship;
4. There is a secondary followership relationship;
5. The bottom-level entities have individual variables;
6. The bottom-level entities have an uncooperative relationship.

We describe the *S20* model by the tri-level programming approach as follows:

$$\min_{x \in X} \quad f^{(1)}(x, y_1, \ldots, y_n, z_{11}, \ldots, z_{1m_1}, \ldots, z_{n1}, \ldots, z_{nm_n})$$

$$\text{s.t.} \quad g^{(1)}(x, y_1, \ldots, y_n, z_{11}, \ldots, z_{1m_1}, \ldots, z_{n1}, \ldots, z_{nm_n}) \le 0,$$

where $y_i, z_{i1}, \ldots, z_{im_i} (i = 1, \ldots, n)$ solve the $i$th middle-level follower's problem and its bottom-level followers' problems:

$$\min_{y_i \in Y_i} \quad f_i^{(2)}(x, y_1, \ldots, y_n, z_{i1}, \ldots, z_{im_i})$$

$$\text{s.t.} \quad g_i^{(2)}(x, y_1, \ldots, y_n, z_{i1}, \ldots, z_{im_i}) \le 0,$$

**Fig. 6.6** The DERD of TLMF decision situation S20

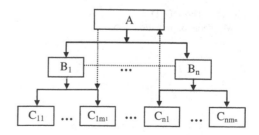

where $z_{ij}(j = 1, \ldots, m_i)$ solves the $i$th middle-level follower's $j$th bottom-level follower's problem:

$$\min_{z_{ij} \in Z_{ij}} \quad f_{ij}^{(3)}(x, y_i, z_{ij})$$

$$\text{s.t.} \quad g_{ij}^{(3)}(x, y_i, z_{ij}) \leq 0, \tag{6.8}$$

where $x \in X \subset R^{l_1}$, $y_i \in Y_i \subset R^{l_{2i}}$, $z_{ij} \in Z_{ij} \subset R^{l_{3ij}}$, $f^{(1)} : X \times \prod_{i=1}^{n} Y_i \times \prod_{i=1}^{n} \prod_{j=1}^{m_i} Z_{ij} \to R$, $f_i^{(2)} : X \times \prod_{i=1}^{n} Y_i \times \prod_{j=1}^{m_i} Z_{ij} \to R$, $f_{ij}^{(3)} : X \times Y_i \times Z_{ij} \to R$, $i = 1, \ldots, n, j = 1, \ldots, m_i$.

In this model, as all decision entities at the middle level have a reference-uncooperative relationship, we have $y_1, \ldots, y_n$ in the objective $f_i^{(2)}$ and constraint $g_i^{(2)}$ for $i = 1, \ldots, n$. While the bottom-level followers attached to the same middle-level follower have an uncooperative relationship, each bottom-level entity's objective function $f_{ij}^{(3)}$ and constraint $g_{ij}^{(3)}$ have no other counterparts' variables for $j = 1, \ldots, m_i, i = 1, \ldots, n$.

### 6. *S25* Model

This model will present a TLMF decision problem which has the following features and is described by DERD in Fig. 6.7:

1. There is a secondary leadership relationship;
2. The middle level entities have individual variables;
3. The middle-level entities have an uncooperative relationship;
4. There is a secondary followership relationship;
5. The bottom-level decision entities have the same variables;
6. The bottom-level neighborhood decision entities have a cooperative relationship.

We describe the *S25* model by the tri-level programming approach as follows:

$$\min_{x \in X} \quad f^{(1)}(x, y_1, \ldots, y_n, z_1, \ldots, z_n)$$

$$\text{s.t.} \quad g^{(1)}(x, y_1, \ldots, y_n, z_1, \ldots, z_n) \leq 0,$$

**Fig. 6.7** The DERD of TLMF decision situation S25

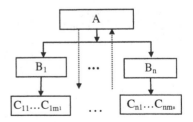

where $y_i, z_i (i = 1, \ldots, n)$ solve the $i$th middle-level follower's problem and its bottom-level followers' problems:

$$\min_{y_i \in Y_i} \quad f_i^{(2)}(x, y_i, z_i)$$

$$\text{s.t.} \quad g_i^{(2)}(x, y_i, z_i) \leq 0,$$

where $z_i$ solves the $i$th middle-level follower's bottom-level follower's problem:

$$\min_{z_i \in Z_i} \quad f_i^{(3)}(x, y_i, z_i)$$

$$\text{s.t.} \quad g_i^{(3)}(x, y_i, z_i) \leq 0, \tag{6.9}$$

where $x \in X \subset R^{l_1}$, $y_i \in Y_i \subset R^{l_{2i}}$, $z_i \in Z_i \subset R^{l_{3ij}}$, $f^{(1)} : X \times \prod_{i=1}^{n} Y_i \times \prod_{i=1}^{n} Z_i \to R, f_i^{(2)} : X \times Y_i \times Z_i \to R, f_{ij}^{(3)} : X \times Y_i \times Z_i \to R, i = 1, \ldots, n, j = 1, \ldots, m_i$.

In this model, as the middle-level decision entities have an uncooperative relationship, each middle-level entity objective function $f_i^{(2)}$ and constraint $g_i^{(2)}$ have no other counterparts' variables for $i = 1, \ldots, n$.

### 7. *S32* Model

This model will present a TLMF decision problem which has the following features and is described by DERD in Fig. 6.8:

1. There is a secondary leadership relationship;
2. The middle-level decision entities have individual variables;
3. The middle-level decision entities have an uncooperative relationship;
4. There is no secondary followership relationship;
5. The bottom-level decision entities have individual variables;
6. The bottom-level decision entities have an uncooperative relationship.

We describe the *S32* model by the tri-level programming approach as follows:

$$\min_{xx \in X} \quad f^{(1)}(x, y_1, \ldots, y_n, z_{11}, \ldots, z_{1m_1}, \ldots, z_{n1}, \ldots, z_{nm_n})$$

$$\text{s.t.} \quad g^{(1)}(x, y_1, \ldots, y_n, z_{11}, \ldots, z_{1m_1}, \ldots, z_{n1}, \ldots, z_{nm_n}) \leq 0,$$

**Fig. 6.8** The DERD of TLMF decision situation S32

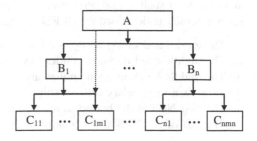

where $y_i, z_{i1}, \ldots, z_{im_i} (i = 1, \ldots, n)$ solve the $i$th middle-level follower's problem and its bottom-level followers' problems:

$$\min_{y_i \in Y_i} \quad f_i^{(2)}(x, y_i, z_{i1}, \ldots, z_{im_i})$$

$$\text{s.t.} \quad g_i^{(2)}(x, y_i, z_{i1}, \ldots, z_{im_i}) \leq 0,$$

where $z_{ij} (j = 1, \ldots, m_i)$ solves the $i$th middle-level follower's $j$th bottom-level follower's problem:

$$\min_{z_{ij} \in Z_{ij}} \quad f_{ij}^{(3)}(y_i, z_{ij})$$

$$\text{s.t.} \quad g_{ij}^{(3)}(y_i, z_{ij}) \leq 0, \tag{6.10}$$

where $x \in X \subset R^{l_1}, y_i \in Y_i \subset R^{l_{2i}}, z_{ij} \in Z_{ij} \subset R^{l_{3ij}}, f^{(1)} : X \times \prod_{i=1}^{n} Y_i \times \prod_{i=1}^{n} \prod_{j=1}^{m_i} Z_{ij} \to R, \ f_i^{(2)} : X \times Y_i \times \prod_{j=1}^{m_i} Z_{ij} \to R, \ f_{ij}^{(3)} : Y_i \times Z_{ij} \to R, \ i = 1, \ldots, n, \ j = 1, \ldots, m_i$.

In this model, $z_{ij}$ are included in the objective functions and constraints of the top-level decision entity to describe the secondary leadership relationship. As there is no secondary followership, however, the top-level variable $x$ is not included in the objectives $f_{ij}^{(3)}(y_i, z_{ij})$ and constraints $g_{ij}^{(3)}(y_i, z_{ij})$ of the bottom level decision problem. The decision entities at both middle and bottom level are uncooperative, so each entity's objective and constraints have only its variables, that is, $f_i^{(2)}$ and $g_i^{(2)}$ have only $y_i$, $f_{ij}^{(3)}$ and $g_{ij}^{(3)}$ have only $z_{ij}$, not other variables of the same level entities.

### 6.4.3 Hybrid TLMF Decision Models

Note that each of the 64 standard situations listed in Table 6.1 supposes that all entities at the same level have the same situations. For example, all the departments in all faculties of the university are uncooperative. However, in some real-world applications, the departments in the Faculty of Science are cooperative, and the departments in the Faculty of Business are uncooperative. We call this a hybrid TLMF decision problem and will describe it by a hybrid TLMF decision model. As an example of such hybrid problems, we present a TLMF decision problem in this section, which is described by DERD in Fig. 6.9:

1. The top-level decision entity is not in a secondary leadership relationship;
2. The middle-level decision entities have individual variables;
3. The middle-level decision entities are uncooperative;
4. There is no secondary followership;
5. The first NES at the bottom level are reference-uncooperative;
6. The rest of the NES at the bottom level have an uncooperative relationship.

**Fig. 6.9** The DERD of a
hybrid TLMF decision
situation

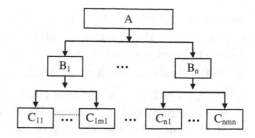

This problem is described by a hybrid model combining $S63$ and $S64$ as follows:

$$\min_{x \in X} \quad f^{(1)}(x, y_1, \ldots, y_n)$$

$$\text{s.t.} \quad g^{(1)}(x, y_1, \ldots, y_n) \leq 0,$$

where $y_i, z_{i1}, \ldots, z_{im_i}\,(i = 1, \ldots, n)$ solve the $i$th middle-level follower's problem and its bottom-level followers' problems:

$$\min_{y_i \in Y_i} \quad f_i^{(2)}(x, y_i, z_{i1}, \ldots, z_{im_i})$$

$$\text{s.t.} \quad g_i^{(2)}(x, y_i, z_{i1}, \ldots, z_{im_i}) \leq 0,$$

where $z_{1j}(j = 1, \ldots, m_1)$ solves the first middle-level follower's $j$th bottom-level follower's problem in a reference-uncooperative situation:

$$\min_{z_{1j} \in Z_{1j}} \quad f_{1j}^{(3)}(y_1, z_{11}, \ldots, z_{1m_1})$$

$$\text{s.t.} \quad g_{1j}^{(3)}(y_1, z_{11}, \ldots, z_{1m_1}) \leq 0,$$

where $z_{ij}(i \neq 1, j = 1, \ldots, m_i)$ solves the $i$th middle-level follower's $j$th bottom-level follower's problem in an uncooperative situation:

$$\min_{z_{ij} \in Z_{ij}} \quad f_{ij}^{(3)}(y_i, z_{ij})$$

$$\text{s.t.} \quad g_{ij}^{(3)}(y_i, z_{ij}) \leq 0, \tag{6.11}$$

where $x \in X \subset R^{l_1}$, $y_i \in Y_i \subset R^{l_{2i}}$, $z_{ij} \in Z_{ij} \subset R^{l_{3ij}}$, $f^{(1)} : X \times \prod_{i=1}^{n} Y_i \to R, f_i^{(2)} : X \times Y_i \times \prod_{j=1}^{m_i} Z_{ij} \to R, f_{1j}^{(3)} : Y_1 \times \prod_{j=1}^{m_1} Z_{1j} \to R, f_{ij}^{(3)} : Y_i \times Z_{ij} \to R, i = 1, \ldots, n, j = 1, \ldots, m_i$.

In this model, as there is no secondary leadership, $z_{ij}$ are not variables in the objective function and constraints of the top-level decision entity. Similarly, as there is no secondary followership, $x$ is not in the objectives and constraints of the bottom level decision entity. The decision entities at the middle level have an uncooperative relationship, so the $i$th entity has only its variables $y_i$ in $f_i^{(2)}$ and constraints $g_i^{(2)}$. The bottom-level entities have two kinds of relationship: the first NES is reference-cooperative (refer to $S63$) and the others are uncooperative (refer to $S64$). Therefore, we have $f_{1j}^{(3)}(y_1, z_{11}, z_{12}, \ldots, z_{1m_1})$ for the first NES and $f_{ij}^{(3)}(y_i, z_{ij})$ $(i = 2, 3, \ldots, n)$ for other NESs at the bottom level.

From the above analysis and discussions, using the standard and hybrid TLMF decision models, we can easily give the rest of the TLMF decision models according to the situations described in Table 6.1, as well as their hybrid models, based on the features of a decision problem.

## 6.5  Case Studies for TLMF Decision Modeling

In this section, we consider four tri-level multi-follower decision cases concerning research development strategy-making within a university to illustrate both DERD and programming approaches for TLMF decision modeling.

### 6.5.1  Case 1: S28 Model

Assume that the university's research strategy involves the university, its three faculties and departments. All three faculties have individual objectives, constraints, variables and do not take each other into consideration. The departments within each faculty are also uncooperative. The university takes the responses of both faculties and departments into account. At the same time, the faculties and departments fully consider the research strategies of the university. This TLMF decision problem is described in Fig. 6.10.

We give the variables, objectives and constraints of these decision entities as follows:

1. **The university (leader):**

Objective $f^{(1)}$ is to maximize research quantum which includes the number of publications (can be transformed to points) and research grant income (can be transformed to points). To achieve this aim, the main strategy of the university is to achieve a good balance between rewarding research performance and building a long-term research development environment. It has

$$\text{Variable } x = (x_1, x_2):$$

$x_1$  How much is used to reward the faculties' research performance, with the aim of encouraging faculties to attract more research grants and generate more publications;

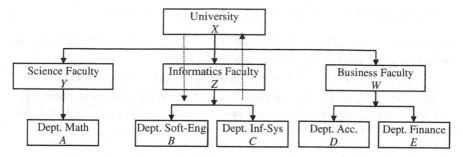

**Fig. 6.10** Case 1 of the university research development strategy-making

$x_2$   How much is used for the university's long-term research investment, such as earlier career researcher development, campus Intranet construction and lab establishment;

Constraints:

$g_1^{(1)} \leq 0$   annual research budget;

$g_2^{(1)} \leq 0$   a fixed number of students;

$g_3^{(1)} \leq 0$   a fix salary budget which is linked to total working hours.

2. *The three faculties (followers)*:

*Science Faculty*: Objective $f_1^{(2)}$ is to maximize the faculty's research budget from the university.

$$\text{Variable } y = (y_1, y_2) :$$

$y_1$   the points granted to reward publication;
$y_2$   the points granted to reward the securing of research grant income;

*Informatics Faculty:* Objective $f_2^{(2)}$ is to maximize the research budget from the university.

Variable:
$z$   how much is used to encourage publication;

*Business Faculty*: Objective $f_3^{(2)}$ is to maximize its research quantum by using the research budget from the university. It is developing a working load policy to reduce the teaching load for researchers who have a high research quantum;

Variable:
$w$   how many points of research quantum per \$ of research budget?

3. *The five departments in the three faculties* (*bottom followers*):

Objectives $f_{ij}^{(3)}, i = 1,2,3, j = 1,2$: all departments have the same objective, that is, to maximize the department's research performance;

Constraints $g_{ij}^{(3)}, i = 1,2,3, j = 1,2$: departments' constraints respectively;

Variables: $a, b, c, d,$ and $e$ are variables of the five departments respectively.

Clearly, this TLMF decision case meets the features of S28 in Table 6.1. We give this case's TLMF model as follows:

$$\max_{x} \quad f^{(1)}(x_1, x_2, y_1, y_2, z, w, a, b, c, d, e) \text{ (University level)}$$

$$\text{s.t.} \quad g^{(1)}(x_1, x_2, y_1, y_2, z, w, a, b, c, d, e) \leq 0,$$

$$\max_{y} \quad f_1^{(2)}(x_1, x_2, y_1, y_2, a) \text{ (Science faculty)}$$

$$\text{s.t.} \quad g_1^{(2)}(x_1, x_2, y_1, y_2, a) \leq 0,$$

$$\max_{a} \quad f_1^{(3)}(x_1, x_2, y_1, y_2, a) \text{ (Mathematics department)}$$

$$\text{s.t.} \quad g_1^{(3)}(x_1, x_2, y_1, y_2, a) \leq 0,$$

$$\max_{z} \quad f_2^{(2)}(x_1, x_2, z, b, c) \text{ (Informatics faculty)}$$

$$\text{s.t.} \quad g_2^{(2)}(x_1, x_2, z, b, c) \leq 0,$$

$$\max_{b} \quad f_{21}^{(3)}(x_1, x_2, z, b)(\text{Soft} - \text{Eng department})$$

$$\text{s.t.} \quad g_{21}^{(3)}(x_1, x_2, z, b) \leq 0,$$

$$\max_{c} \quad f_{22}^{(3)}(x_1, x_2, z, c) \text{ (Inf} - \text{Sys department)}$$

$$\text{s.t.} \quad g_{22}^{(3)}(x_1, x_2, z, c) \leq 0,$$

$$\max_{w} \quad f_3^{(2)}(x_1, x_2, w, d, e) \text{ (Business faculty)}$$

$$\text{s.t.} \quad g_3^{(2)}(x_1, x_2, w, d, e) \leq 0,$$

$$\max_{d} \quad f_{31}^{(3)}(x_1, x_2, w, d) \text{ (Acc.department)}$$

$$\text{s.t.} \quad g_{31}^{(3)}(x_1, x_2, w, d) \leq 0,$$

$$\max_{e} \quad f_{32}^{(3)}(x_1, x_2, w, e) \text{ (Finance department)}$$

$$\text{s.t.} \quad g_{32}^{(3)}(x_1, x_2, w, e) \leq 0,$$

where $x_1, x_2 \in R$ are the decision variables of the university; $y_1, y_2 \in R, z \in R, w \in R$ are of the three faculties respectively, $a, b, c, d, e \in R$ are of the five departments respectively, and $X = \{(x_1, x_2) | x_1 > 0, x_2 > 0\}, Y = \{(y_1, y_2) | y_1 > 0, y_2 > 0\}, Z = \{z | z > 0\}, W = \{w | w > 0\}, A = \{a | a > 0\}, B = \{b | b > 0\}, C = \{c | c > 0\}, D = \{d | d > 0\}, E = \{e | e > 0\}$. As there is a secondary leadership relationship, both objective functions $\max_x f^{(1)}(x_1, x_2, y_1, y_2, z, w, a, b, c, d, e)$ and constraint $g^{(1)}(x_1, x_2, y_1, y_2, z, w, a, b, c, d, e) \leq 0$ of the university include the decision variables of departments $a$, $b$, $c$, $d$, $e$. Similarly, by the secondary followership, $x = (x_1, x_2)$ is included in all departments' objectives and constraints.

## 6.5.2 *Case 2: S27 Model*

In this case, we suppose that all three faculties have uncooperative relationships and all the departments of each faculty have reference-uncooperative relationships. As in Case 1, the university takes into account the reactions of the faculties and of all departments. These departments fully consider both their faculty's and the university's strategies. From the TLMF decision framework in Table 6.1, this case refers to situation *S27* and is described by DERD in Fig. 6.11.

We suppose that the variables, objectives and constraints of decision entities in this case are the same as those of Case 1. This case's TLMF decision model is written as follows:

$$\max_{x} \quad f^{(1)}(x, y, z, w, a, b, c, d, e)$$

$$\text{s.t.} \quad g^{(1)}(x, y, z, w, a, b, c, d, e) \leq 0,$$

$$\max_{y} \quad f_1^{(2)}(x, y, a)$$

$$\text{s.t.} \quad g_1^{(2)}(x, y, a) \leq 0,$$

$$\max_{a} \quad f_1^{(3)}(x, y, a)$$

$$\text{s.t.} \quad g_1^{(3)}(x, y, a) \leq 0,$$

$$\max_{z} \quad f_2^{(2)}(x, z, b, c)$$

$$\text{s.t.} \quad g_2^{(2)}(x, z, b, c) \leq 0,$$

$$\max_{b} \quad f_{21}^{(3)}(x, z, b, c)$$

$$\text{s.t.} \quad g_{21}^{(3)}(x, z, b, c) \leq 0,$$

$$\max_{c} \quad f_{22}^{(3)}(x, z, b, c)$$

$$\text{s.t.} \quad g_{22}^{(3)}(x, z, b, c) \leq 0,$$

$$\max_{w} \quad f_3^{(2)}(x, w, d, e)$$

$$\text{s.t.} \quad g_3^{(2)}(x, w, d, e) \leq 0,$$

$$\max_{d} \quad f_{31}^{(3)}(x, w, d, e)$$

$$\text{s.t.} \quad g_{31}^{(3)}(x, w, d, e) \leq 0,$$

$$\max_{e} \quad f_{32}^{(3)}(x, w, d, e)$$

$$\text{s.t.} \quad g_{32}^{(3)}(x, w, d, e) \leq 0.$$

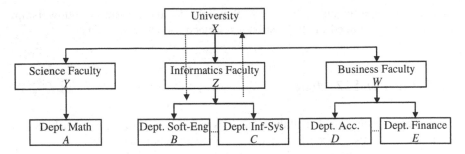

**Fig. 6.11** Case 2 of the university research development strategy-making

As there is a secondary leadership relationship, the university's objective function $f^{(1)}(x,y,z,w,a,b,c,d,e)$ and constraint $g^{(1)}(x,y,z,w,a,b,c,d,e)$ include the variables of departments $a,b,c,d,e$. Similarly, by the secondary followership, $x$ is included in all departments' objectives and constraints. As departments take into account their neighborhood decisions (reference-uncooperative), we have variable $d$ in the objective $f^{(3)}_{32}(x,w,d,e)$ and constraint $g^{(3)}_{32}(x,w,d,e)$ of departments $E$, and $e$ in the objective $f^{(3)}_{31}(x,w,d,e)$ and constraint $g^{(3)}_{31}(x,w,d,e)$ of departments $D$.

### 6.5.3  Case 3: S54 Model

In this case, all three faculties have a reference-uncooperative relationship and all the departments of each faculty have a semi-cooperative relationship. Unlike Case 2, the university does not take the departments' decisions directly into account, nor do all departments directly consider the university's research strategies during their decision process. It can be seen from Table 6.1 that this relates to situation $S54$. This problem's DERD is shown in Fig. 6.12. We use the same notations of

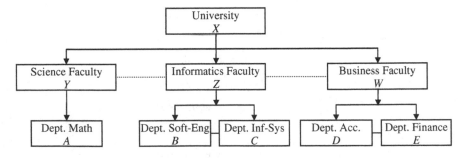

**Fig. 6.12** Case 3 of the university research development strategy-making

variables, objectives and constraints used in Case 1, but the Departments of Soft Eng. and Inf. Sys share variables $b$, and the Departments of Acc. and Finance share variables $e$. We have this case's TLMF decision model as follows:

$$\max_{x} \quad f^{(1)}(x,y,z,w)$$

$$\text{s.t.} \quad g^{(1)}(x,y,z,w) \leq 0,$$

$$\max_{y} \quad f_1^{(2)}(x,y,z,w,a)$$

$$\text{s.t.} \quad g_1^{(2)}(x,y,z,w,a) \leq 0,$$

$$\max_{a} \quad f_1^{(3)}(y,a)$$

$$\text{s.t.} \quad g_1^{(3)}(y,a) \leq 0,$$

$$\max_{z} \quad f_2^{(2)}(x,y,z,w,b)$$

$$\text{s.t.} \quad g_2^{(2)}(x,y,z,w,b) \leq 0,$$

$$\max_{b} \quad f_{21}^{(3)}(z,b)$$

$$\text{s.t.} \quad g_{21}^{(3)}(z,b) \leq 0,$$

$$\max_{c} \quad f_{22}^{(3)}(z,b)$$

$$\text{s.t.} \quad g_{22}^{(3)}(z,b) \leq 0,$$

$$\max_{w} \quad f_3^{(2)}(x,y,z,w,d)$$

$$\text{s.t.} \quad g_3^{(2)}(x,y,z,w,d) \leq 0,$$

$$\max_{d} \quad f_{31}^{(3)}(w,d)$$

$$\text{s.t.} \quad g_{31}^{(3)}(w,d) \leq 0,$$

$$\max_{e} \quad f_{32}^{(3)}(w,e)$$

$$\text{s.t.} \quad g_{32}^{(3)}(w,e) \leq 0.$$

As these faculties have a reference-cooperative relationship, their variables $y$, $z$, $w$ are included in all faculties' objective and constraints such as $f_1^{(2)}(x,y,z,w,a)$ and $g_1^{(2)}(x,y,z,w,a)$. To describe the semi-cooperative relationship between departments, we have $f_{2i}^{(3)}(z,b)$ and $f_{3i}^{(3)}(w,d)$ where variables $b$, $d$ are shared by two departments respectively. This case has no secondary relationships, so $x$ is not included in department functions and $a,b,d$ are not included in the university's objective function.

### 6.5.4 Case 4: Hybrid of S41, S45 and S48 Models

In this case, the three faculties have a semi-cooperative relationship by sharing the same variable $y$. The departments have different relationships in different faculties. In the Science Faculty, the Math department has a second followership relationship with the university. In the Informatics Faculty, the two departments have a cooperative relationship and no secondary relationship. Two departments in the Business Faculty have an uncooperative relationship and no secondary relationship. The three different situations refer to $S41$, $S45$, and $S48$ respectively. This is a hybrid TLMF decision problem. Figure 6.13 describes its DERD. By using the same variables, objectives and constraints of decision entities used in previous cases, we have the following TLMF decision model:

$$\max_{x} \quad f^{(1)}(x,y)$$

$$\text{s.t.} \quad g^{(1)}(x,y) \leq 0,$$

$$\max_{y} \quad f_1^{(2)}(x,y,a)$$

$$\text{s.t.} \quad g_1^{(2)}(x,y,a) \leq 0,$$

$$\max_{a} \quad f_1^{(3)}(x,y,a)$$

$$\text{s.t.} \quad g_1^{(3)}(x,y,a) \leq 0,$$

$$\max_{y} \quad f_2^{(2)}(x,y,b)$$

$$\text{s.t.} \quad g_2^{(2)}(x,y,b) \leq 0,$$

$$\max_{b} \quad f_{21}^{(3)}(y,b)$$

$$\text{s.t.} \quad g_{21}^{(3)}(y,b) \leq 0,$$

$$\max_{y} \quad f_3^{(2)}(x,y,d,e)$$

$$\text{s.t.} \quad g_3^{(2)}(x,y,d,e) \leq 0,$$

$$\max_{d} \quad f_{31}^{(3)}(y,d)$$

$$\text{s.t.} \quad g_{31}^{(3)}(y,d) \leq 0,$$

$$\max_{e} \quad f_{32}^{(3)}(y,e)$$

$$\text{s.t.} \quad g_{32}^{(3)}(y,e) \leq 0.$$

We can see that these three faculties share the same variable $y$ but have individual objectives. To describe the cooperative relationship between the departments in the Informatics Faculty, the two departments share variable $b$, objective function $f_{21}^{(3)}(z,b)$ and constraint $g_{21}^{(3)}(z,b)$. To describe the uncooperative relationship in the

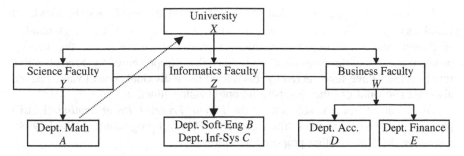

**Fig. 6.13** Case 4 of the university research development strategies making

Faculty of Business, its two departments' objective functions $f_{31}^{(3)}(y,d)$ and $f_{32}^{(3)}(y,e)$, have individual variables. As only the Math Department has a secondary relationship with the university level, $x$ is only included in the Math Department's functions.

Through these four cases, we present a way to model real-world TLMF decision problems by both DERD and programming approaches.

## 6.6 Tri-level Decision Solution Methods

This section focuses on a linear version of tri-level decision problems with a single decision entity at each level.

### 6.6.1 Solution Concepts

According to the basic tri-level decision model (6.1) in a one-level one-entity situation, we present a linear tri-level programming (decision model) as follows.

For $x \in X \subset R^n, y \in Y \subset R^m, z \in Z \subset R^p, f^{(1)}, f^{(2)}, f^{(3)} : X \times Y \times Z \rightarrow R,$

$$
\begin{aligned}
&\min_{x \in X} \quad f^{(1)}(x,y,z) = \alpha_1 x + \beta_1 y + \mu_1 z \\
&\text{s.t.} \quad A_1 x + B_1 y + C_1 z \leq b_1, \\
&\qquad \min_{y \in Y} \quad f^{(2)}(x,y,z) = \alpha_2 x + \beta_2 y + \mu_2 z \\
&\qquad \text{s.t.} \quad A_2 x + B_2 y + C_2 z \leq b_2, \\
&\qquad\qquad \min_{z \in Z} \quad f^{(3)}(x,y,z) = \alpha_3 x + \beta_3 y + \mu_3 z \\
&\qquad\qquad \text{s.t.} \quad A_3 x + B_3 y + C_3 z \leq b_3,
\end{aligned}
\tag{6.12}
$$

where $\alpha_i \in R^n, \beta_i \in R^m, \mu_i \in R^p, b_i \in R^{q_i}, A_i \in R^{q_i \times n}, B_i \in R^{q_i \times m}, C_i \in R^{q_i \times p}, i = 1,2,3.$

The variables $x, y, z$ are called the top-level, middle-level, and bottom-level variables respectively, and $f^{(1)}(x, y, z), f^{(2)}(x, y, z), f^{(3)}(x, y, z)$ the top-level, middle-level, and bottom-level objective functions, respectively. In this model, the decision problem consists of three optimization sub-problems (represented by three objective functions) in a three-level hierarchy. Each level has individual control variables, but also takes account of other levels in its optimization function.

To obtain an optimal solution to the *Linear Tri-level Programming* (LTLP) problem (6.12) based on the solution concept of bi-level programming (Bard 1998), a solution definition is first proposed as follows:

**Definition 6.1**

(a)  Constraint region of the LTLP:

$$S = \{(x, y, z) | x \in X, y \in Y, z \in Z, A_i x + B_i y + C_i z \leq b_i, \, i = 1, 2, 3\}.$$

(b)  Constraint region of the middle level for each fixed $x \in X$:

$$S(x) = \{(y, z) \in Y \times Z | B_i y + C_i z \leq b_i - A_i x, i = 2, 3\}.$$

(c)  Feasible set for the bottom level for each fixed $(x, y) \in X \times Y$:

$$S(x, y) = \{z \in Z | C_3 z \leq b_3 - A_3 x - B_3 y\}.$$

(d)  Projection of $S$ onto the top level's decision space:

$$S(X) = \{x \in X | \exists (y, z) \in Y \times Z, (x, y, z) \in S\}.$$

(e)  Projection of $S$ onto the top and middle levels' decision space:

$$S(X, Y) = \{(x, y) \in X \times Y | \exists z \in Z, (x, y, z) \in S\}.$$

(f)  Rational reaction set of the bottom level for $(x, y) \in S(X, Y)$:

$$P(x, y) = \{z | z \in \mathrm{argmin}[f_3(x, y, \hat{z}) | \hat{z} \in S(x, y)]\}.$$

(g)  Rational reaction set for the middle level for $x \in S(X)$:

$$P(x) = \{(y, z) | (y, z) \in \mathrm{argmin}[f_2(x, \hat{y}, \hat{z}) | (\hat{y}, \hat{z}) \in S(x), \\ \hat{z} \in P(x, \hat{y})]\}.$$

(h)  Inducible region (*IR*):

$$IR = \{(x, y, z) | (x, y, z) \in S, (y, z) \in P(x)\}.$$

Therefore, problem (6.12) is equivalent to the following problem:

$$\min\{f_1(x,y,z)|(x,y,z) \in IR\}. \tag{6.13}$$

## 6.6.2 Theoretical Properties

The three assumptions stated below serve as an introduction to the solution existence theorem.

**Assumption 6.1**

1. $S$ is non-empty and compact.
2. $IR$ is non-empty.
3. $P(x)$ and $P(x,y)$ are point-to-point maps with respect to $x$ and $(x,y)$ respectively.

Three important LTLP theorems are proposed here. Theorem 6.1 proves the existence of an optimal solution of the LTLP model. Theorem 6.2 presents a way to obtain a solution to the LTLP problem. Theorem 6.3 provides the necessary foundations for developing a tri-level $K$th-Best algorithm.

**Theorem 6.1** *If the above assumptions are satisfied, there exists an optimal solution to the linear tri-level decision model* (6.13).

*Proof* Since neither $S$ or $IR$ is empty, there is at least one parameter value $x^* \in S(X)$ and $P(x^*) \neq \emptyset$. Consider a sequence $\{(x^t,y^t,z^t)\}_{t=1}^{\infty} \subseteq IR$ converging to $(x^*,y^*,z^*)$. Then, by the well-known results of linear parametric optimization, $(y^*,z^*) \in P(x^*)$. Hence, $(x^*,y_1^*,\ldots,y_k^*) \in IR$ that shows $IR$ is closed. By Assumption 6.1(1) and $IR \subseteq S$, $IR$ is also bounded. $IR$ is non-empty, so the problem (6.13) consists of minimizing a continuous function over a compact non-empty set, which implies that the problem has an optimal solution.                                        □

**Theorem 6.2** *The inducible region can be written equivalently as a piecewise linear equality constraint comprised of support hyper-planes of S.*

*Proof* Using the notations in the proof of Theorem 6.1, the inducible region $IR$ can be rewritten as follows:

$$IR = \{(x,y,z) \in S|\beta_2 y + \mu_2 z = \min\langle\beta_2\hat{y} + \mu_2\hat{z}|B_i\hat{y} + C_i\hat{z} \leq b_i - A_ix, \hat{y} \geq 0,$$
$$\hat{z} \geq 0, i = 2,3, \mu_3\hat{z} = \min[\mu_3\tilde{z}|C_3\tilde{z} \leq b_3 - A_3x - B_3y, \tilde{z} \geq 0]\rangle\}. \tag{6.14}$$

Let

$$Q(x) = \min[\beta_2 \hat{y} + \mu_2 \hat{z} | (\hat{y}, \hat{z}) \in S(x), \hat{z} \in \text{argmin}[\mu_3 \tilde{z} | \tilde{z} \in S(x, \hat{y})]], \tag{6.15}$$

$$Q(x, y) = \min[\mu_3 \tilde{z} | \tilde{z} \in S(x, y)]. \tag{6.16}$$

It is then necessary to prove that $Q(x)$ is a piecewise linear equality constraint.

According to the expressions for $Q(x)$ and $Q(x, y)$, the first step is to prove that $Q(x, y)$ is a piecewise linear equality constraint for any given $x$ and $y$. Because $Q(x, y)$ can be seen as a linear programming problem with parameters $x$ and $y$, the dual problem of $Q(x, y)$ is

$$\max\{u(A_3 x + B_3 y - b_3) | u C_3 \geq -\mu_3, u \geq 0\}. \tag{6.17}$$

This problem has the same optimal values as $Q(x, y)$ at the solution $u^*$. Let $u^1, \ldots, u^t$ be a listing of all the vertices of the constraint region of the dual problem given by $U = \{u | u C_3 \geq -\mu_3\}$. Because a solution of the dual problem occurs at a vertex of $U$, the equivalent problem is

$$\max\{u(A_3 x + B_3 y - b_3) | u \in \{u^1, \ldots, u^t\}\}. \tag{6.18}$$

This means that $Q(x, y)$ is a piecewise linear function.

Next, it will be proved that $Q(x)$ is a piecewise linear function. Suppose that $z^1, z^2, \ldots, z^s$ are solutions of problem $Q(x, y)$. For each $z^i$, $Q(x)$ becomes a programming problem with parameters $x$ and $z^i$. Therefore, there are $s$ parameterized programming problems, $Q(x)|_{z^1}, \ldots, Q(x)|_{z^s}$. Similarly, each $Q(x)|_{z^i}$ is a piecewise linear function. Hence, the set $IR$ can be rewritten as

$$IR = \bigcup_{i=1}^{s} \{(x, y, z^i) | \beta_2 y = Q(x)|_{z^i} - \mu_2 z^i\} \tag{6.19}$$

which is a piecewise linear equality constraint. $\qquad\qquad\qquad\qquad\qquad \square$

**Corollary 6.1** *A solution to the LTLP problem (6.12) occurs at a vertex of the IR.*

**Theorem 6.3** *The solution $(x^*, y^*, z^*)$ of the linear tri-level programming problem occurs at a vertex of S.*

*Proof* Let $(x^1, y^1, z^1), \ldots, (x^t, y^t, z^t)$ be the distinct vertices of $S$. Because any point in $S$ can be written as a convex combination of these vertices, let $(x^*, y^*, z^*) = \sum_{i=1}^{\bar{t}} \delta_i(x^i, y^i, z^i)$, where $\sum_{i=1}^{\bar{t}} \delta_i = 1, \delta_i > 0, i = 1, \ldots, \bar{t}$ and $\bar{t} \leq t$. It must be shown that $\bar{t} = 1$. Let us write the constraints of (6.12) at $(x^*, y^*, z^*)$ in their piecewise linear form (6.19):

$$0 = Q(x^*)|_{z^*} - \beta_2 y^* - \mu_2 z^*$$

$$= Q\left(\sum_{i=1}^{\bar{i}} \delta_i x^i\right)\bigg|_{z^*} - \beta_2\left(\sum_{i=1}^{\bar{i}} \delta_i y^i\right) - \mu_2\left(\sum_{i=1}^{\bar{i}} \delta_i z^i\right)$$

$$\leq \sum_{i=1}^{\bar{i}} \delta_i Q(x^i)|_{z^*} - \sum_{i=1}^{\bar{i}} \delta_i \beta_2 y^i - \sum_{i=1}^{\bar{i}} \delta_i \mu_2 z^i$$

$$= \sum_{i=1}^{\bar{i}} \delta_i(Q(x^i)|_{z^*} - \beta_2 y^i - \mu_2 z^i)$$

by the convexity of $Q(x)$. However, by definition $Q(x^i)|_{z^*}$,

$$Q(x^i)|_{z^*} = \min_{\substack{(y,z) \in S(x^i) \\ z \in P(x^i, y)}} (\beta_2 y + \mu_2 z) \leq \beta_2 y^i + \mu_2 z^i.$$

Therefore, $Q(x^i)|_{z^*} - \beta_2 y^i - \mu_2 z^i \leq 0, i = 1, \ldots, \bar{i}$. Noting that $\delta_i > 0, i = 1, \ldots, \bar{i}$, the equality in the preceding expression must hold, or else a contradiction would result in the sequence above. Consequently, $Q(x^i)|_{z^*} - \beta_2 y^i - \mu_2 z^i = 0$ for all $i$. These statements imply that $(x^i, y^i, z^i) \in IR, i = 1, \ldots, \bar{i}$, and that $(x^*, y^*, z^*)$ can be written as a convex combination of points in the $IR$. Because $(x^*, y^*, z^*)$ a vertex of the $IR$ by Corollary 6.1 and $P(x)$ and $P(x, y)$ are single-valued, a contradiction results unless $\bar{i} = 1$.                                          □

**Corollary 6.2** *If $(x, y, z)$ is a vertex of IR, then it is also a vertex of S.*

### 6.6.3 Tri-level Kth-Best Algorithm

This section will introduce the tri-level $K$th-Best algorithm for solving the linear tri-level programming problem (6.12).

Theorem 6.3 in Sect. 6.6.2 provides a theoretical foundation and a suitable way to solve problem (6.12). Therefore, it is necessary only to search the extreme points of the constraint region $S$ to find an optimal solution for the LTLP problem (6.12). The main principle of the tri-level $K$th-Best algorithm is shown as follows.

Consider the linear programming problem below:

$$\min\{\alpha_1 x + \beta_1 y + \mu_1 z | (x, y, z) \in S\}. \tag{6.20}$$

The $N$-ranked basic feasible solutions to (6.20) are:

$$\left(x_{[1]}, y_{[1]}, z_{[1]}\right), \left(x_{[2]}, y_{[2]}, z_{[2]}\right), \ldots, \left(x_{[N]}, y_{[N]}, z_{[N]}\right),$$

such that $\alpha_1 x_{[i]} + \beta_1 y_{[i]} + \mu_1 z_{[i]} \le \alpha_1 x_{[i+1]} + \beta_1 y_{[i+1]} + \mu_1 z_{[i+1]}, i = 1, \ldots, N-1$. Then solving the problem (6.12) is equivalent to finding the index

$$K^* = \min\{i \in \{1, \cdots N\} | (x_{[i]}, y_{[i]}, z_{[i]}) \in IR\}.$$

Therefore, a global solution is $(x_{[K^*]}, y_{[K^*]}, z_{[K^*]})$. Similarly, for fixing $x = x_{[i]}$, we have the middle-level and bottom-level problem (6.21) as follows:

$$
\begin{aligned}
\min_{y \in Y} \quad & \beta_2 y + \mu_2 z \\
\text{s.t.} \quad & A_2 x + B_2 y + C_2 z_2, \\
\min_{z \in Z} \quad & \mu_3 z \\
\text{s.t.} \quad & A_3 x + B_3 y + C_3 z_3.
\end{aligned}
\tag{6.21}
$$

Clearly, problem (6.21) is a general bi-level programming scenario which has been discussed in Chap. 3. We can use the $K$th-Best algorithm, the Kuhn-Tucker approach or the Branch-and-bound algorithm to solve this problem.

The procedure of the tri-level $K$th-Best algorithm is described as follows:

---

**Algorithm 6.1: Tri-level $K$th-Best Algorithm**

---

[Begin]

**Step 1:** Set $i \leftarrow 1$. Solve problem (6.20) using the simplex method to obtain the optimal solution, $(x_{[1]}, y_{[1]}, z_{[1]})$. Let $W = \{(x_{[1]}, y_{[1]}, z_{[1]})\}$ and $T = \emptyset$. Go to Step 2.

**Step 2:** Treat the problem as a top- (middle-, bottom-) level problem. This step is equivalent to solving the follower's (middle-, bottom-) decision problem (6.21) for $x = x_{[i]}$. Let $(\tilde{y}, \tilde{z})$ denote the optimal solution to (6.21). If $\tilde{y} = y_{[i]}$ and $\tilde{z} = z_{[i]}$, stop, and $(x_{[i]}, y_{[i]}, z_{[i]})$ is the globally optimal solution of (6.12) with $K^* = i$; otherwise, go to Step 3.

**Step 3:** Let $W_{[i]}$ denote the set of adjacent vertices of $(x_{[i]}, y_{[i]}, z_{[i]})$ such that $(x, y, z) \in W_{[i]}$ implies $\alpha_1 x + \beta_1 y + \mu_1 z \ge \alpha_1 x_{[i]} + \beta_1 y_{[i]} + \mu_1 z_{[i]}$. Let $T = T \cup \{(x_{[i]}, y_{[i]}, z_{[i]})\}$ and $W = (W \cup W_{[i]}) \backslash T$. Go to Step 4.

**Step 4:** Set $i \leftarrow i + 1$ and choose $(x_{[i]}, y_{[i]}, z_{[i]})$ so that

$$\alpha_1 x_{[i]} + \beta_1 y_{[i]} + \mu_1 z_{[i]} = \min\{\alpha_1 x + \beta_1 y + \mu_1 z | (x, y, z) \in W\}.$$

Go to Step 2.

[End]

---

The tri-level $K$th-Best algorithm uses two sub-algorithms: (1) the simplex algorithm, which can obtain an optimal solution for a linear programming problem, and (2) the algorithm for finding the adjacent vertices of a selected vertex. According to the results given by Bard (1984), a vertex is a geometrical interpretation of a feasible solution. Hence, enumerating the adjacent vertices is equivalent to enumerating all the basic feasible solutions for the decision problem.

## 6.6.4 A Numerical Example

We give an example to illustrate how the tri-level $K$th-Best algorithm can be used to solve a tri-level decision problem.

*Example*  6.2  For  $x \in X = \{x | x \geq 0\}, y \in Y = \{y | y \geq 0\}, z \in Z = \{z | z \geq 0\}, f^{(1)},$
$f^{(2)}, f^{(3)} : X \times Y \times Z \to R,$

$$\min_{x \in X} \quad f^{(1)} = x + y + 2z$$
$$\text{s.t.} \quad 2x + y + z \geq 14,$$
$$\min_{y \in Y} \quad f^{(2)} = x + y + 3z$$
$$\text{s.t.} \quad x + y \geq 4,$$
$$y \leq 6,$$
$$\min_{z \in Z} \quad f^{(3)} = x + y - z$$
$$\text{s.t.} \quad y + z \leq 8,$$
$$y + 4z \geq 8,$$
$$y + 2z \leq 13.$$

Now it is possible to use the tri-level $K$th-Best algorithm to obtain a solution for this problem. According to the tri-level $K$th-Best algorithm, solving this problem first requires consideration of the middle level and the bottom level as a whole (middle, bottom) and then solving the problem using the bi-level $K$th-Best algorithm.

From (6.20), let us consider a linear programming problem as follows:

$$\min_{x \in X} \quad f^{(1)} = x + y + 2z$$
$$\text{s.t.} \quad 2x + y + z \geq 14,$$
$$x + y \geq 4,$$
$$y \leq 6,$$
$$y + z \leq 8,$$
$$y + 4z \geq 8,$$
$$y + 2z \leq 13.$$

Now we go through the tri-level $K$th-Best algorithm from Step 1 to Step 4.

Step 1: Set $i \leftarrow 1$. Solve the above problem using the simplex method to obtain the optimal solution, $(x_{[1]}, y_{[1]}, z_{[1]}) = (6, 0, 2)$. Let $W = \{(x_{[1]}, y_{[1]}, z_{[1]})\}$ and $T = \emptyset$. Go to Step 2

Step 2: By the problem (6.21), we have the problem:

$$\min_{y \in Y} \quad f^{(2)} = x + y - z$$

$$\text{s.t.} \quad x + y \geq 4,$$
$$y \leq 6,$$
$$x = 6,$$

$$\min_{z \in Z} \quad f^{(3)} = x + y - z$$

$$\text{s.t.} \quad y + z \leq 8,$$
$$y + 4z \geq 8$$
$$y + 2z \leq 13$$
$$x = 6.$$

Using the bi-level $K$th-Best algorithm, we have $(\tilde{y}_{[1]}, \tilde{z}_{[1]}) = (6, 2) \neq (y_{[1]}, z_{[1]})$ and go to Step 3

Step 3: Find the adjacent vertices of $(x_{[1]}, y_{[1]}, z_{[1]})$ and we have $W_{[1]} = \{(3.75, 6, 0.5), (4, 0, 6)\}$, $T = \{(6, 0, 2)\}$ and $W = \{(3.75, 6, 0.5), (4, 0, 6)\}$. Go to Step 4

Step 4: Update $i = i + 1$, choose $(x_{[2]}, y_{[2]}, z_{[2]}) = (3.75, 6, 0.5)$ and go back to Step 2

Step 2: By the problem (6.21), we have the problem:

$$\min_{y \in Y} \quad f^{(2)} = x + y - z$$

$$\text{s.t.} \quad x + y \geq 4,$$
$$y \leq 6,$$
$$x = 3.75,$$

$$\min_{z \in Z} \quad f^{(3)} = x + y - z$$

$$\text{s.t.} \quad y + z \leq 8,$$
$$y + 4z \geq 8,$$
$$y + 2z \leq 13,$$
$$x = 3.75.$$

Using the bi-level $K$th-Best algorithm, we have $(\tilde{y}_{[2]}, \tilde{z}_{[2]}) = (6, 2) \neq (y_{[2]}, z_{[2]})$ and go to Step 3

Step 3:  Find the adjacent vertices of $\left(x_{[2]}, y_{[2]}, z_{[2]}\right)$ and we have $W_{[2]} = \{(3,6,2)\}$,
$T = \{(6,0,2), (3.75, 6, 0.5)\}$ and $W = \{(4,0,6), (3,6,2)\}$. Go to Step 4

Step 4:  Update $i = i + 1$, choose $\left(x_{[3]}, y_{[3]}, z_{[3]}\right) = (3,6,2)$ and go back to Step 2

Step 2:  By the problem (6.21), we have the problem:

$$\min_{y \in Y} \quad f^{(2)} = x + y - z$$

$$\text{s.t.} \quad x + y \geq 4,$$

$$y \leq 6,$$

$$x = 3,$$

$$\min_{z \in Z} \quad f^{(3)} = x + y - z$$

$$\text{s.t.} \quad y + z \leq 8,$$

$$y + 4z \geq 8,$$

$$y + 2z \leq 13,$$

$$x = 3.$$

Using the bi-level $K$th-Best algorithm, we have $\left(\tilde{y}_{[3]}, \tilde{z}_{[3]}\right) = (6,2)$ $= \left(y_{[3]}, z_{[3]}\right)$. Therefore, $\left(x_{[3]}, y_{[3]}, z_{[3]}\right)$ is an optimal solution of Example 6.2 with $K^* = i = 3$. For the global solution, the objective value of $f_1$ is 13, and the objective function values of $f_2$ and $f_3$ are 15 and 7 respectively. Therefore, the $K$th-Best algorithm provides an useful way to solve the linear tri-level decision problem.

## 6.7 Tri-level Multi-follower Decision Solution Methods

We have proposed 64 kinds of TLMF decision model and this section aims to present solution methods for these models. We take the TLMF decision model $S12$ in its linear version as representative, to illustrate solution concepts and theoretical properties, and describe a TLMF $K$th-Best algorithm for TLMF decision.

### 6.7.1 Solution Concepts

According to the general model $S12$ shown in Sect. 6.4, the model in linear version can be expressed as follows.

For  $x \in X \subset R^k$,  $y \in Y_i \subset R^{k_0}$,  $Y = Y_1 \cap \cdots \cap Y_n$,  $y \in Y$,  $z_{ij} \in Z_{ij} \subset R^{k_{ij}}$, $f^{(1)} : X \times Y \times Z_{11} \times \cdots Z_{1m_1} \times \cdots \times Z_{n1} \times \cdots \times Z_{nm_n} \to R$, $f_i^{(2)} : X \times Y_i \times Z_{i1} \times \cdots \times Z_{im_i} \to R, f_{ij}^{(3)} : X \times Y_i \times Z_{ij} \to R$, and $j = 1, \ldots, m_i, i = 1, \ldots, n$,

$$\min_{x \in X} \quad f^{(1)}(x, y, z_{11}, \ldots, z_{1m_1}, \ldots, z_{n1}, \ldots, z_{nm_n}) = cx + dy + \sum_{i=1}^{n} \sum_{j=1}^{m_i} e_{ij} z_{ij} \quad (6.22a)$$

$$\text{s.t.} \quad Ax + By + \sum_{i=1}^{n} \sum_{j=1}^{m_i} C_{ij} z_{ij} \leq b, \quad (6.22b)$$

where $(y, z_{i1}, \ldots, z_{im_i})(i = 1, \ldots, n)$ is the solution to the $i$th middle-level follower's problem and its bottom-level followers' problems (6.22c–6.22f):

$$\min_{y \in Y_i} \quad f_i^{(2)}(x, y, z_{i1}, \ldots, z_{im_i}) = c_i x + d_i y + \sum_{j=1}^{m_i} g_{ij} z_{ij} \quad (6.22c)$$

$$\text{s.t.} \quad A_i x + B_i y + \sum_{j=1}^{m_i} D_{ij} z_{ij} \leq b_i, \quad (6.22d)$$

where $z_{ij}(j = 1, \ldots, m_i)$ is the solution to the $i$th middle-level follower's $j$th bottom-level follower's problem (6.22e–6.22f):

$$\min_{z_{ij} \in Z_{ij}} \quad f_{ij}^{(3)}(x, y, z_{ij}) = c_{ij} x + d_{ij} y + h_{ij} z_{ij} \quad (6.22e)$$

$$\text{s.t.} \quad A_{ij} x + B_{ij} y + E_{ij} z_{ij} \leq b_{ij}, \quad (6.22f)$$

where $c, c_i, c_{ij} \in R^k, d, d_i, d_{ij} \in R^{k_0}, e_{ij}, g_{ij}, h_{ij} \in R^{k_{ij}}, A \in R^{s \times k}, A_i \in R^{s_i \times k},$ $A_{ij} \in R^{s_{ij} \times k}, B \in R^{s \times k_0}, B_i \in R^{s_i \times k_0}, B_{ij} \in R^{s_{ij} \times k_0}, C_{ij} \in R^{s \times k_{ij}}, D_{ij} \in R^{s_i \times k_{ij}},$ $E_{ij} \in R^{s_{ij} \times k_{ij}}, b \in R^s, b_i \in R^{s_i}, b_{ij} \in R^{s_{ij}}, j = 1, \ldots, m_i, i = 1, \ldots, n.$

To find an optimal solution for the decision model, relevant solution concepts are proposed as follows, based on definitions of bi-level programming and tri-level programming.

**Definition 6.1**

(a) Constraint region of the TLMF decision model:

$$S = \{(x, y, z_{11}, \ldots, z_{1m_1}, \ldots, z_{n1}, \ldots, z_{nm_n}) \in X \times Y \times Z_{11} \times \cdots Z_{1m_1}$$

$$\times \cdots \times Z_{n1} \times \cdots \times Z_{nm_n} | Ax + By + \sum_{i=1}^{n} \sum_{j=1}^{m_i} C_{ij} z_{ij} \leq b,$$

$$A_i x + B_i y + \sum_{j=1}^{m_i} D_{ij} z_{ij} \leq b_i, A_{ij} x + B_{ij} y + E_{ij} z_{ij} \leq b_{ij},$$

$$j = 1, \ldots, m_i, i = 1, \ldots, n\}.$$

(b) Constraint region of the $i$th middle-level follower for each fixed $x \in X$:

$$S_i(x) = \{(y, z_{i1}, \ldots, z_{im_i}) \in Y_i \times Z_{i1} \times \cdots \times Z_{im_i} | A_i x + B_i y$$

$$+ \sum_{j=1}^{m_i} D_{ij} z_{ij} \leq b_i, A_{ij} x + B_{ij} y + E_{ij} z_{ij} \leq b_{ij}, j = 1, \ldots, m_i\}.$$

(c) Feasible set of the $i$th middle-level follower's $j$th bottom-level follower for each fixed $(x, y) \in X \times Y_i$:

$$S_{ij}(x, y) = \{z_{ij} \in Z_{ij} | A_{ij} x + B_{ij} y + E_{ij} z_{ij} \leq b_{ij}\}.$$

(d) Projection of $S$ onto the leader's decision space:

$$S(X) = \{x \in X | \exists (y, z_{11}, \ldots, z_{1m_1}, \ldots, z_{n1}, \ldots, z_{nm_n}),$$
$$(x, y, z_{11}, \ldots, z_{1m_1}, \ldots, z_{n1}, \ldots, z_{nm_n}) \in S\}.$$

(e) Projection of $S$ onto the top-level leader's and the $i$th middle-level follower's decision space:

$$S_i(X, Y) = \{(x, y) | \exists (z_{11}, \ldots, z_{1m_1}, \ldots, z_{n1}, \ldots, z_{nm_n}),$$
$$(x, y, z_{11}, \ldots, z_{1m_1}, \ldots, z_{n1}, \ldots, z_{nm_n}) \in S\}.$$

(f) Rational reaction set of the $i$th middle-level follower's $j$th bottom-level follower for $(x, y) \in S_i(X, Y)$:

$$P_{ij}(x, y) = \{z_{ij} \in Z_{ij} | z_{ij} \in \text{argmin} \left[ f_{ij}^{(3)}(x, y, \hat{z}_{ij}) : \hat{z}_{ij} \in S_{ij}(x, y) \right]\}.$$

(g) Rational reaction set of the $i$th middle-level follower for $x \in S(X)$:

$$P_i(x) = \{(y, z_{i1}, \ldots, z_{im_i}) | (y, z_{i1}, \ldots, z_{im_i}) \in \text{argmin}[f_i^{(2)}(x, \hat{y}, \hat{z}_{i1}, \ldots, \hat{z}_{im_i}) |$$
$$(\hat{y}, \hat{z}_{i1}, \ldots, \hat{z}_{im_i}) \in S_i(x), \hat{z}_{ij} \in P_{ij}(x, \hat{y}), j = 1, \ldots, m_i]\}.$$

(h) Inducible region:

$$IR = \{(x, y, z_{11}, \ldots, z_{1m_1}, \ldots, z_{n1}, \ldots, z_{nm_n}) | (x, y, z_{11}, \ldots, z_{1m_1}, \ldots, z_{n1}, \ldots, z_{nm_n})$$
$$\in S, (y, z_{i1}, \ldots, z_{im_i}) \in P_i(x), i = 1, \ldots, n\}.$$

Therefore, based on the notations, the TLMF decision model (6.22a–6.22f) can be written as:

$$\min_{x, y, z_{11}, \ldots, z_{1m_1}, \ldots, z_{n1}, \ldots, z_{nm_n}} f^{(1)}(x, y, z_{11}, \ldots, z_{1m_1}, \ldots, z_{n1}, \ldots, z_{nm_n})$$
$$\text{s.t.} \quad (x, y, z_{11}, \ldots, z_{1m_1}, \ldots, z_{n1}, \ldots, z_{nm_n}) \in IR.$$

(6.23)

## 6.7.2 Theoretical Properties

For the sake of assuring that an optimal solution to the model (6.22a–6.22f) exists, we give the following assumption.

**Assumption 6.2**

1. $S$ is non-empty and compact.
2. $IR$ is non-empty.
3. $P_i(x)$ and $P_{ij}(x,y)$ are point-to-point maps with respect to $x$ and $(x,y)$ respectively, where $j = 1, \ldots, m_i, i = 1, \ldots, n$.

**Theorem 6.4** *If the TLMF decision model (6.22a–6.22f) meets Assumption 6.2, then there exists an optimal solution.*

*Proof* Let

$$P(x) = \{(y, z_{11}, \ldots, z_{1m_1}, \ldots, z_{n1}, \ldots, z_{nm_n}) : (y, z_{i1}, \ldots, z_{im_i}) \in P_i(x), i = 1, \ldots, n\}.$$

Since neither $S$ nor $IR$ is empty, there is at least one parameter value $x^* \in S(X)$ and $P(x^*) \neq \emptyset$.

Consider a sequence $\left\{(x^t, y^t, z_{11}^t, \ldots, z_{1m_1}^t, \ldots z_{n1}^t, \ldots, z_{nm_n}^t)\right\}_{t=1}^{\infty} \subseteq IR$ converging to $(x^*, y^*, z_{11}^*, \ldots, z_{1m_1}^*, \ldots z_{n1}^*, \ldots, z_{nm_n}^*)$. Then, by the well-known results of linear parametric optimization, we have $(x^*, y^*, z_{11}^*, \ldots, z_{1m_1}^*, \ldots z_{n1}^*, \ldots, z_{nm_n}^*) \in P(x^*)$. Hence, $(x^*, y^*, z_{11}^*, \ldots, z_{1m_1}^*, \ldots z_{n1}^*, \ldots, z_{nm_n}^*) \in IR$ which shows that $IR$ is closed. By Assumption 6.2(1) and $IR \subseteq S$, $IR$ is therefore also bounded, and $IR$ is nonempty, so the problem (6.22a–6.22f) consists of minimizing a continuous function over a compact nonempty set, which implies that the problem has an optimal solution.                                    □

**Theorem 6.5** *The inducible region IR can be expressed equivalently as a piecewise linear equality constraint comprised of supporting hyperplanes of S.*

*Proof* First, denote the optimal value of the $i$th middle-level follower's $j$th bottom-level follower by

$$F_{ij}(x, y) = \min\{h_{ij}\hat{z}_{ij}|\hat{z}_{ij} \in S_{ij}(x, y)\}, \quad j = 1, \ldots, m_i, i = 1, \ldots, n,$$

and define

$$F_i(x) = \min\left\{d_i y + \sum_{j=1}^{m_i} g_{ij}z_{ij}|(y, z_{i1}, \ldots, z_{im_i}) \in S_i(x),\right.$$

$$\left. h_{ij}z_{ij} = F_{ij}(x, y), j = 1, \ldots, m_i\right\}, \ i = 1, \ldots, n.$$

Since $F_{ij}(x,y)$ can be seen as a linear programming problem with parameters $x$ and $y$, the dual problem of $F_{ij}(x,y)$ can be written as

$$\max\left\{(A_{ij}x + B_{ij}y - b_i)u_{ij}|E_{ij}u_{ij} \geq -h_{ij}, u_{ij} \geq 0\right\}. \tag{6.24}$$

If both $F_{ij}(x,y)$ and problem (6.24) have feasible solutions, by the dual theorem of linear programming, both have optimal solutions and the same optimal objective function value. Since a solution to problem (6.24) occurs at a vertex of its constraint region $U_{ij} = \left\{u_{ij}|E_{ij}u_{ij} \geq -h_{ij}, u_{ij} \geq 0\right\}$, adopting $u_{ij}^1, \ldots, u_{ij}^{k_{ij}}$ to express all the vertices of $U_{ij}$, then problem (6.24) can be written as:

$$\max\left\{(A_{ij}x + B_{ij}y - b_i)u_{ij}|u_{ij} \in \{u_{ij}^1, \ldots, u_{ij}^{k_{ij}}\}\right\}. \tag{6.25}$$

Clearly, $F_{ij}(x,y)$ is a piecewise linear function according to problem (6.25).

Next, we prove that $F_i(x)$ is also a piecewise linear function. Assume that $(z_{i1}^1, \ldots, z_{im_i}^1), \ldots, (z_{i1}^{p_i}, \ldots, z_{im_i}^{p_i})$ are solutions to the problem $F_{ij}(x,y)$ for $i = 1, \ldots, n$. For each fixed $i$ and a solution $(z_{i1}^{t_i}, \ldots, z_{im_i}^{t_i})$ where $t_i = 1, \ldots, p_i$, $F_i(x)$ becomes a programming problem with parameters $x$ and $(z_{i1}^{t_i}, \ldots, z_{im_i}^{t_i})$, and there are $p_i$ parameterized programming problems such as $F_i(x)|_{(z_{i1}^1, \ldots, z_{im_i}^1)}, \ldots, F_i(x)|_{(z_{i1}^{p_i}, \ldots, z_{im_i}^{p_i})}$. Considering different combinations of $(z_{i1}^{t_i}, \ldots, z_{im_i}^{t_i})$ for $i = 1, \ldots, n$, there are $\prod_{i=1}^{n} p_i$ parameterized programming problems $F_i(x)|_{(z_{i1}^{t_i}, \ldots, z_{im_i}^{t_i})}$. Therefore, $F_i(x)$ is also a piecewise linear function as $F_{ij}(x,y)$.

Lastly, according to the above definition of $F_i(x)$, the inducible region $IR$ can be rewritten as

$$IR = \left\{(x,y,z_{11}^{t_1}, \ldots, z_{1m_1}^{t_1}, \ldots, z_{n1}^{t_n}, \ldots, z_{nm_n}^{t_n}) \in S|d_iy + \sum_{j=1}^{m_i} g_{ij}z_{ij}\right.$$
$$\left. = F_i(x)|_{(z_{i1}^{t_i}, \ldots, z_{im_i}^{t_i})}, t_i = 1, \ldots, p_i, i = 1, \ldots, n\right\}. \tag{6.26}$$

and it can be seen as a piecewise linear equality constraint.                    □

**Corollary 6.3** *The TLMF decision model (6.22a–6.22f) is equivalent to optimizing $f^{(1)}$ over a feasible region comprised of a piecewise linear equality constraint.*

**Corollary 6.4** *An optimal solution to the TLMF decision model (6.22a–6.22f) occurs at a vertex of IR.*

*Proof* According to the equivalent form (6.23) of the TLMF decision model, and since $f^{(1)}$ is linear, an optimal solution to the problem must occur at a vertex of IR if it exists.                    □

**Theorem 6.6** *The optimal solution $(x^*, y^*, z_{11}^*, \ldots, z_{1m_1}^*, \ldots z_{n1}^*, \ldots, z_{nm_n}^*)$ to the TLMF decision model (6.22a–6.22f) occurs at a vertex of S.*

*Proof* Let $\left(x^1, y^1, z_{11}^1, \ldots, z_{1m_1}^1, \ldots z_{n1}^1, \ldots, z_{nm_n}^1\right), \ldots, \left(x^t, y^t, z_{11}^t, \ldots, z_{1m_1}^t, \ldots z_{n1}^t, \ldots, z_{nm_n}^t\right)$
indicate the distinct vertices of $S$. Since any point in $S$ can be written as a convex combination of these vertices, we have

$$
\left(x^*, y^*, z_{11}^*, \ldots, z_{1m_1}^*, \ldots z_{n1}^*, \ldots, z_{nm_n}^*\right)
$$

$$
= \sum_{r=1}^{\bar{t}} \delta_r \left(x^r, y^r, z_{11}^r, \ldots, z_{1m_1}^r, \ldots z_{n1}^r, \ldots, z_{nm_n}^r\right)
$$

where $\sum_{r=1}^{\bar{t}} \delta_r = 1, \delta_r > 0, r = 1, \ldots, \bar{t}$ and $\bar{t} \le t$.

We can write the constraints of (6.22a–6.22f) in the piecewise linear form (6.26) discussed in Theorem 6.6:

$$
0 = F_i(x^*)|_{(z_{i1}^*, \ldots, z_{im_i}^*)} - d_i y^* - \sum_{j=1}^{m_i} g_{ij} z_{ij}^*
$$

$$
= F_i\left(\sum_{r=1}^{\bar{t}} \delta_r x^r\right)\Big|_{(z_{i1}^*, \ldots, z_{im_i}^*)} - d_i \sum_{r=1}^{\bar{t}} \delta_r y_i^r - \sum_{j=1}^{m_i} g_{ij} \sum_{r=1}^{\bar{t}} \delta_r z_{ij}^r, \quad i = 1, .., n.
$$

Because of the convexity of $F_i(x^*)$, we have

$$
0 \le \sum_{r=1}^{\bar{t}} \delta_r F_i(x^r)|_{(z_{i1}^*, \ldots, z_{im_i}^*)} - \sum_{r=1}^{\bar{t}} \delta_r d_i y_i^r - \sum_{r=1}^{\bar{t}} \delta_r \sum_{j=1}^{m_i} g_{ij} z_{ij}^r
$$

$$
= \sum_{r=1}^{\bar{t}} \delta_r [F_i(x^r)|_{(z_{i1}^*, \ldots, z_{im_i}^*)} - d_i y_i^r - \sum_{j=1}^{m_i} g_{ij} z_{ij}^r], \quad i = 1, .., n.
$$

(6.27)

By the definition of $F_i(x)|_{(z_{i1}^{t_i}, \ldots, z_{im_i}^{t_i})}$, we have

$$
F_i(x^r)|_{(z_{i1}^*, \ldots, z_{im_i}^*)} = \min(d_i y + \sum_{j=1}^{m_i} g_{ij} z_{ij}) \le d_i y_i^r - \sum_{j=1}^{m_i} g_{ij} z_{ij}^r, i = 1, \ldots, n.
$$

Thus, $F_i(x^r)|_{(z_{i1}^*, \ldots, z_{im_i}^*)} - d_i y_i^r - \sum_{j=1}^{m_i} g_{ij} z_{ij}^r \le 0, r = 1, \ldots, \bar{t}, i = 1, .., n.$
Since the above expression (6.27) must be held with $\delta_r > 0, r = 1, \ldots, \bar{t}$, there exist $F_i(x^r)|_{(z_{i1}^*, \ldots, z_{im_i}^*)} - d_i y_i^r - \sum_{j=1}^{m_i} g_{ij} z_{ij}^r \le 0, r = 1, \ldots, \bar{t}, i = 1, \ldots, n$. These statements imply that $\left(x^r, y^r, z_{11}^r, \ldots, z_{1m_1}^r, \ldots z_{n1}^r, \ldots, z_{nm_n}^r\right) \in IR, r = 1, \ldots, \bar{t}$ and that $(x^*, y^*, z_{11}^*, \ldots, z_{1m_1}^*, \ldots z_{n1}^*, \ldots, z_{nm_n}^*)$ can be denoted as a convex combination of the points in the $IR$. Since $(x^*, y^*, z_{11}^*, \ldots, z_{1m_1}^*, \ldots z_{n1}^*, \ldots, z_{nm_n}^*)$ is a vertex of the $IR$ according to Corollary 6.4 and Assumption 6.2(3), there must exist $\bar{t} = 1$, which means $(x^*, y^*, z_{11}^*, \ldots, z_{1m_1}^*, \ldots z_{n1}^*, \ldots, z_{nm_n}^*)$ is a vertex of $S$. $\qquad \square$

**Corollary 6.5** If $\left(x^*, y^*, z_{11}^*, \ldots, z_{1m_1}^*, \ldots z_{n1}^*, \ldots, z_{nm_n}^*\right)$ is a vertex of the IR, it is also a vertex of S.

### 6.7.3 TLMF Kth-Best Algorithm

The above theorems and corollaries provide a theoretical foundation to extend the tri-level $K$th-Best algorithm proposed in Sect. 6.6.3 for solving the TLMF decision problem (6.22a–6.22f). The main principle of the TLMF $K$th-Best algorithm is showed as follows.

First, consider the following linear programming problem:

$$\min_{(x,y,z_{11},\ldots,z_{1m_1},\ldots,z_{n1},\ldots,z_{nm_n})\in S} f^{(1)}(x, y, z_{11}, \ldots, z_{1m_1}, \ldots, z_{n1}, \ldots, z_{nm_n}) \qquad (6.28)$$

and let

$$\left(x^1, y^1, z_{11}^1, \ldots, z_{1m_1}^1, \ldots z_{n1}^1, \ldots, z_{nm_n}^1\right), \ldots, \left(x^N, y^N, z_{11}^N, \ldots, z_{1m_1}^N, \ldots z_{n1}^N, \ldots, z_{nm_n}^N\right)$$

denote the $N$-ranked basic feasible solutions to (6.28), such that

$$f^{(1)}\left(x^K, y^K, z_{11}^K, \ldots, z_{1m_1}^K, \ldots z_{n1}^K, \ldots, z_{nm_n}^K\right)$$
$$\leq f^{(1)}\left(x^{K+1}, y^{K+1}, z_{11}^{K+1}, \ldots, z_{1m_1}^{K+1}, \ldots z_{n1}^{K+1}, \ldots, z_{nm_n}^{K+1}\right), \quad K = 1, \ldots, N-1.$$

Then solving the problem (6.28) is equivalent to searching the index $K^* = \min\left\{K | K \in \{1, \ldots, N\}, \left(x^K, y^K, z_{11}^K, \ldots, z_{1m_1}^K, \ldots z_{n1}^K, \ldots, z_{nm_n}^K\right) \in \text{IR}\right\}$, which ensures that $\left(x^{K^*}, y^{K^*}, z_{11}^{K^*}, \ldots, z_{1m_1}^{K^*}, \ldots z_{n1}^{K^*}, \ldots, z_{nm_n}^{K^*}\right)$ is the global solution to the TLMF problem.

To get $\left(x^{K^*}, y^{K^*}, z_{11}^{K^*}, \ldots, z_{1m_1}^{K^*}, \ldots z_{n1}^{K^*}, \ldots, z_{nm_n}^{K^*}\right)$, we must obtain $\left(y^{K^*}, z_{11}^{K^*}, \ldots, z_{1m_1}^{K^*}, \ldots z_{n1}^{K^*}, \ldots, z_{nm_n}^{K^*}\right)$ by solving a set of uncooperative linear multi-follower bi-level (MFBL) decision problems at the middle and bottom level, so next, for $i = 1, \ldots, n$ and the fixing $x = x^{K^*}$, the middle-level and bottom-level problem becomes:

$$\min_{y \in Y_i} \quad f_i^{(2)}(x, y, z_{i1}, \ldots, z_{im_i}) = c_i x + d_i y + \sum_{j=1}^{m_i} g_{ij} z_{ij}$$

$$\text{s.t.} \quad A_i x + B_i y + \sum_{j=1}^{m_i} D_{ij} z_{ij} \leq b_i,$$

where $z_{ij}(j = 1, \ldots, m_i)$ is the solution to the $i$th middle-level follower's $j$th bottom-level follower's problem:

$$\min_{z_{ij} \in Z_{ij}} f_{ij}^{(3)}(x, y, z_{ij}) = c_{ij}x + d_{ij}y + h_{ij}z_{ij}$$

$$\text{s.t.} \quad A_{ij}x + B_{ij}y + E_{ij}z_{ij} \leq b_{ij}. \tag{6.29}$$

---

### Algorithm 6.2: TLMF $K$th-Best Algorithm

[Begin]

**Step 1:** Set $k = 1$, adopt the simplex method to obtain the optimal solution $\left(x^1, y^1, z_{11}^1, \ldots, z_{1m_1}^1, \ldots z_{n1}^1, \ldots, z_{nm_n}^1\right)$ to the linear programming problem (6.28). Let $W = \left\{\left(x^1, y^1, z_{11}^1, \ldots, z_{1m_1}^1, \ldots z_{n1}^1, \ldots, z_{nm_n}^1\right)\right\}$ and $T = \emptyset$. Go to Step 2.

**Step 2:** Put $x = x^k$ and $i = 1$, solve the uncooperative BLMF decision problems (6.29) and obtain the optimal solution $\left(\hat{y}, \hat{z}_{i1}, \ldots, \hat{z}_{im_i}\right)$ using the BLMF $K$th-Best

algorithm shown as the following subroutine Step 2.1-Step 2.5. Then go to Step 3.

**Step 2.1:** Set $x = x^k$ and $k_i = 1$, adopt the simplex method to obtain the optimal solution $(y^{i1}, z_{i1}^{i1}, \ldots, z_{im_i}^{i1})$ to the linear programming problem (6.30):

$$\min\{f_i^{(2)}(x, y, z_{i1}, \ldots, z_{im_i}) \mid (y, z_{i1}, \ldots, z_{im_i}) \in S_i(x)\}. \tag{6.30}$$

Let $W_i' = \left\{\left(y^{i1}, z_{i1}^{i1}, \ldots, z_{im_i}^{i1}\right)\right\}$ and $T_i = \emptyset$. Go to Step 2.2.

**Step 2.2:** Put $x = x^k, y = y^{ik}$, and $j = 1$. Adopt the simplex method to solve the problem (6.31):

$$\min\{f_{ij}^{(3)}(x, y, z_{ij}) \mid z_{ij} \in S_{ij}(x, y)\}. \tag{6.31}$$

and obtain the optimal solution $\tilde{z}_{ij}$.

**Step 2.3:** If $\tilde{z}_{ij} \neq z_{ij}^{ik_i}$, go to Step 2.4. If $\tilde{z}_{ij} = z_{ij}^{ik_i}$ and $j \neq m_i$, set $j = j + 1$ and go to Step 2.2. If $\tilde{z}_{ij} = z_{ij}^{ik_i}$ and $j = m_i$, stop the subroutine, $K_i^* = k_i$ and go to Step 2 with $\left(\hat{y}, \hat{z}_{i1}, \ldots, \hat{z}_{im_i}\right) = \left(y^{ik_i}, z_{i1}^{ik_i}, \ldots, z_{im_i}^{ik_i}\right)$.

**Step 2.4:** Let $W_{k_i}$ denote the set of adjacent vertices of $\left(y^{ik_i}, z_{i1}^{ik_i}, \ldots, z_{im_i}^{ik_i}\right)$ such that $\left(y, z_{i1}, \ldots, z_{im_i}\right) \in W_{k_i}$ implies

$$\{f_i^{(2)}(x^k, y, z_{i1}, \ldots, z_{im_i}) \geq f_i^{(2)}(x^k, y^{ik_i}, z_{i1}^{ik_i}, \ldots, z_{im_i}^{ik_i}).$$

Let $T_i = T_i \cup \left\{\left(y^{ik_i}, z_{i1}^{ik_i}, \ldots, z_{im_i}^{ik_i}\right)\right\}$ and $W_i' = W_i' \cup W_{k_i}/T_i$. Go to Step 2.5.

**Step 2.5:** Set $k_i = k_i + 1$ and choose $(y^{ik_i}, z_{i1}^{ik_i}, \ldots, z_{im_i}^{ik_i})$ such that

$$f_i^{(2)}(x^k, y^{ik_i}, z_{i1}^{ik_i}, \ldots, z_{im_i}^{ik_i})$$
$$= \min\{f_i^{(2)}(x^k, y, z_{i1}, \ldots, z_{im_i}) \mid (y, z_{i1}, \ldots, z_{im_i}) \in W_i'\}.$$

Go to Step 2.2.

**Step   3:**  If  $(\hat{y}, \hat{z}_{i1}, \dots, \hat{z}_{im_i}) \neq (y^k, z_{i1}^k, \dots, z_{im_i}^k)$,    go   to   Step   4.   If $(\hat{y}, \hat{z}_{i1}, \dots, \hat{z}_{im_i}) = (y^k, z_{i1}^k, \dots, z_{im_i}^k)$ and $i \neq n$, set $i = i + 1$ and go to Step 2. If $(\hat{y}, \hat{z}_{i1}, \dots, \hat{z}_{im_i}) = (y^k, z_{i1}^k, \dots, z_{im_i}^k)$ and $i = n$, stop and $(x^k, y^k, z_{11}^k, \dots, z_{1m_1}^k,$ $\dots z_{n1}^k, \dots, z_{nm_n}^k)$ is the optimal solution to the TLMF decision problem (6.22) and $K^* = k$.

**Step   4:**  Let  $W_k$  denote   the   set   of   adjacent   vertices   of $(x^k, y^k, z_{11}^k, \dots, z_{1m_1}^k, \dots z_{n1}^k, \dots, z_{nm_n}^k)$   such   that   $(x, y, z_{11}, \dots, z_{1m_1}, \dots, z_{n1},$ $\dots, z_{nm_n}) \in W_k$ implies

$$f^{(1)}(x, y, z_{11}, \dots, z_{1m_1}, \dots, z_{n1}, \dots, z_{nm_n})$$
$$\geq f^{(1)}(x^k, y^k, z_{11}^k, \dots, z_{1m_1}^k, \dots z_{n1}^k, \dots, z_{nm_n}^k).$$

Let $T = T \cup \{(x^k, y^k, z_{11}^k, \dots, z_{1m_1}^k, \dots z_{n1}^k, \dots, z_{nm_n}^k)\}$ and $W = W \cup W_k/T$. Go to Step 5.

**Step   5:**  Set $k = j + k$ and  choose $(x^k, y^k, z_{11}^k, \dots, z_{1m_1}^k, \dots z_{n1}^k, \dots, z_{nm_n}^k)$ such that

$$f^{(1)}(x^k, y^k, z_{11}^k, \dots, z_{1m_1}^k, \dots z_{n1}^k, \dots, z_{nm_n}^k)$$

$$= \min_{(x, y, z_{11}, \dots, z_{1m_1}, \dots, z_{n1}, \dots, z_{nm_n}) \in W} f^{(1)}(x, y, z_{11}, \dots, z_{1m_1}, \dots, z_{n1}, \dots, z_{nm_n}).$$

Go to Step 2.
[End]

---

Clearly, problem (6.29) is an uncooperative MFBL decision problem. It can be solved by the multi-follower (uncooperative) $K$th-Best algorithm given in Sect. 4.4.3, or the multi-follower (uncooperative) Kuhn-Tucker approach given in Sect. 4.4.4.

Detailed procedures of the TLMF $K$th-Best algorithm are presented as follows:

## 6.7.4  A Numerical Example

A numerical example is adopted to illustrate how the TLMF $K$th-Best algorithm works.

*Example 6.3* Consider a TLMF decision problem in a linear version shown as follows   with   $x \in R$,   $y \in R$,   $z_{ij} \in R$   and   $X = \{x | x \geq 0\}$,   $Y_i = \{y | y \geq 0\}$, $Z_{ij} = \{z_{ij} | z_{ij} \geq 0\}, i = 1, 2, m_i = 2, j = 1, \dots, m_i$.

$$\min_{x \in X} \quad f^{(1)}(x, y, z_{11}, z_{12}, z_{21}, z_{22}) = -1.5x - y + 2z_{11} + z_{12} - z_{21} - 1.5z_{22}$$

s.t.   $x + y + z_{11} + z_{12} + z_{21} + z_{22} \geq 10,$

$x \leq 1.5,$

$$\min_{y \in Y_1} \quad f_1^{(2)}(x, y, z_{11}, z_{12}) = x + y + z_{11} + z_{12}$$

s.t.   $x + y + z_{11} + z_{12} \geq 6.5,$

$$\min_{z_{11} \in Z_{11}} \quad f_{11}^{(3)}(x, y, z_{11}) = x + y + 3z_{11}$$

s.t.   $x + y + z_{11} \geq 3.5,$

$z_{11} \leq 2,$

$$\min_{z_{12} \in Z_{12}} \quad f_{12}^{(3)}(x, y, z_{12}) = x + y + 2z_{12}$$

s.t.   $x + y + z_{12} \geq 5,$

$z_{12} \leq 4,$

$$\min_{y \in Y_2} \quad f_2^{(2)}(x, y, z_{21}, z_{22}) = x - y + 2z_{21} + 3z_{22}$$

s.t.   $x + y + z_{21} + z_{22} \geq 5.5,$

$x + y \leq 2,$

$$\min_{z_{21} \in Z_{21}} \quad f_{21}^{(3)}(x, y, z_{21}) = x + y + 2z_{21}$$

s.t.   $x + y + z_{21} \geq 3,$

$z_{21} \leq 2,$

$$\min_{z_{22} \in Z_{22}} \quad f_{22}^{(3)}(x, y, z_{22}) = x + y + z_{22}$$

s.t.   $x + y + z_{22} \geq 4.5,$

$z_{22} \leq 3.$

We can adopt the TLMF $K$th-Best algorithm to solve the linear semi-cooperative decision problem. First, we have to solve a linear programming problem in the format (6.28) of the leader.

Step 1:  Set $k = 1$ and adopt the simplex method to obtain the optimal solution to the problem (6.28). The optimal solution to (6.28) is $(x^1, y^1, z_{11}^1, z_{12}^1, z_{21}^1, z_{22}^1) = (1.5, 0.5, 1.5, 3, 2, 3)$ and now $W = \{(1.5, 0.5, 1.5, 3, 2, 3)\}$ and $T = \emptyset$. Go to Step 2 and iteration 1 will start

Step 2:  Put $x = 1.5$ and $i = 1$, and solve the BLMF decision problem in the form of (6.29). We can obtain the optimal solution $(\hat{y}, \hat{z}_{11}, \hat{z}_{12}) = (0.5, 1.5, 3)$ to (6.29) and go to Step 3

Step 3:  Evidently, $(\hat{y}, \hat{z}_{i1}, \ldots, \hat{z}_{im_i}) = \left(y^k, z_{i1}^k, \ldots, z_{im_i}^k\right)$, $i = 1$ and $n = 2$, so $i \neq n$, set $i = 2$ and go to Step 2

**Table 6.2** The detailed computing process of the TLMF $K$th-Best algorithm

| Iteration $k$ | $(x^k, y^k, z^k_{11}, z^k_{12}, z^k_{21}, z^k_{22})$ | $W_k$ | $T$ | $W$ |
|---|---|---|---|---|
| 2 | (0,2,1.5,3,2,3) | {(0,2,1.5,3,1,3), (0,2,1.5,3,2,2.5)} | {(1.5,0.5,1.5,3,1,3), (0,2,1.5,3,2,3)} | {(1.5,0.5,1.5,3,1,3), (1.5,0.5,1.5,3,2,2.5), (0,2,1.5,3,1,3), (0,2,1.5,3,2,2.5)} |
| 3 | (1.5,0.5,1.5,3,2,2.5) | {(1.5,0.5,1.5,3,1,2.5)} | {(1.5,0.5,1.5,3,1,3), (0,2,1.5,3,2,3), (1.5,0.5,1.5,3,2,2.5)} | {(1.5,0.5,1.5,3,1,3), (0,2,1.5,3,1,3), (0,2,1.5,3,2,2.5), (1.5,0.5,1.5,3,1,2.5)} |
| 4 | (1.5,0.5,1.5,3,1,3) | ∅ | {(1.5,0.5,1.5,3,1,3), (0,2,1.5,3,2,3), (1.5,0.5,1.5,3,2,2.5), (1.5,0.5,1.5,3,1,3)} | {(0,2,1.5,3,1,3), (0,2,1.5,3,2,2.5), (1.5,0.5,1.5,3,1,2.5)} |
| 5 | (0,2,1.5,3,2,2.5) | {(0,2,1.5,3,1,2.5)} | {(1.5,0.5,1.5,3,1,3), (0,2,1.5,3,2,3), (1.5,0.5,1.5,3,2,2.5), (1.5,0.5,1.5,3,1,3), (0,2,1.5,3,2,2.5)} | {(0,2,1.5,3,1,3), (1.5,0.5,1.5,3,1,2.5), (0,2,1.5,3,1,2.5)} |
| 6 | (0,2,1.5,3,1,3) | ∅ | {(1.5,0.5,1.5,3,1,3), (0,2,1.5,3,2,3), (1.5,0.5,1.5,3,2,2.5), (1.5,0.5,1.5,3,1,3), (0,2,1.5,3,2,2.5), (0,2,1.5,3,1,3)} | {(1.5,0.5,1.5,3,1,2.5), (0,2,1.5,3,1,2.5)} |
| 7 | (1.5,0.5,1.5,3,1,2.5) | | | |

Step 2: Put $x = 1.5$ and $i = 2$, and solve the BLMF decision problem (6.29). We can obtain the optimal solution $(\hat{y}, \hat{z}_{21}, \hat{z}_{22}) = (0.5, 1, 2.5)$ to (6.29) and go to Step 3

Step 3: Now, $(\hat{y}, \hat{z}_{i1}, \ldots, \hat{z}_{im_i}) \neq (y^k, z_{i1}^k, \ldots, z_{im_i}^k)$ and go to Step 4

Step 4: Find the adjacent vertices of $(x^1, y^1, z_{11}^1, z_{12}^1, z_{21}^1, z_{22}^1)$ and the set of adjacent vertices $W_1 = \{(0, 2, 1.5, 3, 2, 3), (1.5, 0.5, 1.5, 3, 1, 3), (1.5, 0.5, 1.5, 3, 2, 2.5)\}$, $T = \{(1.5, 0.5, 1.5, 3, 2, 3)\}$, $W = \{(0, 2, 1.5, 3, 2, 3), (1.5, 0.5, 1.5, 3, 1, 3), (1.5, 0.5, 1.5, 3, 2, 2.5)\}$. Go to Step 5

Step 5: Set $k = 2$ and choose $(x^2, y^2, z_{11}^2, z_{12}^2, z_{21}^2, z_{22}^2) = (0, 2, 1.5, 3, 2, 3)$ and go to Step 2. This step means that iteration 1 has stopped and we cannot obtain an optimal solution through the iteration. The next iteration will be then executed.

In this way, we ultimately achieve the optimal solution through seven iterations. The searched vertices and the detailed computing process of iterations 2–7 are shown as Table 6.2.

In iteration 7, $(x^7, y^7, z_{11}^7, z_{12}^7, z_{21}^7, z_{22}^7) = (1.5, 0.5, 1.5, 3, 1, 2.5)$ is the optimal solution to the TLMF decision problem and the objective function values of all decision entities are $f^{(1)} = -1.5, f_1^{(2)} = 6.5, f_2^{(2)} = 10.5, f_{11}^{(3)} = 6.5, f_{12}^{(3)} = 8, f_{21}^{(3)} = 4, f_{22}^{(3)} = 4.5$.

It is worthwhile to note that $W_4 = \emptyset$ and $W_6 = \emptyset$ in Table 6.1 do not mean that adjacent vertices of $(x^4, y^4, z_{11}^4, z_{12}^4, z_{21}^4, z_{22}^4)$ and $(x^6, y^6, z_{11}^6, z_{12}^6, z_{21}^6, z_{22}^6)$ do not exist but may imply that their adjacent vertices have been found in previous iterations and have been involved in $W$.

The results show that the TLMF $K$th-Best algorithm provides a practical way of solving the proposed TLMF decision problem. However, the computational load of the algorithm may grow steeply with the number of variables and constraints. Therefore, the execution efficiency of the TLMF $K$th-Best algorithm is needed to explore sufficient numeric experiments.

## 6.8 Summary

In a hierarchical organization, interactive decision entities exist within a predominantly hierarchical structure and the execution of decisions is sequential, from the top to the middle and then to the bottom levels. Each entity independently maximizes its own objective, but is affected by the actions of other entities at the same or different levels through externalities. Multiple followers commonly appear in both middle and bottom levels and have various relationships with each other, which results in the complication of this problem.

This chapter presents four main issues in the area: (1) it establishes a TLMF decision framework which identifies 64 standard situations and their possible combinations of TLMF decision problems; (2) it develops a DERD approach to effectively model various TLMF decision problems; (3) it gives a general and standard set of models using both DERD and programming modeling approaches, as well as hybrid TLMF decision models; (4) it presents solution concepts, theoretical properties and related algorithms for a TLMF decision problem.

# Part III
# Fuzzy Multi-level Decision Making

# Chapter 7
# Fuzzy Bi-level Decision Making

Various uncertain issues naturally appear in organizational bi-level decision problems. Fuzzy sets and fuzzy systems can be used to handle uncertainties. This chapter introduces related definitions, theorems and models of *fuzzy bi-level decision-making* (FBLDM) and develops related algorithms to solve the uncertain issues in bi-level decision-making.

This chapter is organized as follows. Section 7.1 discusses uncertain issues in bi-level decision-making and presents a framework for *fuzzy bi-level multi-follower* (FBLMF) decision-making. Section 7.2 provides some concepts in related to fuzzy sets and systems. Section 7.3 gives a fuzzy bi-level decision-making model and a set of fuzzy bi-level multi-follower decision models. Based on the FBLDM model, a definition of the optimal solution for the FBLDM problem and a fuzzy approximation $K$th-Best algorithm for this problem are proposed in Sect. 7.4. Section 7.5 presents a solution concept and method for solving the FBLMF decision problem. Conclusions are discussed in Sect. 7.6.

## 7.1 Problem Identification

Although bi-level decision-making theory and technology have been well studied and applied with remarkable success in different domains, most of the existing solution approaches assume that the objective functions and constraints are characterized with precise parameters. Accordingly, these parameters are required to be fixed at some values in an experimental and/or subjective manner through the experts' understanding of the nature of them in the problem-formulation process. It has been observed that, in most real-world situations, the possible values of these parameters are often only imprecisely or ambiguously known to the experts who establish this model. This may result in a difficulty to set parameters in the objective functions or constraints of the used model.

*Example 7.1* A logistics channel of a company concerns the movement of goods from the source to the place of consumption. Decisions are made at two main stages, supplier and distributor. The two stages are interrelated in a way that a

© Springer-Verlag Berlin Heidelberg 2015
G. Zhang et al., *Multi-Level Decision Making*,
Intelligent Systems Reference Library 82, DOI 10.1007/978-3-662-46059-7_7

decision, such as price or commission, made at supplier stage affects the decision making in its distributor stage. Also, the distributor executes its policies after, and in view of, decisions made at the supplier level. The supplier and its distributor independently maximize their net benefits or minimize their costs but the net benefits or costs are always affected by the actions of the other. To model the bi-level decision model, we need to establish the objective functions for both the leader and follower. However, the parameters of the objective functions (as well as the constraint conditions) are often imprecise and uncertain since they are obtained from experience or experiments which often contain uncertain information such as "about $100". Therefore the bi-level decision model will have uncertain parameters.

In terms of this observation and analysis, it would be certainly more appropriate to interpret uncertain parameters as fuzzy numerical data which can be represented by means of fuzzy sets (Zadeh 1965). A bi-level programming problem in which the parameters, either in the objective functions or in the constraints, are described by fuzzy values is called a *fuzzy bi-level programming* (FBLP) or a *fuzzy bi-level decision-making* (FBLDM) problem.

Multiple followers may be involved in a fuzzy bi-level decision. As we discussed in Chap. 4, these followers may have different objectives and/or different constraints, but could work cooperatively through sharing variables involved in their objective and/or constraints or they could work uncooperatively. Therefore the leader's decision will be affected not only by the followers' individual reactions but also by the relationships among the followers. This chapter deals with both fuzzy parameter and multi-follower situations by developing fuzzy bi-level multi-follower decision models and solution methods. Referring to Table 7.1, the FBLMF has nine situations.

## 7.2 Fuzzy Sets and Systems

*Fuzzy sets*, introduced by Zadeh in 1965, provide us with a new mathematical tool to deal with uncertainty of information. Since then, fuzzy set theory has been rapidly developed and many successful real applications of fuzzy sets in wide-ranging fields

Table 7.1 A framework for fuzzy bi-level multi-follower decision-making

| Relationships among followers | Relationship factor | | | Situation ($F$-$Si$) |
|---|---|---|---|---|
| | Decision variables | Objectives | Constraints | |
| Uncooperative | Individual | Individual | Individual | $F$-$S1$ |
| Cooperative | Sharing | Sharing | Sharing | $F$-$S2$ |
| Semi-cooperative | Sharing | Sharing | Individual | $F$-$S3$ |
| | | Individual | Sharing | $F$-$S4$ |
| | | | Individual | $F$-$S5$ |
| Reference-uncooperative | Individual, but taking other followers' variables as references | Sharing | Sharing | $F$-$S6$ |
| | | | Individual | $F$-$S7$ |
| | | Individual | Sharing | $F$-$S8$ |
| | | | Individual | $F$-$S9$ |

have been appearing. In this section, we will review basic concepts of fuzzy sets, fuzzy relations, fuzzy numbers, linguistic variables, and fuzzy linear programming, which will be used in the rest of the chapters in the book.

## 7.2.1 Fuzzy Sets

**Definition 7.1** (*Fuzzy set*) Let $X$ be a universal set. Then a fuzzy set $\tilde{A}$ of $X$ is defined by its membership function:

$$
\begin{aligned}
\mu_{\tilde{A}} &: X \to [0,1], \\
x &\longmapsto \mu_{\tilde{A}}(x) \in [0,1].
\end{aligned}
\tag{7.1}
$$

The value of $\mu_{\tilde{A}}(x)$ represents the grade of membership of $x$ in $X$ and is interpreted as the degree to which $x$ belongs to $\tilde{A}$, therefore the closer the value of $\mu_{\tilde{A}}(x)$ is 1, the more belongs to $\tilde{A}$.

A crisp or ordinary set $A$ of $X$ can also be viewed as a fuzzy set in $X$ with a membership function as its characteristic function, i.e.,

$$
\mu_{\tilde{A}} = \begin{cases} 1, & x \in A, \\ 0, & x \notin A, \end{cases}
\tag{7.2}
$$

A fuzzy set $\tilde{A}$ can be characterised as a set of ordered pairs of elements $x$ and its grade $\mu_{\tilde{A}}(x)$ and is noted

$$
\tilde{A} = \{(x, \mu_{\tilde{A}}(x)) | x \in X\}.
\tag{7.3}
$$

where each pair $(x, \mu_{\tilde{A}}(x))$ is called a *singleton*.

When $X$ is a countable set, a fuzzy set $\tilde{A}$ on $X$ is expressed as

$$
\tilde{A} = \sum_{x_i \in X} \frac{\mu(x_i)}{x_i}.
\tag{7.4}
$$

When $X$ is a finite set whose elements are $x_1, x_2, \ldots, x_n$, a fuzzy set $\tilde{A}$ on $X$ is expressed as

$$
\tilde{A} = \{(x_1, \mu_{\tilde{A}}(x_1)), (x_2, \mu_{\tilde{A}}(x_2)), \ldots, (x_n, \mu_{\tilde{A}}(x_n))\}.
\tag{7.5}
$$

When $X$ is an infinite and uncountable set, a fuzzy set $\tilde{A}$ on $X$ is expressed as

$$
\tilde{A} = \int_X \frac{\mu(x)}{x}.
\tag{7.6}
$$

These expressions mean that the grade of membership of $x$ is $\mu_{\tilde{A}}(x)$ and the operations '$+$,' '$\sum$,' and '$\int$' do not refer to ordinary addition and integral but they are union, and '$/$' does not indicate an ordinary division but it is merely a marker.

The following basic notions are defined for fuzzy sets.

**Definition 7.2** (*Support of a fuzzy set*) Let $\tilde{A}$ be a fuzzy set on $X$. Then the support of $\tilde{A}$, denoted by $\mathrm{supp}(\tilde{A})$, is the crisp set given by

$$\mathrm{supp}\,(\tilde{A}) = \{x \in X | \mu_{\tilde{A}}(x) > 0\}. \tag{7.7}$$

**Definition 7.3** (*Normal fuzzy set*) Let $\tilde{A}$ be a fuzzy set on $X$. The *height* of $\tilde{A}$, denoted by $hgt(\tilde{A})$, is defined as

$$\mathrm{hgt}(\tilde{A}) = \sup_{x \in X} \mu_{\tilde{A}}(x). \tag{7.8}$$

If $\mathrm{hgt}(\tilde{A}) = 1$, then the fuzzy set $\tilde{A}$ is called a *normal fuzzy set*; otherwise it is called *subnormal*.

**Definition 7.4** (*Empty fuzzy set*) A fuzzy set $\tilde{A}$ is empty, denoted by $\emptyset$, if $\mu_{\tilde{A}}(x) = 0$ for all $x \in X$.

**Definition 7.5** (*α-cut*) Let $\tilde{A}$ be a fuzzy set on $X$ and $\alpha \in [0, 1]$. The *α-cut* of the fuzzy set $\tilde{A}$ is the crisp set $A_\alpha$ given by

$$A_\alpha = \{x \in X | \mu_{\tilde{A}}(x) \geq \alpha\}. \tag{7.9}$$

### 7.2.2 Fuzzy Numbers

**Definition 7.6** (*Fuzzy number*) A fuzzy set $\tilde{a}$ on $R$ is called a *fuzzy number* if it satisfies the following conditions:

1. $\tilde{a}$ is normal, i.e., there exists an $x_0 \in R$ such that $\mu_{\tilde{a}}(x_0) = 1$;
2. $a_\alpha$ is a closed interval for every $\alpha \in [0, 1]$, noted by $[a_\alpha^L, a_\alpha^R]$;
3. The support of $\tilde{a}$ is bounded.

Let $F(R)$ be the set of all fuzzy numbers on $R$. By the decomposition theorem of fuzzy sets, we have

$$\tilde{a} = \bigcup_{\alpha \in [0,1]} \alpha [a_\alpha^L, a_\alpha^R], \tag{7.10}$$

for every $\tilde{a} \in F(R)$. For any real number $\lambda \in R$, we define $\mu_\lambda(x)$ by

$$\mu_\lambda(x) = \begin{cases} 1, & x = \lambda, \\ 0, & x \neq \lambda. \end{cases}$$

Then $\lambda \in F(R)$.

**Definition 7.7** A triangular fuzzy number $\tilde{a}$ can be defined by a triplet $\left(a_0^L, a, a_0^R\right)$ and the membership function $\mu_{\tilde{a}}(x)$ is defined as:

$$\mu_{\tilde{a}}(x) = \begin{cases} 0, & x < a_0^L \\ \frac{x - a_0^L}{a - a_0^L}, & a_0^L \leq x \leq a \\ \frac{a_0^R - x}{a_0^R - a}, & a \leq x \leq a_0^R \\ 0, & a_0^R < x \end{cases} \tag{7.11}$$

where $a = a_1^L = a_1^R$.

**Definition 7.8** If $\tilde{a}$ is a fuzzy number and $a_\lambda^L > 0$ for any $\lambda \in (0, 1]$, then $\tilde{a}$ is called a *positive fuzzy number*. Let $F_+^*(R)$ be the set of all finite positive fuzzy numbers on $R$.

**Definition 7.9** For any $\tilde{a}, \tilde{b} \in F(R)$ and $0 < \alpha \in R$, the sum $(\tilde{a} + \tilde{b})$, difference $(\tilde{a} - \tilde{b})$, scalar product $(\alpha\tilde{a})$ and product $(\tilde{a} \times \tilde{b})$ of two fuzzy numbers are defined by their membership functions:

$$\mu_{\tilde{a}+\tilde{b}}(t) = \sup_{t=u+v} \min\left\{\mu_{\tilde{a}}(u), \mu_{\tilde{b}}(v)\right\}, \tag{7.12}$$

$$\mu_{\tilde{a}-\tilde{b}}(t) = \sup_{t=u-v} \min\left\{\mu_{\tilde{a}}(u), \mu_{\tilde{b}}(v)\right\}, \tag{7.13}$$

$$\mu_{\alpha\tilde{b}}(t) = \min_{t=\alpha u}\{\mu_{\tilde{a}}(u)\}, \tag{7.14}$$

$$\mu_{\tilde{a}\times\tilde{b}}(t) = \sup_{t=u\times v} \min\left\{\mu_{\tilde{a}}(u), \mu_{\tilde{b}}(v)\right\}, \tag{7.15}$$

where we set $\sup\{\emptyset\} = -\infty$.

**Theorem 7.1** *For any* $\tilde{a}, \tilde{b} \in F(R)$, *and* $0 < \alpha \in R$,

$$\tilde{a} + \tilde{b} = \bigcup_{\lambda \in (0,1]} \lambda\left[a_\lambda^L + b_\lambda^L, a_\lambda^R + b_\lambda^R\right],$$

$$\tilde{a} - \tilde{b} = \bigcup_{\lambda \in (0,1]} \lambda\left[a_\lambda^L - b_\lambda^R, a_\lambda^R - b_\lambda^L\right] = \tilde{a} + (-\tilde{b})$$

$$= \bigcup_{\lambda \in (0,1]} \lambda\left[a_\lambda^L + (-b_\lambda^R), a_\lambda^R + (-b_\lambda^L)\right],$$

$$\alpha\tilde{a} = \bigcup_{\lambda \in (0,1]} \lambda\left[\alpha\tilde{a}_\lambda^L, \alpha\tilde{a}_\lambda^R\right],$$

$$\tilde{a} \times \tilde{b} = \bigcup_{\lambda \in (0,1]} \lambda\left[a_\lambda^L \times b_\lambda^L, a_\lambda^R \times b_\lambda^R\right].$$

**Definition 7.10** For any $\tilde{a} \in F_+^*(R)$ and $0 < \alpha \in Q_+$ ($Q_+$ is a set of all positive rational numbers), the positive fuzzy power $\alpha$ of $\tilde{a}$ is defined by the membership function:

$$\mu_{\tilde{a}^\alpha}(t) = \sup_{t=u^\alpha} \min\{\mu_{\tilde{a}}(u)\} \tag{7.16}$$

where we set $\sup\{\emptyset\} = -\infty$.

**Theorem 7.2** *For any $\tilde{a} \in F_+^*(R)$ and $0 < \alpha \in Q_+$,*

$$\tilde{a}^\alpha = \bigcup_{\lambda \in (0,1]} \lambda \left[ \left(a_\lambda^L\right)^\alpha, \left(a_\lambda^R\right)^\alpha \right].$$

**Definition 7.11** Let $\tilde{a}$ and $\tilde{b}$ be two fuzzy numbers. We define:

1. $\tilde{a} \succcurlyeq \tilde{b}$ iff $a_\lambda^L \geq b_\lambda^L$ and $a_\lambda^R \geq b_\lambda^R$, $\lambda \in (0,1]$;
2. $\tilde{a} = \tilde{b}$ iff $\tilde{a} \succcurlyeq \tilde{b}$ and $\tilde{b} \succcurlyeq \tilde{a}$;
3. $\tilde{a} \succ \tilde{b}$ iff $\tilde{a} \succcurlyeq \tilde{b}$ and there exists a $\lambda_0 \in (0,1]$ such that $a_{\lambda_0}^L > b_{\lambda_0}^L$ or $a_{\lambda_0}^R > b_{\lambda_0}^R$.

**Definition 7.12** If $\tilde{a}$ is a fuzzy number and $0 < a_\lambda^L \leq a_\lambda^R \leq 1$, for any $\lambda \in (0,1]$, then $\tilde{a}$ is called a *normalized positive fuzzy number*. Let $F_N^*(R)$ be the set of all normalised positive triangular fuzzy numbers on $R$.

**Definition 7.13** Let $\tilde{a}, \tilde{b} \in F(R)$. Then the quasi-distance function of $\tilde{a}$ and $\tilde{b}$ is defined as

$$d(\tilde{a}, \tilde{b}) \left( \int_0^1 \frac{1}{2} \left[ \left(a_\lambda^L - b_\lambda^L\right)^2 + \left(a_\lambda^R - b_\lambda^R\right)^2 \right] d\lambda \right)^{\frac{1}{2}}.$$

**Definition 7.14** Let $\tilde{a}, \tilde{b} \in F(R)$. Then the fuzzy number $\tilde{a}$ is said to closer to the fuzzy number $\tilde{b}$ as $d(\tilde{a}, \tilde{b})$ approaches 0.

**Proposition 7.1** *If both $\tilde{a}$ and $\tilde{b}$ are real numbers, then the quasi-distance measurement $d(\tilde{a}, \tilde{b})$ is identical to the Euclidean distance.*

**Proposition 7.2** *Let $\tilde{a}, \tilde{b} \in F(R)$.*

1. *If they are identical, then $d(\tilde{a}, \tilde{b}) = 0$.*
2. *If $\tilde{a}$ is a real number or $\tilde{b}$ is a real number and $d(\tilde{a}, \tilde{b}) = 0$, then $\tilde{a} = \tilde{b}$.*

**Proposition 7.3** *Let $\tilde{a}, \tilde{b}, \tilde{c} \in F(R)$. Then $\tilde{b}$ is closer to $\tilde{a}$ than $\tilde{c}$ if and only if $d(\tilde{b}, \tilde{a}) < d(\tilde{c}, \tilde{a})$.*

**Proposition 7.4** *Let $\tilde{a}, \tilde{b} \in F(R)$. If $d(\tilde{a}, 0) < d(\tilde{b}, 0)$, then $\tilde{a}$ is closer to 0 than $\tilde{b}$.*

**Definition 7.15** Let $\tilde{a}_i \in F(R)$, $i = 1, \ldots, n$. We define $\tilde{a} = (\tilde{a}_1, \ldots, \tilde{a}_n)$ as follows:

$$\mu_{\tilde{a}} : R^n \to [0, 1],$$

$$x \longmapsto \bigwedge_{i=1}^{n} \mu_{\tilde{a}_i}(x_i),$$

where $x = (x_1, \ldots, x_n)^T \in R^n$, and $\tilde{a}$ is called an $n$-dimensional fuzzy number on $R^n$. Let $F(R^n)$ be the set of all $n$-dimensional fuzzy numbers on $R^n$.

**Proposition 7.5** *For every $\tilde{a} \in F(R^n)$, $\tilde{a}$ is normal.*

**Proposition 7.6** *For every $\tilde{a} \in F(R^n)$, the $\lambda$-cut of $\tilde{a}$ is an $n$-dimensional closed rectangular region for any $\lambda \in (0, 1]$.*

**Proposition 7.7** *For every $\tilde{a} \in F(R^n)$, $\tilde{a}$ is a convex fuzzy set, i.e.,*

$$\mu_{\tilde{a}}(\lambda x + (1 - \lambda)y) \geqq \mu_{\tilde{a}}(x) \wedge \mu_{\tilde{a}}(y),$$

*whenever $\lambda \in [0, 1]$, $x = (x_1, \ldots, x_n)$, $y = (y_1, \ldots, y_n) \in R^n$.*

**Proposition 7.8** *For every $\tilde{a} \in F(R^n)$, and $\lambda_1, \lambda_2 \in (0, 1]$, if $\lambda_1 \leq \lambda_2$, then $\tilde{a}_{\lambda_2} \subset \tilde{a}_{\lambda_1}$.*

**Definition 7.16** For any $\tilde{a}, \tilde{b} \in F(R^n)$, and $0 < \alpha \in R$, the sum $(\tilde{a} + \tilde{b})$, difference $(\tilde{a} - \tilde{b})$, scalar product $(\alpha\tilde{a})$ and product $(\tilde{a} \times \tilde{b})$ are defined by the membership functions:

$$\mu_{\tilde{a}+\tilde{b}}(x) = \bigwedge_{i=1}^{n} \mu_{\tilde{a}_i+\tilde{b}_i}(x_i), \tag{7.17}$$

$$\mu_{\tilde{a}-\tilde{b}}(x) = \bigwedge_{i=1}^{n} \mu_{\tilde{a}_i-\tilde{b}_i}(x_i), \tag{7.18}$$

$$\mu_{\alpha\tilde{a}}(x) = \bigwedge_{i=1}^{n} \mu_{\alpha\tilde{a}_i}(x_i), \tag{7.19}$$

$$\mu_{\tilde{a}\times\tilde{b}}(x) = \bigwedge_{i=1}^{n} \mu_{\tilde{a}_i\times\tilde{b}_i}(x_i). \tag{7.20}$$

**Definition 7.17** For any $\tilde{a} = (\tilde{a}_1, \ldots, \tilde{a}_n)$, $\tilde{a}_i \in F_+^*(R)$ $(i = 1, \ldots, n)$ and $0 < \alpha \in Q_+$,

$$\mu_{\tilde{a}^\alpha}(x) = \bigwedge_{i=1}^{n} \mu_{\tilde{a}_i^\alpha}(x_i). \tag{7.21}$$

**Definition 7.18** For any $n$-dimensional fuzzy numbers $\tilde{a}, \tilde{b} \in F(R^n)$, and $\alpha \in (0, 1]$ we define:

1. $\tilde{a} \succeq_\alpha \tilde{b}$ iff $a_\lambda^L \geqq b_\lambda^L$ and $a_\lambda^R \geqq b_\lambda^R$, $\lambda \in [\alpha, 1]$;
2. $\tilde{a} \succ_\alpha \tilde{b}$ iff $a_\lambda^L \geq b_\lambda^L$ and $a_\lambda^R \geq b_\lambda^R$, $\lambda \in [\alpha, 1]$;
3. $\tilde{a} \succ_\alpha \tilde{b}$ iff $a_\lambda^L > b_\lambda^L$ and $a_\lambda^R > b_\lambda^R$, $\lambda \in [\alpha, 1]$.

We call the binary relations $\succeq$, $\succeq$, and $\succ$ a fuzzy greater-than order, a strict fuzzy greater-than order, and a strong fuzzy greater-than order, respectively.

**Definition 7.19** Let $\tilde{a}, \tilde{b} \in F(R)$ be two fuzzy numbers, the *ranking* of two fuzzy numbers are defined as:

$$\tilde{a} \preccurlyeq \tilde{b} \quad \text{if } m(\tilde{a}) < m(\tilde{b}), \tag{7.22}$$

or

$$m(\tilde{a}) = m(\tilde{b}) \quad \text{and} \quad \sigma(\tilde{a}) \geq \sigma(\tilde{b}), \tag{7.23}$$

where the mean $m(\tilde{a})$ and the standard deviation $\sigma(\tilde{a})$ are defined as:

$$m(\tilde{a}) = \frac{\int_{s(\tilde{a})} x\tilde{a}(x)dx}{\int_{s(\tilde{a})} \tilde{a}(x)dx}, \tag{7.24}$$

$$\sigma(\tilde{a}) = \left(\frac{\int_{s(\tilde{a})} x^2\tilde{a}(x)dx}{\int_{s(\tilde{a})} \tilde{a}(x)dx} - (m(\tilde{a}))^2\right)^{\frac{1}{2}}, \tag{7.25}$$

where $S(\tilde{a}) = \{x | \tilde{a}(x) > 0\}$ is the support of fuzzy number $\tilde{a}$.

For triangular fuzzy number $\tilde{a} = (l, m, n)$, we have

$$m(\tilde{a}) = \frac{1}{3}(l + m + n), \tag{7.26}$$

$$\sigma(\tilde{a}) = \frac{1}{18}(l^2 + m^2 + n^2 - lm - ln - mn). \tag{7.27}$$

## 7.3 Fuzzy Bi-level Decision Models

In this section, we will present a set of fuzzy bi-level decision-making models. We first propose a FBLDM model and a set of FBLMF decision-making models.

For $x \in X \subset R^n$, the leader's decision variable, $y \in Y \subset R^m$, the follower's decision variable, $F : X \times Y \to F(R)$, the leader's objective function, and $f : X \times Y \to F(R)$, the follower's objective function, in general, a linear FBLDM model can be written as follows:

$$\min_{x \in X} \quad F(x,y) = \tilde{c}_1 x + \tilde{d}_1 y \qquad (7.28a)$$

$$\text{s.t.} \quad \tilde{A}_1 x + \tilde{B}_1 y \preceq \tilde{b}_1, \qquad (7.28b)$$

where $y$ is the solution of the lower level problem

$$\min_{y \in Y} \quad f(x,y) = \tilde{c}_2 x + \tilde{d}_2 y \qquad (7.28c)$$

$$\text{s.t.} \quad \tilde{A}_2 x + \tilde{B}_2 y \preceq \tilde{b}_2, \qquad (7.28d)$$

where $\tilde{c}_1, \tilde{c}_2 \in F(R^n)$, $\tilde{d}_1, \tilde{d}_2 \in F(R^m)$, $\tilde{b}_1 \in F(R^p)$, $\tilde{b}_2 \in F(R^q)$, $\tilde{A}_1 \in F(R^{p \times n})$, $\tilde{B}_1 \in F(R^{p \times m})$, $\tilde{A}_2 \in F(R^{q \times n})$, $\tilde{B}_2 \in F(R^{q \times n})$.

Based on the normal bi-level decision-making model and the general model of the bi-level multi-follower decision-making model discussed in Chap. 4, we give a set of *fuzzy bi-level multi-follower* (FBLMF) decision models:

### 1. *F-S1 Model for linear FBLMF decision problems*
For $x \in X \subset R^n, y_i \in Y_i \subset R^{m_i}, F : X \times Y_1 \times \cdots \times Y_k \to F(R), f_i : X \times Y_i \to F(R)$, and $i = 1, \ldots, k$, a linear FBLMF decision model, in which $k(\geq 2)$ followers are involved and have individual decision variables, objective functions and constraints, is defined as follows (*this is called an uncooperative FBLMF decision model*).

$$\min_{x \in X} \quad F(x, y_1, \ldots, y_k) = \tilde{c}x + \sum_{i=1}^{k} \tilde{d}_i y_i$$

$$\text{s.t.} \quad \tilde{A}x + \sum_{i=1}^{k} \tilde{B}_i y_i \preceq \tilde{b}, \qquad (7.29)$$

where $y_i$ is the solution to the $i$th follower's problem:

$$\min_{y_i \in Y_i} \quad f_i(x, y_1, \ldots, y_k) = \tilde{c}_i x + \tilde{e}_i y_i$$

$$\text{s.t.} \quad \tilde{A}_i x + \tilde{C}_i y_i \preceq \tilde{b}_i,$$

where $\tilde{c}, \tilde{c}_i \in F(R^n), \tilde{d}_i, \tilde{e}_i \in F(R^{m_i}), \tilde{A} \in F(R^{p \times n}), \tilde{B}_i \in F(R^{p \times m_i}),$ $\tilde{b} \in F(R^p),$ $\tilde{A}_i \in F(R^{q_i \times n}), \tilde{C}_i \in F(R^{q_i \times m_i}), \tilde{b}_i \in F(R^{q_i}), i = 1, \dots, k.$

2. **F-S2 Model for linear FBLMF decision problems**

For $x \in X \subset R^n, y \in Y \subset R^m, F : X \times Y \to F(R), f_i : X \times Y \to F(R), i = 1, \dots, k,$ a linear FBLMF decision model, in which $k(\geq 2)$ followers are involved and have shared decision variables, objective functions and constraints, is defined as follows (*this is called a cooperative FBLMF decision model*).

$$\min_{x \in X} \quad F(x, y) = \tilde{c}x + \tilde{d}y$$

$$\text{s.t.} \quad \tilde{A}x + \tilde{B}y \preceq \tilde{b}, \tag{7.30}$$

where $y_i$ is the solution to the $i$th follower's problem:

$$\min_{y \in Y} \quad f_i(x, y) = \tilde{c}'x + \tilde{d}'y_i$$

$$\text{s.t.} \quad \tilde{A}'x + \tilde{B}'y \preceq \tilde{b}',$$

where $\tilde{c}, \tilde{c}' \in F(R^n), \tilde{d}, \tilde{d}' \in F(R^m), \tilde{A} \in F(R^{p \times n}), \tilde{B} \in F(R^{p \times m}), \tilde{b} \in F(R^p), \tilde{A}' \in F(R^{q \times n}), \tilde{B}' \in F(R^{q \times m}), \tilde{b}' \in F(R^q).$

3. **F-S3 Model for linear FBLMF decision problems**

For $x \in X \subset R^n, y \in Y_i \subset R^m, Y = Y_1 \cap \cdots \cap Y_k, F : X \times Y \to F(R), f_i : X \times Y_i \to F(R), i = 1, \dots, k,$ a linear FBLMF decision model, in which $k(\geq 2)$ followers are involved and have shared decision variables and objective functions but different constraint, is defined as follows (*this is called a semi-cooperative FBLMF decision model*).

$$\min_{x \in X} \quad F(x, y) = \tilde{c}x + \tilde{d}y$$

$$\text{s.t.} \quad \tilde{A}x + \tilde{B}y \preceq \tilde{b}, \tag{7.31}$$

where $y$ is the solution to the $i$th follower's problem:

$$\min_{y \in Y_i} \quad f_i(x, y) = \tilde{c}'x + \tilde{d}'y_i$$

$$\text{s.t.} \quad \tilde{A}_i x + \tilde{C}_i y_i \preceq \tilde{b}_i,$$

where $\tilde{c}, \tilde{c}' \in F(R^n), \tilde{d}, \tilde{d}' \in F(R^m), \tilde{A} \in F(R^{p \times n}), \tilde{B} \in F(R^{p \times m}), \tilde{b} \in F(R^p), \tilde{A}_i \in F(R^{q_i \times n}), \tilde{C}_i \in F(R^{q_i \times m}), \tilde{b}_i \in F(R^{q_i}), i = 1, \dots, k.$

4. **F-S4 Model for linear FBLMF decision problems**

For $x \in X \subset R^n, y \in Y \subset R^m, F : X \times Y \to F(R), f_i : X \times Y \to F(R), i = 1, \dots, k,$ a linear FBLMF decision model, in which $k(\geq 2)$ followers are involved and have shared decision variables and constraints but different objective functions, is defined as follows (*this is also called a semi-cooperative FBLMF decision model*).

$$\min_{x \in X} \quad F(x,y) = \tilde{c}x + \tilde{d}y$$
$$\text{s.t.} \quad \tilde{A}x + \tilde{B}y \preceq \tilde{b}, \tag{7.32}$$

where $y$ is the solution to the $i$th follower's problem:

$$\min_{y \in Y_i} \quad f_i(x,y) = \tilde{c}_i x + \tilde{e}_i y_i$$
$$\text{s.t.} \quad \tilde{A}'x + \tilde{B}'y \preceq \tilde{b}',$$

where $\tilde{c}, \tilde{c}_i \in F(R^n), \tilde{d}, \tilde{e}_i \in F(R^m), \tilde{A} \in F(R^{p \times n}), \tilde{B} \in F(R^{p \times m}), \tilde{b} \in F(R^p), \tilde{A}' \in F(R^{q \times n}), \tilde{B}' \in F(R^{q \times m}), \tilde{b}' \in F(R^q)$, $i = 1, \ldots, k$.

### 5. *F-S5 Model for linear FBLMF decision problems*

For $x \in X \subset R^n, y \in Y_i \subset R^m, Y = Y_1 \cap \cdots \cap Y_k, F : X \times Y \to F(R), f_i : X \times Y_i \to F(R)$, $i = 1, \ldots, k$, a linear FBLMF decision model, in which $k(\geq 2)$ followers are involved and have shared decision variables but different objective functions and constraints, is defined as follows (*this is also called a semi-cooperative FBLMF decision model*).

$$\min_{x \in X} \quad F(x,y) = \tilde{c}x + \tilde{d}y$$
$$\text{s.t.} \quad \tilde{A}x + \tilde{B}y \preceq \tilde{b}, \tag{7.33}$$

where $y$ is the solution to the $i$th follower's problem:

$$\min_{y \in Y_i} \quad f_i(x,y) = \tilde{c}_i x + \tilde{e}_i y_i$$
$$\text{s.t.} \quad \tilde{A}_i x + \tilde{C}_i y_i \preceq \tilde{b}_i,$$

where $\tilde{c}, \tilde{c}_i \in F(R^n), \tilde{d}, \tilde{e}_i \in F(R^m), \tilde{A} \in F(R^{p \times n}), \tilde{B} \in F(R^{p \times m}), \tilde{b} \in F(R^p), \tilde{A}_i \in F(R^{q_i \times n}), \tilde{C}_i \in F(R^{q_i \times m}), \tilde{b}_i \in F(R^{q_i})$, $i = 1, \ldots, k$.

### 6. *F-S6 Model for linear FBLMF decision problems*

For $x \in X \subset R^n, y_i \in Y_i \subset R^{m_i}, F : X \times Y_1 \times \cdots \times Y_k \to F(R), f_i : X \times Y_1 \times \cdots \times Y_k \to F(R)$, $i = 1, \ldots, k$, a linear FBLMF decision model, in which $k(\geq 2)$ followers are involved and there are individual decision variables in the shared objective functions and constraints among them, but the followers take other followers' decision variables as references, is defined as follows (*this is called a reference-uncooperative FBLMF decision model*).

$$\min_{x \in X} \quad F(x, y_1, \ldots, y_k) = \tilde{c}x + \sum_{i=1}^{k} \tilde{d}_i y_i$$

$$\text{s.t.} \quad \tilde{A}x + \sum_{i=1}^{k} \tilde{B}_i y_i \preceq \tilde{b}, \tag{7.34}$$

where $y_i$ is the solution to the $i$th follower's problem:

$$\min_{y_i \in Y_i} \quad f_i(x, y_1, \ldots, y_k) = \tilde{c}'x + \sum_{i=1}^{k} \tilde{e}_i y_i$$

$$\text{s.t.} \quad \tilde{A}'x + \sum_{i=1}^{k} \tilde{C}_i y_i \preceq \tilde{b}',$$

where $\tilde{c}, \tilde{c}' \in F(R^n), \tilde{d}_i, \tilde{e}_i \in F(R^{m_i}), \tilde{A} \in F(R^{p \times n}), \tilde{B}_i \in F(R^{p \times m_i}), \tilde{b} \in F(R^p), \tilde{A}' \in F(R^{q \times n}), \tilde{C}_i \in F(R^{q \times m_i}), \tilde{b}' \in F(R^q), i = 1, \ldots, k.$

7. *F-S7 Model for linear FBLMF decision problems*
For  $x \in X \subset R^n, y_i \in Y_i \subset R^{m_i}, F : X \times Y_1 \times \cdots \times Y_k \to F(R), f_i : X \times Y_1 \times \cdots \times Y_k \to F(R), i = 1, \ldots, k,$ a linear FBLMF decision model, in which $k(\geq 2)$ followers are involved and there are individual decision variables in the shared objective functions and separate constraints among them, but the followers take other followers' decision variables as references, is defined as follows (*this is also called a reference-uncooperative FBLMF decision model*).

$$\min_{x \in X} \quad F(x, y_1, \ldots, y_k) = \tilde{c}x + \sum_{i=1}^{k} \tilde{d}_i y_i$$

$$\text{s.t.} \quad \tilde{A}x + \sum_{i=1}^{k} \tilde{B}_i y_i \preceq \tilde{b}, \tag{7.35}$$

where $y_i$ is the solution to the $i$th follower's problem:

$$\min_{y_i \in Y_i} \quad f_i(x, y_1, \ldots, y_k) = \tilde{c}'x + \sum_{i=1}^{k} \tilde{e}_i y_i$$

$$\text{s.t.} \quad \tilde{A}_i x + \sum_{s=1}^{k} \tilde{C}_{is} y_s \preceq \tilde{b}_i,$$

where $\tilde{c}, \tilde{c}' \in F(R^n), \tilde{d}_i, \tilde{e}_i \in F(R^{m_i}), \tilde{A} \in F(R^{p \times n}), \tilde{B}_i \in F(R^{p \times m_i}), \tilde{b} \in F(R^p), \tilde{A}_i \in F(R^{q_i \times n}), \tilde{C}_{is} \in F(R^{q_i \times m_i}), \tilde{b}_i \in F(R^{q_i}), i, s = 1, \ldots, k.$

## 8. F-S8 Model for linear FBLMF decision problems

For $x \in X \subset R^n, y_i \in Y_i \subset R^{m_i}, F : X \times Y_1 \times \cdots \times Y_k \to F(R), f_i : X \times Y_1 \times \cdots \times Y_k \to F(R), i = 1, \ldots, k,$ a linear FBLMF decision model, in which $k(\geq 2)$ followers are involved and there are individual decision variables in the separate objective functions and shared constraints among them, but the followers take other followers' decision variables as references, is defined as follows (*this is also called a reference-uncooperative FBLMF decision model*).

$$\min_{x \in X} \quad F(x, y_1, \ldots, y_k) = \tilde{c}x + \sum_{i=1}^{k} \tilde{d}_i y_i$$

$$\text{s.t.} \quad \tilde{A}x + \sum_{i=1}^{k} \tilde{B}_i y_i \preceq \tilde{b},$$

where $y_i$ is the solution to the $i$th follower's problem:

$$\min_{y_i \in Y_i} \quad f_i(x, y_1, \ldots, y_k) = \tilde{c}_i x + \sum_{s=1}^{k} \tilde{e}_{is} y_s$$

$$\text{s.t.} \quad \tilde{A}'x + \sum_{i=1}^{k} \tilde{C}_i y_i \preceq \tilde{b}',$$

where $\tilde{c}, \tilde{c}_i \in F(R^n), \tilde{d}_i, \tilde{e}_{is} \in F(R^{m_i}), \tilde{A} \in F(R^{p \times n}), \tilde{B}_i \in F(R^{p \times m_i}), \tilde{b} \in (R^p), \tilde{A}' \in F(R^{q \times n}), \tilde{C}_i \in F(R^{q \times m_i}), \tilde{b}' \in F(R^q), i, s = 1, \ldots, k.$

## 9. F-S9 Model for linear FBLMF decision problems

For $x \in X \subset R^n, y_i \in Y_i \subset R^{m_i}, F : X \times Y_1 \times \cdots \times Y_k \to F(R), f_i : X \times Y_1 \times \cdots \times Y_k \to F(R), i = 1, \ldots, k,$ a linear FBLMF decision model, in which $k(\geq 2)$ followers are involved and there are individual decision variables in the separate objective functions and constraints among them, but the followers take other followers' decision variables as references, is defined as follows (*this is also called a reference-uncooperative FBLMF decision model*).

$$\min_{x \in X} \quad F(x, y_1, \ldots, y_k) = \tilde{c}x + \sum_{i=1}^{k} \tilde{d}_i y_i$$

$$\text{s.t.} \quad \tilde{A}x + \sum_{i=1}^{k} \tilde{B}_i y_i \preceq \tilde{b},$$

where $y_i$ is the solution to the $i$th follower's problem:

$$\min_{y_i \in Y_i} \quad f_i(x, y_1, \ldots, y_k) = \tilde{c}_i x + \sum_{s=1}^{k} \tilde{e}_{is} y_s$$

$$\text{s.t.} \quad \tilde{A}_i x + \sum_{i=1}^{k} \tilde{C}_{is} y_s \preceq \tilde{b}_i,$$

where $\tilde{c}, \tilde{c}_i \in F(R^n), \tilde{d}_i, \tilde{e}_{is} \in F(R^{m_i}), \tilde{A} \in F(R^{p \times n}), \tilde{B}_i \in F(R^{p \times m_i}), \tilde{b} \in F(R^p), \tilde{A}_i \in F(R^{q_i \times n}), \tilde{C}_{is} \in F(R^{q_i \times m_i}), \tilde{b}_i \in F(R^{q_i}), i, s = 1, \ldots, k.$

We will discuss related solution concepts and solution methods in Sect. 7.4.


## 7.4 Fuzzy Approximation $K$th-Best Algorithm

### 7.4.1 Property and Algorithm

Based on the FBLDM model proposed in Sect. 7.3, we integrate the classic $K$th-Best algorithm and fuzzy number techniques (Zadeh 1975) into our proposed *fuzzy approximation $K$th-Best* (FA-$K$th-Best) algorithm to deal with the FBLDM problem.

Turning to solve the FBLDM problem, we first need to transfer the FBLDM problem into a non-fuzzy bi-level decision problem. In related to the FBLDM problem, we now consider the following multi-objective linear bi-level programming problem:

For $x \in X \subset R^n$, $y \in Y \subset R^m$, $F : X \times Y \to F(R)$, and $f : X \times Y \to F(R)$,

$$\min_{x \in X} \quad (F(x, y))_\lambda^L = c_{1\lambda}^L x + d_{1\lambda}^L y, \quad \lambda \in [0, 1]$$

$$\min_{x \in X} \quad (F(x, y))_\lambda^R = c_{1\lambda}^R x + d_{1\lambda}^R y, \quad \lambda \in [0, 1] \tag{7.36a}$$

$$\text{s.t.} \quad A_{1\lambda}^L x + B_{1\lambda}^L y \leq b_{1\lambda}^L, A_{1\lambda}^R x + B_{1\lambda}^R y \leq b_{1\lambda}^R, \quad \lambda \in [0, 1] \tag{7.36b}$$

$$\min_{y \in Y} \quad (f(x, y))_\lambda^L = c_{2\lambda}^L x + d_{2\lambda}^L y, \quad \lambda \in [0, 1]$$

$$\min_{y \in Y} \quad (f(x, y))_\lambda^R = c_{2\lambda}^R x + d_{2\lambda}^R y, \quad \lambda \in [0, 1] \tag{7.36c}$$

$$\text{s.t.} \quad A_{2\lambda}^L x + B_{2\lambda}^L y \leq b_{2\lambda}^L, A_{2\lambda}^R x + B_{2\lambda}^R y \leq b_{2\lambda}^R, \quad \lambda \in [0, 1] \tag{7.36d}$$

where $c_{1\lambda}^L$, $c_{1\lambda}^R$, $c_{2\lambda}^L$, $c_{2\lambda}^R \in R^n$, $d_{1\lambda}^L$, $d_{1\lambda}^R$, $d_{2\lambda}^L$, $d_{2\lambda}^R \in R^m$, $b_{1\lambda}^L$, $b_{1\lambda}^R \in R^p$, $b_{2\lambda}^L$, $b_{2\lambda}^R \in R^q$, $A_{1\lambda}^L = (a_{ij\lambda}^L)$, $A_{1\lambda}^R = (a_{ij\lambda}^R) \in R^{p \times n}$, $B_{1\lambda}^L = (b_{ij\lambda}^L)$, $B_{1\lambda}^R = (b_{ij\lambda}^R) \in R^{p \times m}$, $A_{2\lambda}^L = (e_{ij\lambda}^L)$, $A_{2\lambda}^R = (e_{ij\lambda}^R) \in R^{q \times n}$, $B_{2\lambda}^L = (s_{ij\lambda}^L)$, $B_{2\lambda}^R = (s_{ij\lambda}^R) \in R^{q \times m}$.

For simplicity, we denote the constraint region of problem (7.36a–7.36d) by

$$S = \{(x,y) | A_{1\lambda}^L x + B_{1\lambda}^L y \leq b_{1\lambda}^L, A_{1\lambda}^R x + B_{1\lambda}^R y \leq b_{1\lambda}^R, A_{2\lambda}^L x$$
$$+ B_{2\lambda}^L y \leq b_{2\lambda}^L, A_{2\lambda}^R x + B_{2\lambda}^R y \leq b_{2\lambda}^L, \quad \lambda \in [0,1]\}.$$

Note that problem (7.36a–7.36d) is a deterministic multi-objective linear bi-level programming problem. It is difficult to obtain the complete optimal solution of the lower level problem, even though it does not exist in some practical problems, since the lower level is a multi-objective optimization problem. To overcome this complication, we define the solutions of the lower level problem by means of Pareto optimality in the multi-objective optimization problem.

We also give some definitions of the problem (7.36a–7.36d).

**Definition 7.20** For each fixed $x$, the constraint region of the lower level problem is:

$$S(x) = \{y | A_{2\lambda}^L x + B_{2\lambda}^L y \leq b_{2\lambda}^L, A_{2\lambda}^R x + B_{2\lambda}^R y \leq b_{2\lambda}^R, \lambda \in [0,1]\}.$$

**Definition 7.21** A point $y^* \in S(x^*)$ is called a Pareto optimal solution to the lower level problem in the problem (7.36a–7.36d) if there is no $y \in S(x^*)$ such that

$$c_{2\lambda}^L x + d_{2\lambda}^L y \leq c_{2\lambda}^L x^* + d_{2\lambda}^L y^*,$$
$$c_{2\lambda}^R x + d_{2\lambda}^R y \leq c_{2\lambda}^R x^* + d_{2\lambda}^R y^*.$$

with at least one strong inequality, and $\lambda \in [0,1]$.

Denote the set of Pareto optimal solutions of the lower level problem by $P(x)$.

**Definition 7.22** The feasible region of the problem (7.36a–7.36d) is defined as:

$$IR = \{(x,y) | (x,y) \in S, y \in P(x)\}.$$

Thus, the problem (7.36a–7.36d) can be equivalently written as:

$$\min_{x,y} (c_{1\lambda}^L x + d_{1\lambda}^L y, c_{1\lambda}^R x + d_{1\lambda}^R y)$$

$$\text{s.t.} (x,y) \in IR.$$

We then introduce the definitions of the solutions of the problem (7.36a–7.36d).

**Definition 7.23** A point $(x^*, y^*) \in IR$ is said to be a complete optimal solution to the problem (7.36a–7.36d) if it holds that

$$c_{1\lambda}^L x^* + d_{1\lambda}^L y^* \le c_{1\lambda}^L x + d_{1\lambda}^L y,$$
$$c_{1\lambda}^R x^* + d_{1\lambda}^R y^* \le c_{1\lambda}^R x + d_{1\lambda}^R y,$$

for $\lambda \in [0, 1]$ and all $(x, y) \in IR$.

**Definition 7.24** A point $(x^*, y^*) \in IR$ is said to be a Pareto optimal solution to the problem (7.36a–7.36d) if there is no $(x, y) \in IR$ such that

$$c_{1\lambda}^L x + d_{1\lambda}^L y \le c_{1\lambda}^L x^* + d_{1\lambda}^L y^*,$$
$$c_{1\lambda}^R x + d_{1\lambda}^R y \le c_{1\lambda}^R x^* + d_{1\lambda}^R y^*,$$

with at least one strong inequality, and $\lambda \in [0, 1]$.

**Definition 7.25** A point $(x^*, y^*) \in IR$ is said to be a weak Pareto optimal solution to the problem (7.36a–7.36d) if there is no $(x, y) \in IR$ such that

$$c_{1\lambda}^L x + d_{1\lambda}^L y < c_{1\lambda}^L x^* + d_{1\lambda}^L y^*,$$
$$c_{1\lambda}^R x + d_{1\lambda}^R y < c_{1\lambda}^R x^* + d_{1\lambda}^R y^*,$$

with at least one strong inequality, and $\lambda \in [0, 1]$.

**Theorem 7.3** *Problems* (7.29) *and* (7.36a–7.36d) *are equivalent for* $\lambda \in [0, 1]$.

*Proof* It follows from Definition (7.10) that the objective functions of both problems are equal. From Definition (7.11), it is easy to check that the constraint regions of the two problems are the same. This completes the proof.  □

**Lemma 7.1** *If there is* $(x^*, y^*)$ *such that* $c_{\alpha}^L x + d_{\alpha}^L y \ge c_{\alpha}^L x^* + d_{\alpha}^L y^*$, $c_{\beta}^L x + d_{\beta}^L y \ge c_{\beta}^L x^* + d_{\beta}^L y^*$, $c_{\alpha}^R x + d_{\alpha}^R y \ge c_{\alpha}^R x^* + d_{\alpha}^R y^*$, *and* $c_{\beta}^R x + d_{\beta}^R y \ge c_{\beta}^R x^* + d_{\beta}^R y^*$ *for any* $(x, y) \in R^n \times R^m (0 \le \beta < \alpha \le 1)$ *and fuzzy sets* $\tilde{c}$ *and* $\tilde{d}$ *have a trapezoidal membership function* (Fig. 7.1):

**Fig. 7.1** The trapezoidal membership function

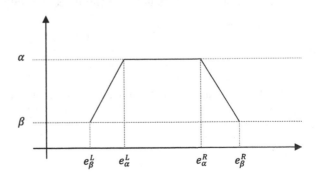

$$\mu_{\tilde{e}}(x) = \begin{cases} 0 & x < e_{\beta}^{L} \\ \frac{\alpha-\beta}{e_{\alpha}^{L}-e_{\beta}^{L}}\left(x - c_{\beta}^{L}\right) + \beta & e_{\beta}^{L} \leq x < e_{\alpha}^{L} \\ \alpha & e_{\alpha}^{L} \leq x \leq e_{\alpha}^{R} \\ \frac{\alpha-\beta}{e_{\alpha}^{R}-e_{\beta}^{R}}\left(x - e_{\beta}^{R}\right) + \beta & e_{\alpha}^{R} < x \leq e_{\beta}^{R} \\ 0 & e_{\alpha}^{R} < x \leq e_{\beta}^{R} \end{cases}$$

*then*

$$c_{\lambda}^{L}x + d_{\lambda}^{L}y \geq c_{\lambda}^{L}x^{*} + d_{\lambda}^{L}y^{*},$$
$$c_{\lambda}^{R}x + d_{\lambda}^{R}y \geq c_{\lambda}^{R}x^{*} + d_{\lambda}^{R}y^{*},$$

*for any $\lambda \in [\beta, \alpha]$.*

**Theorem 7.4** *For $x \in X \subset R^{n}$, $y \in Y \subset R^{m}$, if all the fuzzy parameters $\tilde{a}_{ij}$, $\tilde{b}_{ij}$, $\tilde{e}_{ij}$, $\tilde{s}_{ij}$, $\tilde{c}_{i}$, $\tilde{b}_{1}$, $\tilde{b}_{2}$, and $\tilde{d}_{i}$ have trapezoidal membership functions in the FBLDM problem (7.29),*

$$\mu_{\tilde{z}}(t) = \begin{cases} 0 & t < z_{\beta}^{L} \\ \frac{\alpha-\beta}{z_{\alpha}^{L}-z_{\beta}^{L}}\left(t - z_{\beta}^{L}\right) + \beta & z_{\beta}^{L} \leq t \leq z_{\alpha}^{L} \\ \alpha & z_{\alpha}^{L} \leq t < z_{\alpha}^{R} \\ \frac{\alpha-\beta}{z_{\beta}^{R}-z_{\alpha}^{R}}\left(-t + z_{\beta}^{R}\right) + \beta & z_{\alpha}^{R} \leq t \leq z_{\beta}^{R} \\ 0 & z_{\beta}^{R} < t \end{cases} \tag{7.37}$$

*where $\tilde{z}$ denotes $\tilde{a}_{ij}$, $\tilde{b}_{ij}$, $\tilde{e}_{ij}$, $\tilde{s}_{ij}$, $\tilde{c}_{i}$, $\tilde{b}_{1}$, $\tilde{b}_{2}$, and $\tilde{d}_{i}$ respectively, then $(x^{*}, y^{*})$ is a complete optimal solution to the problem (7.36a–7.36d) if and only if $(x^{*}, y^{*})$ is a complete optimal solution to the multi-objective linear bi-level programming (MOL-BLP) problem (7.38a–7.38d):*

$$\min_{x \in X} \quad (F(x,y))_{\alpha}^{L} = c_{1\alpha}^{L}x + d_{1\alpha}^{L}y,$$
$$\min_{x \in X} \quad (F(x,y))_{\alpha}^{R} = c_{1\alpha}^{R}x + d_{1\alpha}^{R}y,$$
$$\min_{x \in X} \quad (F(x,y))_{\beta}^{L} = c_{1\beta}^{L}x + d_{1\beta}^{L}y, \tag{7.38a}$$
$$\min_{x \in X} \quad (F(x,y))_{\beta}^{R} = c_{1\beta}^{R}x + d_{1\beta}^{R}y,$$

$$\text{s.t.} \quad A_{1\,\alpha}^{L}x + B_{1\,\alpha}^{L}y \leq b_{1\,\alpha}^{L},$$
$$A_{1\,\alpha}^{R}x + B_{1\,\alpha}^{R}y \leq b_{1\,\alpha}^{R},$$
$$A_{1\,\beta}^{L}x + B_{1\,\beta}^{L}y \leq b_{1\,\beta}^{L}, \tag{7.38b}$$
$$A_{1\,\beta}^{R}x + B_{1\,\beta}^{R}y \leq b_{1\,\beta}^{R},$$

$$\min_{y \in Y} \quad (f(x,y))_\alpha^L = c_{2\alpha}^L x + d_{2\alpha}^L y,$$

$$\min_{y \in Y} \quad (f(x,y))_\alpha^R = c_{2\alpha}^R x + d_{2\alpha}^R y,$$

$$\min_{y \in Y} \quad (f(x,y))_\beta^L = c_{2\beta}^L x + d_{2\beta}^L y,$$ (7.38c)

$$\min_{y \in Y} \quad (f(x,y))_\beta^R = c_{2\beta}^R x + d_{2\beta}^R y,$$

$$\text{s.t.} \quad A_{2\alpha}^L x + B_{2\alpha}^L y \leq b_{2\alpha}^L,$$

$$A_{2\alpha}^R x + B_{2\alpha}^R y \leq b_{2\alpha}^R,$$

$$A_{2\beta}^L x + B_{2\beta}^L y \leq b_{2\beta}^L,$$ (7.38d)

$$A_{2\beta}^R x + B_{2\beta}^R y \leq b_{2\beta}^R.$$

*Proof* Denote the constraint regions of (7.36a–7.36d) and (7.38a–7.38d) by $S_1$ and $S_2$, the feasible regions by $IR_1$ and $IR_2$ respectively. Now in order to show that the sets $IR_1$ and $IR_2$ are the same, we first prove $S_1$ and $S_2$ are equal.

For any $\lambda \in [\beta, \alpha]$, it is easy to check that if $(x,y)$ satisfies (7.36b) and (7.36d), then it satisfies (7.38b) and (7.38d). Now we prove, if $(x,y)$ satisfies (7.38b) and (7.38d), then it satisfies (7.36b) and (7.36d). In fact, for any $\lambda \in [\beta, \alpha]$,

$$a_{ij\lambda}^L = \frac{\lambda - \beta}{\alpha - \beta}\left(a_{ij\alpha}^L - a_{ij\beta}^L\right) + a_{ij\beta}^L,$$

$$b_{ij\lambda}^L = \frac{\lambda - \beta}{\alpha - \beta}\left(b_{ij\alpha}^L - b_{ij\beta}^L\right) + b_{ij\beta}^L,$$

$$b_{1\lambda}^R = \frac{\lambda - \beta}{\alpha - \beta}\left(b_{1\alpha}^R - b_{1\beta}^R\right) + b_{1\beta}^R,$$

we have

$$A_{1\lambda}^L x + B_{1\lambda}^L y = (a_{ij\lambda}^L)x + (b_{ij\lambda}^L)y$$

$$= \left(\frac{\lambda - \beta}{\alpha - \beta}\left(a_{ij\alpha}^L - a_{ij\beta}^L\right) + a_{ij\beta}^L\right)x + \left(\frac{\lambda - \beta}{\alpha - \beta}\left(b_{ij\alpha}^L - b_{ij\beta}^L\right) + b_{ij\beta}^L\right)y$$

$$= \frac{\lambda - \beta}{\alpha - \beta}(a_{ij\alpha}^L)x + \left(1 - \frac{\lambda - \beta}{\alpha - \beta}\right)(a_{ij\beta}^L)x + \frac{\lambda - \beta}{\alpha - \beta}(b_{ij\alpha}^L)y + \left(1 - \frac{\lambda - \beta}{\alpha - \beta}\right)(b_{ij\beta}^L)y$$

$$= \frac{\lambda - \beta}{\alpha - \beta}\left((a_{ij\alpha}^L)x + (b_{ij\alpha}^L)y\right) + \left(1 - \frac{\lambda - \beta}{\alpha - \beta}\right)\left((a_{ij\beta}^L)x + (b_{ij\beta}^L)y\right)$$

$$= \frac{\lambda - \beta}{\alpha - \beta}\left((A_{1\alpha}^L)x + (B_{1\alpha}^L)y\right) + \left(1 - \frac{\lambda - \beta}{\alpha - \beta}\right)\left((A_{1\beta}^L)x + (B_{1\beta}^L)y\right)$$

$$\leq \frac{\lambda - \beta}{\alpha - \beta}b_{1\alpha}^L + \left(1 - \frac{\lambda - \beta}{\alpha - \beta}\right)b_{1\beta}^L = b_{1\lambda}^L$$

from (7.38b). Similarly, we can prove

$$A_{1\lambda}^R x + B_{1\lambda}^R y \le b_{1\lambda}^R,$$
$$A_{2\lambda}^L x + B_{2\lambda}^L y \le b_{2\lambda}^L,$$
$$A_{2\lambda}^R x + B_{2\lambda}^R y \le b_{2\lambda}^R,$$

for any $\lambda \in [\beta, \alpha]$ from (7.38b) and (7.38d).

Thus, $S_1 = S_2$.

In order to show $IR_1 = IR_2$, we consider the following two cases.

**Case 1** For any $(x, y) \in IR_2$, we prove $(x, y) \in IR_1$.

It follows from $(x, y) \in IR_2$ that $(x, y) \in S_2$, and then $(x, y) \in S_1$. We prove this result by contradiction, and assume that $(x, y)$ is not a feasible point of problem (7.36a–7.36d). Thus, there exists $y'$ such that $(x, y') \in S_1$, and

$$(f(x, y'))_\lambda^L \le (f(x, y))_\lambda^L \quad \text{and} \quad (f(x, y'))_\lambda^R \le (f(x, y))_\lambda^R,$$

which imply that

$$(f(x, y'))_\alpha^L \le (f(x, y))_\alpha^L,$$
$$(f(x, y'))_\alpha^R \le (f(x, y))_\alpha^R,$$
$$(f(x, y'))_\beta^L \le (f(x, y))_\beta^L,$$
$$(f(x, y'))_\beta^R \le (f(x, y))_\beta^R.$$

That is, $(x, y)$ is not a feasible point of the problem (7.38a–7.38d), which contradicts the definition of $(x, y)$. Hence, $(x, y) \in IR_1$, and then $IR_1 \supseteq IR_2$.

**Case 2** For any $(x, y) \in IR_1$, we prove $(x, y) \in IR_2$.

It follows from $(x, y) \in IR_1$ that $(x, y) \in S_1$, and then $(x, y) \in S_2$. We prove this result by contradiction, and assume that $(x, y)$ is not a feasible point of the problem (7.38a–7.38d). Thus, there exists $y'$ such that $(x, y') \in S_2$, and

$$(f(x, y'))_\alpha^L \le (f(x, y))_\alpha^L,$$
$$(f(x, y'))_\alpha^R \le (f(x, y))_\alpha^R,$$
$$(f(x, y'))_\beta^L \le (f(x, y))_\beta^L,$$
$$(f(x, y'))_\beta^R \le (f(x, y))_\beta^R,$$

It follows from Lemma 7.1 that

$$(f(x,y'))_\lambda^L \leq (f(x,y))_\lambda^L \quad \text{and} \quad (f(x,y'))_\lambda^R \leq (f(x,y))_\lambda^R,$$

which contradicts $(x,y) \in IR_1$. Hence, $(x,y) \in IR_2$, and then $IR_2 \supseteq IR_1$.
   Therefore, $IR_2 = IR_1$.
   Now, we prove this theorem. If $(x^*, y^*)$ is a complete optimal solution to the problem (7.38a–7.38d), then for any $(x,y) \in IR_2$, we have

$$(F(x^*,y^*))_\alpha^L \leq (F(x,y))_\alpha^L,$$
$$(F(x^*,y^*))_\alpha^R \leq (F(x,y))_\alpha^R,$$
$$(F(x^*,y^*))_\beta^L \leq (F(x,y))_\beta^L,$$
$$(F(x^*,y^*))_\beta^R \leq (F(x,y))_\beta^R.$$

It follows from Lemma 7.1 that

$$(F(x^*,y^*))_\lambda^L \leq (F(x,y))_\lambda^L \quad \text{and} \quad (F(x^*,y^*))_\lambda^R \leq (F(x,y))_\lambda^R.$$

Thus, $(x^*, y^*)$ is a complete optimal solution to the problem (7.36a–7.36d).
   Furthermore, if $(x^*, y^*)$ is a complete optimal solution to the problem (7.36a–7.36d), then for any $(x,y) \in IR_1$, we find

$$(F(x^*,y^*))_\lambda^L \leq (F(x,y))_\lambda^L \quad \text{and} \quad (F(x^*,y^*))_\lambda^R \leq (F(x,y))_\lambda^R.$$

Obviously, $(x^*, y^*)$ is a complete optimal solution to the problem (7.38a–7.38d).
The proof is complete.                                                                          □

**Theorem 7.5** *For $x \in X \subset R^n$, $y \in Y \subset R^m$, if all the fuzzy parameters $\tilde{a}_{ij}$, $\tilde{b}_{ij}$, $\tilde{e}_{ij}$, $\tilde{s}_{ij}$, $\tilde{c}_i$, $\tilde{b}_1$, $\tilde{b}_2$, and $\tilde{d}_i$ have trapezoidal membership functions (7.37) in the FBLDM problem (7.29), then $(x^*, y^*)$ is a Pareto optimal solution to the problem (7.36a–7.36d) if and only if $(x^*, y^*)$ is a Pareto optimal solution to the problem (7.38a–7.38d).*

*Proof* Clearly, it follows from the proof of Theorem 7.4 that the feasible regions of problems (7.36a–7.36d) and (7.38a–7.38d) are the same. Let $(x^*, y^*)$ be a Pareto optimal solution to the problem (7.36a–7.36d). We prove it is also a Pareto optimal solution to the problem (7.38a–7.38d) by contradiction, and suppose that there exists a $(\bar{x}, \bar{y}) \in IR_2$ such that, for $\lambda = \alpha, \beta$,

$$(F(\bar{x},\bar{y}))_{\lambda}^{L} \leq (F(x^{*},y^{*}))_{\lambda}^{L},$$
$$(F(\bar{x},\bar{y}))_{\lambda}^{R} \leq (F(x^{*},y^{*}))_{\lambda}^{R},$$

By using Lemma 7.1, for any $\lambda \in [\beta, \alpha]$, we have

$$(F(\bar{x},\bar{y}))_{\lambda}^{L} \leq (F(x^{*},y^{*}))_{\lambda}^{L} \quad \text{and} \quad (F(\bar{x},\bar{y}))_{\lambda}^{R} \leq (F(x^{*},y^{*}))_{\lambda}^{R},$$

which contradicts the assumption that $(x^{*}, y^{*})$ is a Pareto optimal solution to the problem (7.36a–7.36d). Thus, $(x^{*}, y^{*})$ is a Pareto optimal solution to the problem (7.38a–7.38d).

Let $(x^{*}, y^{*})$ be a Pareto optimal solution to the problem (7.38a–7.38d). If $(x^{*}, y^{*})$ is not a Pareto optimal solution to the problem (7.36a–7.36d), then there exist $(\bar{x}, \bar{y}) \in IR_{1}$ such that

$$(F(\bar{x},\bar{y}))_{\lambda}^{L} \leq (F(x^{*},y^{*}))_{\lambda}^{L},$$
$$(F(\bar{x},\bar{y}))_{\lambda}^{R} \leq (F(x^{*},y^{*}))_{\lambda}^{R},$$

Hence, for $\lambda = \alpha$ and $\lambda = \beta$, we have

$$(F(\bar{x},\bar{y}))_{\alpha}^{L} \leq (F(x^{*},y^{*}))_{\alpha}^{L},$$
$$(F(\bar{x},\bar{y}))_{\alpha}^{R} \leq (F(x^{*},y^{*}))_{\alpha}^{R},$$
$$(F(\bar{x},\bar{y}))_{\beta}^{L} \leq (F(x^{*},y^{*}))_{\beta}^{L},$$
$$(F(\bar{x},\bar{y}))_{\beta}^{R} \leq (F(x^{*},y^{*}))_{\beta}^{R},$$

which contradicts the assumption that $(x^{*}, y^{*})$ is a Pareto optimal solution to the problem (7.38a–7.38d), and then $(x^{*}, y^{*})$ is a Pareto optimal solution to the problem (7.36a–7.36d). This completes the proof. □

**Theorem 7.6** *For $x \in X \subset R^{n}$, $y \in Y \subset R^{m}$, if all the fuzzy parameters $\tilde{a}_{ij}$, $\tilde{b}_{ij}$, $\tilde{e}_{ij}$, $\tilde{s}_{ij}$, $\tilde{c}_{i}$, $\tilde{b}_{1}$, $\tilde{b}_{2}$, and $\tilde{d}_{i}$ have trapezoidal membership functions (7.37) in the problem (7.29), then $(x^{*}, y^{*})$ is a weak Pareto optimal solution to the problem (7.36a–7.36d) if and only if $(x^{*}, y^{*})$ is a weak Pareto optimal solution to the problem (7.38a–7.38d).*

*Proof* The proof is similar to that of Theorem 7.5. □

---

**Algorithm 7.1: Fuzzy Approximation *K*th-Best Algorithm for the FBLDM Problem**

[Begin]

**Step 0:** Set $l=1$.

**Step 1:** Give their weights for fuzzy objectives of every leader and follower and $\sum_{j=1}^{2^l+1} w_{j1} = 1$ *and* $\sum_{j=1}^{2^l+1} w_{j2} = 1$, respectively.

**Step 2:** Transfer the problem (7.29) to the problem (7.38) by using Theorem 7.5.

**Step 3:** Let the interval $[0,1]$ be decomposed into $2^l$ mean sub-intervals with $\left(2^l+1\right)$ nodes $\lambda_i \left(i=0,1,\ldots,2^l\right)$, which are arranged in the order of $0 = \lambda_0 < \lambda_1 < \cdots < \lambda_{2^l} = 1$ and a range of errors $\varepsilon > 0$.

**Step 4:** Solve $(\text{MOL}-\text{BLP})_2^l$, i.e. (7.38) by using the weighting method and *K*th-Best algorithm when $\beta=0$ and $\alpha=1$, we obtain an optimal solution $(\mathbf{x},\ \mathbf{y})_{2^l}$.

**Step 5:** Solve $(\text{MOL}-\text{BLP})_2^{l+1}$, and obtain an optimal solution $(x,y)_{2^{l+1}}$.

**Step 6:** If $\left\|(x,y)_{2^{l+1}} - (x,y)_{2^l}\right\| < \varepsilon$, then $(x,y)_{2^{l+1}}$ is an approximation solution to the FBLDM problem, otherwise, update $l$ to $2l$ and go back to Step 1.

**Step 7:** Show the approximation solution.

[End]

---

To obtain a solution of the FBLDM problem (7.29), we only need to solve the problem (7.38a–7.38d). However, the lower level problem of the problem (7.38a–7.38d) is a multi-objective programming scenario. It is not true if we assume the follower's rational reaction set $P(x)$ is a point-to-point map with respect to $x$. To deal with this difficulty, there are at least three approaches, e.g. the optimistic formulation, pessimistic formulation (Dempe 2002), and selection function approach (Dempe and Starostina 2006). Since the follower has preferences (weighting) in related to the objectives, we adopt the third approach, i.e. the selection function approach, for solving the problem (7.38a–7.38d). Using the weighting method (Sakawa 1993), we can transform the problem (7.38a–7.38d) into a linear bi-level programming problem, and the resulting problem can be solved by the *K*th-Best algorithm. Now, this section will introduce a fuzzy approximation *K*th-Best algorithm to solve the FBLDM problem and a case-based example is used to illustrate the proposed approach.

A case-based example will be given to illustrate the proposed algorithm in the next section.

### 7.4.2 Illustrative Examples

We give examples here to illustrate how to use the proposed FBLDM decision model and the Fuzzy Approximation *K*th-Best algorithm for solving the FBLDM problems in practice.

*Example 7.2* In a logistic system, consider $\min_{x \in X} F(x,y) = \tilde{1}x - \tilde{2}y$ is a simplified objective function of a supplier, and $\min_{y \in Y} f_1(x,y) = \tilde{1}x + \tilde{1}y$ is a simplified objective function of its distributor, with $x \in R^1$, $y \in R^1$, and $X = \{x|x \geq 0\}$, $Y = \{y|y \geq 0\}$,

$$\min_{x \in X} \quad F(x,y) = \tilde{1}x - \tilde{2}y$$
$$\text{s.t.} \quad -\tilde{1}x + \tilde{3}y \preceq \tilde{4},$$
$$\min_{y \in Y} \quad f(x,y) = \tilde{1}x + \tilde{1}y \qquad\qquad (7.39)$$
$$\text{s.t.} \quad \tilde{1}x - \tilde{1}y \preceq \tilde{0},$$
$$-\tilde{1}x - \tilde{1}y \preceq \tilde{0},$$

where

$$\mu_{\tilde{1}}(t) = \begin{cases} 0, & t < 0 \\ t, & 0 \leq t < 1 \\ 2-t, & 1 \leq t < 2 \\ 0, & 2 \leq t \end{cases}, \quad \mu_{\tilde{2}}(t) = \begin{cases} 0, & t < 1 \\ t-1, & 1 \leq t < 2 \\ 3-t, & 2 \leq t < 3 \\ 0, & 3 \leq t \end{cases}$$

$$\mu_{\tilde{3}}(t) = \begin{cases} 0, & t < 2 \\ t-2, & 2 \leq t < 3 \\ 4-t, & 3 \leq t < 4 \\ 0, & 4 \leq t \end{cases}, \quad \mu_{\tilde{4}}(t) = \begin{cases} 0, & t < 3 \\ t-3, & 3 \leq t < 4 \\ 5-t, & 4 \leq t < 5 \\ 0, & 5 \leq t \end{cases}$$

$$\mu_{\tilde{0}}(t) = \begin{cases} 0, & t < -1 \\ t+1, & 1 \leq t < 0 \\ 1-t, & 0 \leq t < 1 \\ 0, & 1 \leq t \end{cases}$$

The proposed Fuzzy Approximation *K*th-Best algorithm is used for solving the problem (7.39) by the following steps.

Step 0:  Set $l = 1$.

Step 1:  Give their weights for fuzzy objectives of every leader and follower and $\sum_{j=1}^{2^l+1} w_{j1} = 1$ and $\sum_{j=1}^{2^l+1} w_{j2} = 1$, respectively.

Step 2: By using Theorem 7.5, transfer the problem (7.39) to the following problem:

$$\min_{x \in X} \quad (f(x,y))_1^{L(R)} = x - 2y$$

$$\min_{x \in X} \quad (F(x,y))_0^L = -3y$$

$$\min_{x \in X} \quad (F(x,y))_0^R = 2x - y$$

$$\text{s.t.} \quad -1x + 3y \le 4,$$

$$-2x + 2y \le 3,$$

$$0x + 4y \le 5,$$

$$\min_{y \in Y} \quad (f(x,y))_1^{L(R)} = 1x + 1y$$

$$\min_{y \in Y} \quad (f(x,y))_0^L = 0x + 0y$$

$$\min_{y \in Y} \quad (f(x,y))_0^R = 2x + 2y$$

$$\text{s.t.} \quad 1x - 1y \le 0,$$

$$0x - 2y \le -1,$$

$$2x - 0y \le 1,$$

$$-1x - 1y \le 0,$$

$$0x - 0y \le 0,$$

$$-2x - 2y \le -1.$$

Step 3: Let the interval $[0, 1]$ be decomposed into mean sub-intervals with $(2^l + 1)$ nodes $\lambda_i (i = 0, 1, \ldots, 2^l)$, which are arranged in the order of $0 = \lambda_0 < \lambda_1 < \cdots < \lambda_{2^l} = 1$ and a range of errors $\varepsilon > 0$.

Step 4: The problem is then transferred to the following linear bi-level programming problem by using the weighting method.

$$\min_{x \in X} \quad (F(x,y)) = 3x - 6y$$

$$\text{s.t.} \quad -1x + 3y \le 4,$$

$$-2x + 2y \le 3,$$

$$0x + 4y \le 5,$$

$$\min_{y \in Y} \quad (f(x,y)) = 3x + 3y$$

$$\text{s.t.} \quad 1x - 1y \le 0,$$

$$0x - 2y \le -1,$$

$$2x - 0y \le 1,$$

$$-1x - 1y \le 0,$$

$$0x - 0y \le 0,$$

$$-2x - 2y \le 1.$$

Step 5: The result is $(x,y)_{2^1} = (0, 0.5)$, and

$$(F(x,y))_1^{L(R)} = -1,$$
$$(F(x,y))_0^L = -1.5,$$
$$(F(x,y))_0^R = -0.5,$$
$$(f(x,y))_1^{L(R)} = 0.5,$$
$$(f(x,y))_0^L = 0,$$
$$(f(x,y))_0^R = 1.$$

Step 2: When $l = 2$, we solve the following problem:

$$\min_{x \in X} \quad (F(x,y))_1^{L(R)} = x - 2y$$
$$\min_{x \in X} \quad (F(x,y))_{\frac{1}{2}}^L = 0.5x - 2.5y$$
$$\min_{x \in X} \quad (F(x,y))_0^L = -3y$$
$$\min_{x \in X} \quad (F(x,y))_{\frac{1}{2}}^R = 1.5x - 1.5y$$
$$\min_{x \in X} \quad (F(x,y))_0^R = 2x - y$$
$$\text{s.t.} \quad -1x + 3y \leq 4,$$
$$-1.5x + 2.5y \leq 3.5,$$
$$-2x + 2y \leq 3,$$
$$-0.5x + 3.5y \leq 4.5,$$
$$0x + 4y \leq 5,$$
$$\min_{y \in Y} \quad (f(x,y))_1^{L(R)} = 1x + 1y$$
$$\min_{y \in Y} \quad (f(x,y))_{\frac{1}{2}}^L = 0.5x + 0.5y$$
$$\min_{y \in Y} \quad (f(x,y))_0^L = 0x + 0y$$
$$\min_{y \in Y} \quad (f(x,y))_{\frac{1}{2}}^R = 1.5x + 1.5y$$
$$\min_{y \in Y} \quad (f(x,y))_0^R = 2x + 2y$$

s.t.   $1x - 1y \leq 0,$

$0.5x - 1.5y \leq -0.5,$

$0x - 2y \leq -1,$

$1.5x - 0.5y \leq 0.5,$

$2x - 0y \leq 1,$

$-1.5x - 1.5y \leq -0.5,$

$-1x - 1y \leq 0,$

$-2x - 2y \leq -1,$

$0x - 0y \leq 0,$

$-0.5x - 0.5y \leq 0.5.$

**Step 4:** The problem is then transferred to the following linear bi-level programming problem by using method of weighting

$$\min_{x \in X} \ (F(x,y)) = 5x - 10y$$

s.t.   $-1x + 3y \leq 4,$

$-1.5x + 2.5y \leq 3.5,$

$-2x + 2y \leq 3,$

$-0.5x + 3.5y \leq 4.5,$

$0x + 4y \leq 5,$

$$\min_{y \in Y} \ (f(x,y)) = 5x + 5y$$

s.t.   $1x - 1y \leq 0,$

$0.5x - 1.5y \leq -0.5,$

$0x - 2y \leq -1,$

$1.5x - 0.5y \leq 0.5,$

$2x - 0y \leq 1,$

$-1.5x - 1.5y \leq -0.5,$

$-1x - 1y \leq 0,$

$-2x - 2y \leq -1,$

$0x - 0y \leq 0,$

$-0.5x - 0.5y \leq 0.5.$

Step 5: The result is $(x, y)_{2^2} = (0, 0.5)$, and

$$(F(x, y))_1^{L(R)} = -1,$$
$$(F(x, y))_{\frac{1}{2}}^{L} = -1.25,$$
$$(F(x, y))_0^{L} = -1.5,$$
$$(F(x, y))_{\frac{1}{2}}^{R} = -0.75,$$
$$(F(x, y))_0^{R} = -0.5,$$
$$(f(x, y))_1^{L(R)} = 0.5,$$
$$(f(x, y))_{\frac{1}{2}}^{L} = 0.25,$$
$$(f(x, y))_0^{L} = 0,$$
$$(f(x, y))_{\frac{1}{2}}^{R} = 0.75,$$
$$(f(x, y))_0^{R} = 1.$$

Step 6: Since $\| (x, y)_{2^2} - (x, y)_{2^1} \| = 0$, $(x, y) = (0, 0.5)$ is the optimal solution for this example. We have the solution $x = 0, y = 0.5$ of the problem (7.39) with $F(x, y) = -0.5 \times \tilde{2}$ and $f(x, y) = 0.5 \times \tilde{1}$.

In summary, since the logistics planning problem involves two decision entities at a hierarchical structure, we have handled it by the bi-level decision techniques. When modeling this problem in practice, uncertain factors are often involved. That is, the parameters of the bi-level decision model for logistics planning, involved in the objective functions and constraints, are uncertain values. It therefore becomes a fuzzy bi-level decision problem. We apply the developed fuzzy-number-based $K$th-Best algorithm to find an optimal solution for this problem.

## 7.5 Fuzzy Multi-Follower Approximation $K$th-Best Algorithm

To model a real-world bi-level decision problem, some uncertain parameters often appear in the objective functions and/or constraints of the leader and/or the follower. Also, multiple followers may be involved in a decision problem and work cooperatively according to each of the possible decisions made by the leader, but they may have different objectives and/or constraints. Following previous sections and the BLMF decision problem presented in Chap. 4, this section proposes a fuzzy multi-follower approximation $K$th-Best algorithm to solve the FBLMF problem.

Associated with the FBLMF decision problem $F$-$S9$, we now consider the following linear bi-level multi-follower decision problem:

For $x \in X \subset R^n, y_i \in Y_i \subset R^{m_i}, F : X \times Y_1 \times \cdots \times Y_k \to F(R), f_i : X \times Y_1 \times \cdots \times Y_k \to F(R)$, and $i = 1, \ldots, k$,

$$\min_{x \in X} \ (F(x, y_1, \ldots, y_k))_\lambda^{L(R)} = c_\lambda^{L(R)} x + \sum_{j=1}^{k} d_{j\lambda}^{L(R)} y_j, \quad \lambda \in [0, 1] \qquad (7.40a)$$

$$\text{s.t.} \ \ A_\lambda^{L(R)} x + \sum_{j=1}^{k} B_{j\lambda}^{L(R)} y_j \le b_\lambda^{L(R)}, \quad \lambda \in [0, 1] \qquad (7.40b)$$

where $y_i$ is the solution to the $i$th follower's problem:

$$\min_{y_i \in Y_i} \ (f_i(x, y_1, \ldots, y_k))_\lambda^{L(R)} = c_{i\lambda}^{L(R)} x + \sum_{s=1}^{k} e_{is\lambda}^{L(R)} y_s, \quad \lambda \in [0, 1] \qquad (7.40c)$$

$$A_{i\lambda}^{L(R)} x + \sum_{s=1}^{k} C_{is\lambda}^{L(R)} y_s \le b_{i\lambda}^{L(R)}, \quad \lambda \in [0, 1] \qquad (7.40d)$$

where $c_\lambda^{L(R)}, c_{i\lambda}^{L(R)} \in R^n, \ d_{j\lambda}^{L(R)} \in R^{m_j}, e_{is\lambda}^{L(R)} \in R^{m_s}, A_\lambda^{L(R)} \in R^{p \times n}, \ B_{j\lambda}^{L(R)} \in R^{p \times m_j},$
$b_\lambda^{L(R)} \in R^p, A_{i\lambda}^{L(R)} \in R^{q_i \times n}, C_{is\lambda}^{L(R)} \in R^{q_i \times m_s}, b_{i\lambda}^{L(R)} \in R^{q_i}$.

By using Definition 7.18, we have the following results.

**Theorem 7.7** *Let* $(x^*, y_1^*, \ldots, y_k^*)$ *be the optimal solution of the BLMF problem defined by (7.40a–7.40d). Then it is also an optimal solution of the FBLMF decision problem defined by F-S9.*

We can prove the following theorem similar to the proof of Theorem 7.4.

**Theorem 7.8** *If all the fuzzy parameters of F-S9 have triangular membership functions:*

$$\mu_{\tilde{z}_i}(t) = \begin{cases} 0, & t_i < z_{0i}^L \\ \frac{t_i - z_{0i}^L}{z_i - z_{0i}^L}, & z_{0i}^L \le t_i < z_i \\ \frac{z_{0i}^R - t_i}{z_{0i}^R - z_i}, & z_i \le t_i < z_{0i}^R \\ 0, & t_i \ge z_{0i}^R \end{cases}$$

*then* $(x^*, y_1^*, \ldots, y_k^*)$ *is a complete optimal solution to the problem (7.40a–7.40d) if and only if* $(x^*, y_1^*, \ldots, y_k^*)$ *is a complete optimal solution to the multi-objective bi-level multi-follower (MO-BLMF) problem:*

$$\min_{x \in X} \ (F(x, y_1, \ldots, y_k))_c = cx + \sum_{i=1}^{k} d_i y_i$$

$$\min_{x \in X} \ (F(x, y_1, \ldots, y_k))_0^L = c_0^L x + \sum_{i=1}^{k} d_{i0}^L y_i$$

$$\min_{x \in X} \ (F(x, y_1, \ldots, y_k))_0^R = c_0^R x + \sum_{i=1}^{k} d_{i0}^R y_i$$

$$\text{s.t.} \quad Ax + \sum_{(i=0)}^{k} B_i y_i \leq b,$$

$$A_0^L x + \sum_{i=0}^{k} B_{i0}^L y_i \leq b_0^L,$$

$$A_0^R x + \sum_{i=0}^{k} B_{i0}^R y_i \leq b_0^R,$$

$$\min_{y_i \in Y_i} \ (f_i(x, y_1, \ldots, y_k))_c = c_i x + \sum_{j=1}^{k} e_{ij} y_j \qquad (7.41)$$

$$\min_{y_i \in Y_i} \ (f_i(x, y_1, \ldots, y_k))_0^L = c_{i0}^L x + \sum_{j=1}^{k} e_{ij0}^L y_j$$

$$\min_{y_i \in Y_i} \ (f_i(x, y_1, \ldots, y_k))_0^R = c_{i0}^R x + \sum_{j=1}^{k} e_{ij0}^R y_j$$

$$\text{s.t.} \quad A_i x + \sum_{j=0}^{k} C_{ij} y_j \leq b_i,$$

$$A_{i0}^L x + \sum_{j=0}^{k} C_{ij0}^L y_j \leq b_{i0}^L,$$

$$A_{i0}^R x + \sum_{j=0}^{k} C_{ij0}^R y_j \leq b_{i0}^R.$$

To solve the problem (7.41), we can transform it into the following problem by employing the weighting method (Sakawa 1993):

$$\min_{x \in X} \ (F(x, y_1, \ldots, y_k))_a = cx + \sum_{i=1}^{k} d_i y_i + c_0^L x + \sum_{i=1}^{k} d_{i0}^L y_i + c_0^R x + \sum_{i=1}^{k} d_{i0}^R y_i$$

$$\text{s.t.} \quad Ax + \sum_{i=0}^{k} B_i y_i \leq b,$$

$$A_0^L x + \sum_{i=0}^{k} B_{i0}^L y_i \leq b_0^L, \qquad (7.42)$$

$$A_0^R x + \sum_{i=0}^{k} B_{i0}^R y_i \leq b_0^R,$$

$$\min_{y_i \in Y_i} \quad f_i(x, y_1, \ldots, y_k) = c_i x + \sum_{j=1}^{k} e_{ij} y_j + c_{i0}^L x + \sum_{j=1}^{k} e_{ij0}^L y_j + c_{i0}^R x + \sum_{j=1}^{k} e_{ij0}^R y_j$$

$$\text{s.t.} \quad A_i x + \sum_{j=0}^{k} C_{ij} y_j \leq b_i,$$

$$A_{i0}^L x + \sum_{j=0}^{k} C_{ij0}^L y_j \leq b_{i0}^L,$$

$$A_{i0}^R x + \sum_{j=0}^{k} C_{ij0}^R y_j \leq b_{i0}^R.$$

Note that the problem (7.42) is a general linear bi-level multi-follower problem which has been discussed in Sect. 4.6. Denoting the constraint region of the problem (7.42) by $S$, we can easily obtain the following result by Theorem 4.11.

**Theorem 7.9** *The solution of the linear BLMF problem (7.42) occurs at a vertex of S.*

Theorem 7.9 has provided the theoretical foundation for our new approach to solve FBLMF decision problems. According to the objective function of the upper level, we apply the descendent order to all the extreme points on $S$, and select the first extreme point to check if it is on the inducible region $IR$. If yes, the current extreme point is the optimal solution. Otherwise, we select the next one and check it.

Based on the above results, we propose a fuzzy multi-follower approximation $K$th-Best algorithm to solve the FBLMF decision problem $F$-$S9$ as follows:

---

**Algorithm 7.2: Fuzzy Approximation $K$th-Best Algorithm for the FBLMF Decision Problem $F$-$S9$**

[Begin]

**Step 1:** Given the weight $w_j (j = 1, \ldots, k)$ for the followers' problem respectively and let $\sum_{j=1}^{k} w_j = 1$.

**Step 2:** Transform the FBLMF decision problem to the problem (7.42).

**Step 3:** Solve the problem (7.42) by using the MF-$K$th-Best algorithm proposed in Section 4.6.3.

**Step 4:** Show the approximation solution of the FBLMF problem $F$-$S9$.

[End]

---

## 7.6  Summary

This chapter first discusses uncertain issue in bi-level decision-making. After introducing fuzzy sets and related concepts, it presents a fuzzy bi-level decision-making model and a set of fuzzy bi-level multi-follower decision-making models. It then presents a set of algorithms to solve these problems. The proposed fuzzy approximation $K$th-Best algorithm and fuzzy multi-follower approximation $K$th-Best algorithm will have a wide range of applications in terms of dealing with various fuzzy bi-level decision-making problems.

# Chapter 8
# Fuzzy Multi-objective Bi-level Decision Making

In Chap. 7, we presented a set of solution approaches and related algorithms to solve a fuzzy bi-level programming problem. This chapter extends the results given in Chap. 7 by adding the capability to handle the multi-objective issue, that is, the leader, or the follower, or both have multiple objectives. We call it *fuzzy multi-objective bi-level programming* (FMO-BLP) or *fuzzy multi-objective bi-level decision-making* (FMO-BLD). Obviously, the fuzzy bi-level programming developed in Chap. 7 is a special case of the FMO-BLP. To obtain a solution for the FMO-BLP problem, we develop a fuzzy approximation Kuhn-Tucker approach. The approach can deal with various complex FMO-BLP problems and allow the parameters in the FMO-BLP model to be any form of membership functions.

The rest of this chapter is organized as follows. Section 8.1 identifies the FMO-BLP problem. Section 8.2 defines the FMO-BLP model and reviews related theorems and properties of the FMO-BLP problem. A fuzzy approximation Kuhn-Tucker approach and related examples for solving the FMO-BLP problem is presented in Sect. 8.3. A summary is given in Sect. 8.4.

## 8.1 Problem Identification

In some situations of bi-level decision problems, multiple conflicting objectives may need to be considered simultaneously by the leader, or the follower, or both. For example, a coordinator of a multi-division firm considers three objectives in making an aggregate production plan: to maximize net profits, quality of products, and workers' satisfaction. The three objectives are in conflict with each other, but must be considered simultaneously. Any improvement in one objective may be achieved only at the expense of others. The multi-objective optimization problem has been well researched by many researchers. However, in a bi-level model, the selection of a solution for the leader is not only a compromised solution for the three objective functions of the leader, but also is affected by the follower's optimal reactions. Therefore, how to find an optimal solution for a multiple objectives bi-level program is a new issue.

© Springer-Verlag Berlin Heidelberg 2015
G. Zhang et al., *Multi-Level Decision Making*,
Intelligent Systems Reference Library 82, DOI 10.1007/978-3-662-46059-7_8

Furthermore, fuzzy parameters may be in the objective functions or the constraints or both since uncertain values are often involved in a bi-level decision problem in practice as we mentioned in Chap. 7. This results in a FMO-BLP problem. In addition, there may be multiple followers involved in the FMO-BLP problem, that is, the *fuzzy multi-objective bi-level multi-follower* (FMO-BLMF) decision problem.

In the following sections, we will mainly consider the linear version of the FMO-BLP problem.

## 8.2  Fuzzy Multi-objective Bi-level Decision Model

In this section, we present a FMO-BLP model and review related definitions, theorems and properties for the FMO-BLP problem.

Consider the following FMO-BLP problem:

For $x \in X \subset R^n$, $y \in Y \subset R^m$, $F : X \times Y \to (F^*(R))^s$, and $f : X \times Y \to (F^*(R))^t$,

$$\min_{x \in X} \quad F(x,y) = \left( \tilde{c}_{11}x + \tilde{d}_{11}y, \tilde{c}_{21}x + \tilde{d}_{21}y, \ldots, \tilde{c}_{s1}x + \tilde{d}_{s1}y \right) \qquad (8.1a)$$

$$\text{s.t.} \quad \tilde{A}_1 x + \tilde{B}_1 y \leqq \tilde{b}_1, \qquad (8.1b)$$

$$\min_{y \in Y} \quad f(x,y) = \left( \tilde{c}_{12}x + \tilde{d}_{12}y, \tilde{c}_{22}x + \tilde{d}_{22}y, \ldots, \tilde{c}_{t2}x + \tilde{d}_{t2}y \right) \qquad (8.1c)$$

$$\text{s.t.} \quad \tilde{A}_2 x + \tilde{B}_2 y \leqq \tilde{b}_2, \qquad (8.1d)$$

where $\tilde{c}_{i1}, \tilde{c}_{j2} \in (F^*(R))^n, \tilde{d}_{i1}, \tilde{d}_{j2} \in (F^*(R))^m, \tilde{b}_1 \in (F^*(R))^p, \tilde{b}_2 \in (F^*(R))^q, \tilde{A}_1 = \left( \tilde{a}_{ij} \right)_{p \times n}, \tilde{a}_{ij} \in F^*(R), \tilde{B}_1 = \left( \tilde{b}_{ij} \right)_{p \times m}, \tilde{b}_{ij} \in F^*(R), \tilde{A}_2 = \left( \tilde{e}_{ij} \right)_{q \times n}, \tilde{e}_{ij} \in F^*(R), \tilde{B}_2 = \left( \tilde{s}_{ij} \right)_{q \times m}, \tilde{s}_{ij} \in F^*(R), i = 1,2,\ldots,s, j = 1,2,\ldots,t.$

For the sake of simplicity, denote the constraint region of problem (8.1a)–(8.1d) by

$$\tilde{S} = \left\{ (x,y) \,|\, \tilde{A}_1 x + \tilde{B}_1 y \leqq \tilde{b}_1, \tilde{A}_2 x + \tilde{B}_2 y \leqq \tilde{b}_2 \right\}.$$

We assume that $\tilde{S}$ is compact.

In the FMO-BLP problem (8.1a)–(8.1d), $F(x,y) = (F_1(x,y), \ldots, F_s(x,y))$ and $f(x,y) = (f_1(x,y), \ldots, f_t(x,y))$ of the leader and the follower are $s$-dimensional and $t$-dimensional fuzzy numbers, respectively.

Since problem (8.1a)–(8.1d) has fuzzy variables, we first use the $\lambda$-level sets of fuzzy variables to describe the objective functions and the constraints. The FMO-BLP problem (8.1a)–(8.1d) can be reformulated as the following multi-objective bi-level programming problem:

$$\min_{x \in X} \ (F(x,y))_\lambda^{L(R)} = \left((F_1(x,y))_\lambda^L, (F_1(x,y))_\lambda^R, \dots, (F_s(x,y))_\lambda^L, (F_s(x,y))_\lambda^R\right) \quad (8.2a)$$

$$\text{s.t.} \quad A_1{}_\lambda^L x + B_1{}_\lambda^L y \le b_1{}_\lambda^L, A_1{}_\lambda^R x + B_1{}_\lambda^R y \le b_1{}_\lambda^R, \quad \lambda \in [0,1] \quad\quad (8.2b)$$

$$\min_{y \in Y} \ (f(x,y))_\lambda^{L(R)} = \left((f_1(x,y))_\lambda^L, (f_1(x,y))_\lambda^R, \dots, (f_t(x,y))_\lambda^L, (f_t(x,y))_\lambda^R\right) \quad (8.2c)$$

$$\text{s.t.} \quad A_2{}_\lambda^L x + B_2{}_\lambda^L y \le b_2{}_\lambda^L, A_2{}_\lambda^R x + B_2{}_\lambda^R y \le b_2{}_\lambda^R, \quad \lambda \in [0,1] \quad\quad (8.2d)$$

where $(F_i(x,y))_\lambda^L = c_{i1}{}_\lambda^L x + d_{i1}{}_\lambda^L y$, $c_{i1}{}_\lambda^L, c_{i1}{}_\lambda^R, c_{j2}{}_\lambda^L, c_{j2}{}_\lambda^R \in R^n$, $d_{i1}{}_\lambda^L, d_{i1}{}_\lambda^R, d_{j2}{}_\lambda^L, d_{j2}{}_\lambda^R \in R^m$, $b_1{}_\lambda^L, b_1{}_\lambda^R \in R^p$, $b_2{}_\lambda^L, b_2{}_\lambda^R \in R^q$, $A_1{}_\lambda^L = \left(a_{ij}{}_\lambda^L\right)$, $A_1{}_\lambda^R = \left(a_{ij}{}_\lambda^R\right) \in R^{p \times n}$, $(f_i(x,y))_\lambda^L = c_{j2}{}_\lambda^L x + d_{j2}{}_\lambda^L y$, $A_2{}_\lambda^L = \left(e_{ij}{}_\lambda^L\right), A_2{}_\lambda^R = \left(e_{ij}{}_\lambda^R\right) \in R^{q \times n}$, $B_1{}_\lambda^L = \left(b_{ij}{}_\lambda^L\right), B_1{}_\lambda^R = \left(b_{ij}{}_\lambda^R\right) \in R^{p \times m}$, $B_2{}_\lambda^L = \left(s_{ij}{}_\lambda^L\right), B_2{}_\lambda^R = \left(s_{ij}{}_\lambda^R\right) \in R^{q \times n}$, $\lambda \in [0,1], i = 1, \dots, s, j = 1, \dots, t$.

For simplicity, we denote the constraint region of problem (8.2a)–(8.2d) by $S = \{(x,y)|A_1{}_\lambda^L x + B_1{}_\lambda^L y \le b_1{}_\lambda^L, A_1{}_\lambda^R x + B_1{}_\lambda^R y \le b_1{}_\lambda^R, A_2{}_\lambda^L x + B_2{}_\lambda^L y \le b_2{}_\lambda^L, A_2{}_\lambda^R x + B_2{}_\lambda^R y\} \le b_2{}_\lambda^R\}$, and assume that $S$ is compact.

Clearly, problem (8.2a)–(8.2d) is a deterministic multi-objective linear bi-level programming problem. It is difficult to obtain the complete optimal solution of the lower level problem; such a solution does not exist to some practical problems, since the lower level is a multi-objective optimization problem. To overcome this difficulty, we define the solutions of the lower level problem by means of Pareto optimality in multi-objective optimization problem.

Next, we give some definitions of the problem (8.2a)–(8.2d).

**Definition 8.1** For each fixed $x$, the constraint region of the lower level problem is:

$$S(x) = \left\{y \,|\, A_2{}_\lambda^L x + B_2{}_\lambda^L y \le b_2{}_\lambda^L, A_2{}_\lambda^R x + B_2{}_\lambda^R y \le b_2{}_\lambda^R, \lambda \in [0,1]\right\}.$$

**Definition 8.2** A point $y^* \in S(x^*)$ is called a Pareto optimal solution to the lower level problem in problem (8.2a)–(8.2d) if there is no $y \in S(x^*)$ such that

$$(f_1(x,y))_\lambda^L \le (f_1(x^*,y^*))_\lambda^L,$$
$$(f_1(x,y))_\lambda^R \le (f_1(x^*,y^*))_\lambda^R,$$
$$\vdots$$
$$(f_t(x,y))_\lambda^L \le (f_t(x^*,y^*))_\lambda^L,$$
$$(f_t(x,y))_\lambda^R \le (f_t(x^*,y^*))_\lambda^R,$$

with at least one strong inequality, and $\lambda \in [0,1]$.

Denote the set of Pareto optimal solutions of the lower level problem by $P(x)$.

**Definition 8.3** The feasible region of the problem (8.2a)–(8.2d) is defined as:

$$IR = \{(x,y) \mid (x,y) \in S, y \in P(x)\}.$$

Thus, problem (8.2a)–(8.2d) can be equivalently written as:

$$\min_{x,y} \; (F(x,y))_\lambda^{L(R)} = ((F_1(x,y))_\lambda^L, (F_1(x,y))_\lambda^R, \ldots, (F_s(x,y))_\lambda^L, (F_s(x,y))_\lambda^R)$$
$$\text{s.t.} \quad (x,y) \in IR.$$

We then introduce the definitions of the solutions of problem (8.2a)–(8.2d).

**Definition 8.4** A point $(x^*,y^*) \in IR$ is called a complete optimal solution to problem (8.2a)–(8.2d) if it holds that

$$(F_1(x^*,y^*))_\lambda^L \le (F_1(x,y))_\lambda^L,$$
$$(F_1(x^*,y^*))_\lambda^R \le (F_1(x,y))_\lambda^R,$$

$$\vdots$$

$$(F_s(x^*,y^*))_\lambda^L \le (F_s(x,y))_\lambda^L,$$
$$(F_s(x^*,y^*))_\lambda^R \le (F_s(x,y))_\lambda^R,$$

for $\lambda \in [0,1]$ and all $(x,y) \in IR$.

**Definition 8.5** A point $(x^*,y^*) \in IR$ is called a Pareto optimal solution to problem (8.2a)–(8.2d) if there is no $(x,y) \in IR$ such that

$$(F_1(x,y))_\lambda^L \le (F_1(x^*,y^*))_\lambda^L,$$
$$(F_1(x,y))_\lambda^R \le (F_1(x^*,y^*))_\lambda^R,$$

$$\vdots$$

$$(F_s(x,y))_\lambda^L \le (F_s(x^*,y^*))_\lambda^L,$$
$$(F_s(x,y))_\lambda^R \le (F_s(x^*,y^*))_\lambda^R,$$

with at least one strong inequality, and $\lambda \in [0,1]$.

**Fig. 8.1** The trapezoidal membership function

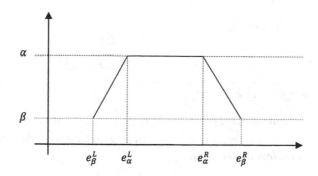

**Definition 8.6** A point $(x^*, y^*) \in IR$ is called a weak Pareto optimal solution to problem (8.2a)–(8.2d) if there is no $(x, y) \in IR$ such that

$$(F_1(x,y))_\lambda^L < (F_1(x^*,y^*))_\lambda^L,$$
$$(F_1(x,y))_\lambda^R < (F_1(x^*,y^*))_\lambda^R,$$

$$\vdots$$

$$(F_s(x,y))_\lambda^L < (F_s(x^*,y^*))_\lambda^L,$$
$$(F_s(x,y))_\lambda^R < (F_s(x^*,y^*))_\lambda^R,$$

with at least one strong inequality, and $\lambda \in [0,1]$.

**Theorem 8.1** *Problems* (8.1a)–(8.1d) *and* (8.2a)–(8.2d) *are equivalent for* $\lambda \in [0,1]$.

*Proof* It follows from Definition (7.10) that the objective functions of both problems are equal. From Definition (7.11), it is easy to check that the constraint regions of the two problems are the same. This completes the proof. □

**Lemma 8.1** *If there is* $(x^*, y^*)$ *such that* $c_\alpha^L x + d_\alpha^L y \geq c_\alpha^L x^* + d_\alpha^L y^*$, $c_\beta^L x + d_\beta^L y \geq c_\beta^L x^* + d_\beta^L y^*$, $c_\alpha^R x + d_\alpha^R y \geq c_\alpha^R x^* + d_\alpha^R y^*$, *and* $c_\beta^R x + d_\beta^R y \geq c_\beta^R x^* + d_\beta^R y^*$ *for any* $(x, y) \in R^n \times R^m (0 \leq \beta < \alpha \leq 1)$ *and fuzzy sets* $\tilde{c}$ *and* $\tilde{d}$ *have a trapezoidal membership function which is also shown in Fig. 8.1:*

$$\mu_{\tilde{e}}(x) = \begin{cases} 0, & x < e_\beta^L \\ \frac{\alpha-\beta}{e_\alpha^L - e_\beta^L}\left(x - c_\beta^L\right) + \beta, & e_\beta^L \leq x < e_\alpha^L \\ \alpha, & e_\alpha^L \leq x \leq e_\alpha^R \\ \frac{\alpha-\beta}{e_\alpha^R - e_\beta^R}\left(x - e_\beta^R\right) + \beta, & e_\alpha^R < x \leq e_\beta^R \\ 0, & e_\beta^R < x \end{cases}$$

*then we have*

$$c_\lambda^L x + d_\lambda^L y \geq c_\lambda^L x^* + d_\lambda^L y^*,$$
$$c_\lambda^R x + d_\lambda^R y \geq c_\lambda^R x^* + d_\lambda^R y^*,$$

*for any $\lambda \in [\beta, \alpha]$.*

**Theorem 8.2** *For $x \in X \subset R^n$, $y \in Y \subset R^m$, if all the fuzzy parameters $\tilde{a}_{ij}$, $\tilde{b}_{ij}$, $\tilde{e}_{ij}$, $\tilde{s}_{ij}$, $\tilde{c}_{ij}$, $\tilde{b}_1$, $\tilde{b}_2$, and $\tilde{d}_{ij}$ have trapezoidal membership functions in the FMO-BLP problem (8.1a)–(8.1d),*

$$\mu_{\tilde{z}}(t) = \begin{cases} 0, & t < z_\beta^L \\ \frac{\alpha - \beta}{z_\alpha^L - z_\beta^L}\left(t - z_\beta^L\right) + \beta, & z_\beta^L \leq t \leq z_\alpha^L \\ \alpha, & z_\alpha^L \leq t < z_\alpha^R \\ \frac{\alpha - \beta}{z_\beta^R - z_\alpha^R}\left(-t + z_\beta^R\right) + \beta, & z_\alpha^R \leq t \leq z_\beta^R \\ 0, & z_\beta^R < t \end{cases} \tag{8.3}$$

*where $\tilde{z}$ denotes $\tilde{a}_{ij}$, $\tilde{b}_{ij}$, $\tilde{e}_{ij}$, $\tilde{s}_{ij}$, $\tilde{c}_{ij}$, $\tilde{b}_1$, $\tilde{b}_2$, and $\tilde{d}_{ij}$ respectively, then $(x^*, y^*)$ is a complete optimal solution to the problem (8.2a)–(8.2d) if and only if $(x^*, y^*)$ is a complete optimal solution to the multi-objective linear bi-level programming (MOL-BLP) problem:*

$$\min_{x \in X} \quad (F_1(x,y))_\alpha^L = c_{11}{}_\alpha^L x + d_{11}{}_\alpha^L y$$
$$\min_{x \in X} \quad (F_1(x,y))_\alpha^R = c_{11}{}_\alpha^R x + d_{11}{}_\alpha^R y$$
$$\min_{x \in X} \quad (F_1(x,y))_\beta^L = c_{11}{}_\beta^L x + d_{11}{}_\beta^L y$$
$$\min_{x \in X} \quad (F_1(x,y))_\beta^R = c_{11}{}_\beta^R x + d_{11}{}_\beta^R y$$

$$\vdots \tag{8.4a}$$

$$\min_{x \in X} \quad (F_s(x,y))_\alpha^L = c_{s1}{}_\alpha^L x + d_{s1}{}_\alpha^L y$$
$$\min_{x \in X} \quad (F_s(x,y))_\alpha^R = c_{s1}{}_\alpha^R x + d_{s1}{}_\alpha^R y$$
$$\min_{x \in X} \quad (F_s(x,y))_\beta^L = c_{s1}{}_\beta^L x + d_{s1}{}_\beta^L y$$
$$\min_{x \in X} \quad (F_s(x,y))_\beta^R = c_{s1}{}_\beta^R x + d_{s1}{}_\beta^R y$$

$$\text{s.t.} \quad A_1{}_\alpha^L x + B_1{}_\alpha^L y \leq b_1{}_\alpha^L,$$
$$A_1{}_\alpha^R x + B_1{}_\alpha^R y \leq b_1{}_\alpha^R,$$
$$A_1{}_\beta^L x + B_1{}_\beta^L y \leq b_1{}_\beta^L,$$
$$A_1{}_\beta^R x + B_1{}_\beta^R y \leq b_1{}_\beta^R, \tag{8.4b}$$

$$\min_{y \in Y} \ (f_1(x,y))_\alpha^L = c_{12}\,{}_\alpha^L x + d_{12}\,{}_\alpha^L y$$

$$\min_{y \in Y} \ (f_1(x,y))_\alpha^R = c_{12}\,{}_\alpha^R x + d_{12}\,{}_\alpha^R y$$

$$\min_{y \in Y} \ (f_1(x,y))_\beta^L = c_{12}\,{}_\beta^L x + d_{12}\,{}_\beta^L y$$

$$\min_{y \in Y} \ (f_1(x,y))_\beta^R = c_{12}\,{}_\beta^R x + d_{12}\,{}_\beta^R y$$

$$\vdots \tag{8.4c}$$

$$\min_{y \in Y} \ (f_t(x,y))_\alpha^L = c_{t2}\,{}_\alpha^L x + d_{t2}\,{}_\alpha^L y$$

$$\min_{y \in Y} \ (f_t(x,y))_\alpha^R = c_{t2}\,{}_\alpha^R x + d_{t2}\,{}_\alpha^R y$$

$$\min_{y \in Y} \ (f_t(x,y))_\beta^L = c_{t2}\,{}_\beta^L x + d_{t2}\,{}_\beta^L y$$

$$\min_{y \in Y} \ (f_t(x,y))_\beta^R = c_{t2}\,{}_\beta^R x + d_{t2}\,{}_\beta^R y$$

$$\text{s.t.} \quad
\begin{aligned}
A_2\,{}_\alpha^L x + B_2\,{}_\alpha^L y &\le b_2\,{}_\alpha^L, \\
A_2\,{}_\alpha^R x + B_2\,{}_\alpha^R y &\le b_2\,{}_\alpha^R, \\
A_2\,{}_\beta^L x + B_2\,{}_\beta^L y &\le b_2\,{}_\beta^L, \\
A_2\,{}_\beta^R x + B_2\,{}_\beta^R y &\le b_2\,{}_\beta^R.
\end{aligned} \tag{8.4d}$$

*Proof* Denote the constraint regions of (8.2a)–(8.2d) and (8.4a)–(8.4d) by $S_1$ and $S_2$, the feasible regions by $IR_1$ and $IR_2$ respectively. Now in order to show that the sets $IR_1$ and $IR_2$ are the same, we first prove that $S_1$ and $S_2$ are equal.

For any $\lambda \in [\beta,\alpha]$, it is easy to verify that if $(x,y)$ satisfies (8.2b) and (8.2d), then it satisfies (8.4b) and (8.4d). Now we prove, if $(x,y)$ satisfies (8.4b) and (8.4d), then it satisfies (8.2b) and (8.2d). In fact, for any $\lambda \in [\beta,\alpha]$,

$$a_{ij}\,{}_\lambda^L = \frac{\lambda - \beta}{\alpha - \beta}\left(a_{ij}\,{}_\alpha^L - a_{ij}\,{}_\beta^L\right) + a_{ij}\,{}_\beta^L,$$

$$b_{ij}\,{}_\lambda^L = \frac{\lambda - \beta}{\alpha - \beta}\left(b_{ij}\,{}_\alpha^L - b_{ij}\,{}_\beta^L\right) + b_{ij}\,{}_\beta^L,$$

$$b_1\,{}_\lambda^R = \frac{\lambda - \beta}{\alpha - \beta}\left(b_1\,{}_\alpha^R - b_1\,{}_\beta^R\right) + b_1\,{}_\beta^R,$$

we have

$$
\begin{aligned}
A_1 {}_\lambda^L x + B_1 {}_\lambda^L y &= (a_{ij} {}_\lambda^L)x + (b_{ij} {}_\lambda^L)y \\
&= \left( \frac{\lambda - \beta}{\alpha - \beta} \left( a_{ij} {}_\alpha^L - a_{ij} {}_\beta^L \right) + a_{ij} {}_\beta^L \right)x \\
&\quad + \left( \frac{\lambda - \beta}{\alpha - \beta} \left( b_{ij} {}_\alpha^L - b_{ij} {}_\beta^L \right) + b_{ij} {}_\beta^L \right)y \\
&= \frac{\lambda - \beta}{\alpha - \beta}(a_{ij} {}_\alpha^L)x + \left( 1 - \frac{\lambda - \beta}{\alpha - \beta} \right) \left( a_{ij} {}_\beta^L \right)x \\
&\quad + \frac{\lambda - \beta}{\alpha - \beta}(b_{ij} {}_\alpha^L)y + \left( 1 - \frac{\lambda - \beta}{\alpha - \beta} \right) \left( b_{ij} {}_\beta^L \right)y \\
&= \frac{\lambda - \beta}{\alpha - \beta}\left( (a_{ij} {}_\alpha^L)x + (b_{ij} {}_\alpha^L)y \right) \\
&\quad + \left( 1 - \frac{\lambda - \beta}{\alpha - \beta} \right) \left( \left( a_{ij} {}_\beta^L \right)x + \left( b_{ij} {}_\beta^L \right)y \right) \\
&= \frac{\lambda - \beta}{\alpha - \beta}\left( (A_1 {}_\alpha^L)x + (B_1 {}_\alpha^L)y \right) \\
&\quad + \left( 1 - \frac{\lambda - \beta}{\alpha - \beta} \right) \left( \left( A_1 {}_\beta^L \right)x + \left( B_1 {}_\beta^L \right)y \right) \\
&\leq \frac{\lambda - \beta}{\alpha - \beta} b_1 {}_\alpha^L + \left( 1 - \frac{\lambda - \beta}{\alpha - \beta} \right) b_1 {}_\beta^L = b_1 {}_\lambda^L
\end{aligned}
$$

from (8.4b). Similarly, we can obtain that

$$
\begin{aligned}
A_1 {}_\lambda^R x + B_1 {}_\lambda^R y &\leq b_1 {}_\lambda^R, \\
A_2 {}_\lambda^L x + B_2 {}_\lambda^L y &\leq b_2 {}_\lambda^L, \\
A_2 {}_\lambda^R x + B_2 {}_\lambda^R y &\leq b_2 {}_\lambda^R,
\end{aligned}
$$

for any $\lambda \in [\beta, \alpha]$ from (8.4b) and (8.4d).

Thus, $S_1 = S_2$.

Next, we will prove that $IR_1 = IR_2$, and consider the following two cases.

**Case 1** For any $(x, y) \in IR_2$, we prove $(x, y) \in IR_1$.

It follows from $(x, y) \in IR_2$ that $(x, y) \in S_2$, and then $(x, y) \in S_1$. We prove this result by contradiction, and assume that $(x, y)$ is not a feasible point of problem (8.2a)–(8.2d). Thus, there exists $y'$ such that $(x, y') \in S_1$, and

$$
\left( f_j(x, y') \right)_\lambda^L \leq \left( f_j(x, y) \right)_\lambda^L \text{ and } \left( f_j(x, y') \right)_\lambda^R \leq \left( f_j(x, y) \right)_\lambda^R, \quad j = 1, 2, \ldots, t,
$$

which imply that

$$
\begin{aligned}
\left( f_j(x, y') \right)_\alpha^L &\leq \left( f_j(x, y) \right)_\alpha^L, \\
\left( f_j(x, y') \right)_\alpha^R &\leq \left( f_j(x, y) \right)_\alpha^R, \\
\left( f_j(x, y') \right)_\beta^L &\leq \left( f_j(x, y) \right)_\beta^L, \\
\left( f_j(x, y') \right)_\beta^R &\leq \left( f_j(x, y) \right)_\beta^R, \quad j = 1, 2, \ldots, t
\end{aligned}
$$

That is, $(x, y)$ is not a feasible point of problem (8.4a)–(8.4d), which contradicts the definition of $(x, y)$. Hence, $(x, y) \in IR_1$, and then $IR_1 \supseteq IR_2$.

**Case 2** For any $(x, y) \in IR_1$, we prove $(x, y) \in IR_2$.
It follows from $(x, y) \in IR_1$ that $(x, y) \in S_1$, and then $(x, y) \in S_2$. We prove this result by contradiction, and assume that $(x, y)$ is not a feasible point of problem (8.4a)–(8.4d). Thus, there exists $y'$ such that $(x, y') \in S_2$, and

$$\left(f_j(x, y')\right)_\alpha^L \leq \left(f_j(x, y)\right)_\alpha^L,$$
$$\left(f_j(x, y')\right)_\alpha^R \leq \left(f_j(x, y)\right)_\alpha^R,$$
$$\left(f_j(x, y')\right)_\beta^L \leq \left(f_j(x, y)\right)_\beta^L,$$
$$\left(f_j(x, y')\right)_\beta^R \leq \left(f_j(x, y)\right)_\beta^R, \quad j = 1, 2, \ldots, t.$$

It follows from Lemma 8.1 that

$$\left(f_j(x, y')\right)_\lambda^L \leq \left(f_j(x, y)\right)_\lambda^L \quad \text{and} \quad \left(f_j(x, y')\right)_\lambda^R \leq \left(f_j(x, y)\right)_\lambda^R, \quad j = 1, 2, \ldots, t,$$

which contradicts $(x, y) \in IR_1$. Hence, $(x, y) \in IR_2$, and then $IR_2 \supseteq IR_1$.
Therefore, $IR_2 = IR_1$.
Now, if $(x^*, y^*)$ is a complete optimal solution to problem (8.4a)–(8.4d), then for any $(x, y) \in IR_2$, we have

$$\left(F_j(x^*, y^*)\right)_\alpha^L \leq \left(F_j(x, y)\right)_\alpha^L,$$
$$\left(F_j(x^*, y^*)\right)_\alpha^R \leq \left(F_j(x, y)\right)_\alpha^R,$$
$$\left(F_j(x^*, y^*)\right)_\beta^L \leq \left(F_j(x, y)\right)_\beta^L,$$
$$\left(F_j(x^*, y^*)\right)_\beta^R \leq \left(F_j(x, y)\right)_\beta^R, \quad j = 1, 2, \ldots, t.$$

It follows from Lemma 8.1 that

$$\left(F_j(x^*, y^*)\right)_\lambda^L \leq \left(F_j(x, y)\right)_\lambda^L \quad \text{and} \quad \left(F_j(x^*, y^*)\right)_\lambda^R \leq \left(F_j(x, y)\right)_\lambda^R, \quad j = 1, 2, \ldots, t.$$

Thus, $(x^*, y^*)$ is a complete optimal solution to problem (8.2a)–(8.2d).
Furthermore, if $(x^*, y^*)$ is a complete optimal solution to problem (8.2a)–(8.2d), then for any $(x, y) \in IR_1$, we find that

$$\left(F_j(x^*, y^*)\right)_\lambda^L \leq \left(F_j(x, y)\right)_\lambda^L \quad \text{and} \quad \left(F_j(x^*, y^*)\right)_\lambda^R \leq \left(F_j(x, y)\right)_\lambda^R, \quad j = 1, 2, \ldots, t.$$

Obviously, $(x^*, y^*)$ is a complete optimal solution to problem (8.4a)–(8.4d).
The proof is complete. □

**Corollary 8.1** *For $x \in X \subset R^n$, $y \in Y \subset R^m$, if all the fuzzy parameters $\tilde{a}_{ij}$, $\tilde{b}_{ij}$, $\tilde{e}_{ij}$, $\tilde{s}_{ij}$, $\tilde{c}_{ij}$, $\tilde{b}_1$, $\tilde{b}_2$, and $\tilde{d}_{ij}$ have piecewise trapezoidal membership functions in the FMO-BLP problem (8.1a)–(8.1d),*

$$
\mu_{\tilde{z}}(t) =
\begin{cases}
0, & t < z_{\alpha_0}^L \\[4pt]
\frac{\alpha_1 - \alpha_0}{z_{\alpha_1}^L - z_{\alpha_0}^L}\left(t - z_{\alpha_0}^L\right) + \alpha_0, & z_{\alpha_0}^L \le t < z_{\alpha_1}^L \\[4pt]
\frac{\alpha_1 - \alpha_0}{z_{\alpha_2}^L - z_{\alpha_1}^L}\left(t - z_{\alpha_1}^L\right) + \alpha_1, & z_{\alpha_1}^L \le t < z_{\alpha_2}^L \\[2pt]
\vdots & \vdots \\[2pt]
\alpha, & z_{\alpha_{n-1}}^L \le t < z_{\alpha_n}^L \\[4pt]
\frac{\alpha_n - \alpha_{n-1}}{z_{\alpha_{n-1}}^R - z_{\alpha_n}^R}\left(-t + z_{\alpha_{n-1}}^R\right) + \alpha_{n-1}, & z_{\alpha_n}^R \le t < z_{\alpha_{n-1}}^R \\[2pt]
\vdots & \vdots \\[4pt]
\frac{\alpha_n - \alpha_{n-1}}{z_{\alpha_{n-1}}^R - z_{\alpha_n}^R}\left(-t + z_{\alpha_{n-1}}^R\right) + \alpha_{n-1}, & z_{\alpha_1}^R \le t < z_{\alpha_0}^R \\[4pt]
0, & z_{\alpha_0}^R < t
\end{cases}
\tag{8.5}
$$

*where $\tilde{z}$ denotes $\tilde{a}_{ij}$, $\tilde{b}_{ij}$, $\tilde{e}_{ij}$, $\tilde{s}_{ij}$, $\tilde{c}_{ij}$, $\tilde{b}_1$, $\tilde{b}_2$, and $\tilde{d}_{ij}$ respectively, then $(x^*, y^*)$ is a complete optimal solution to the problem (8.2a)–(8.2d) if and only if $(x^*, y^*)$ is a complete optimal solution to the following multi-objective linear bi-level programming (MOL-BLP) problem:*

$$
\min_{x \in X} \quad (F_1(x, y))_{\alpha_0}^L = c_{11}{}_{\alpha_0}^L x + d_{11}{}_{\alpha_0}^L y
$$

$$\vdots$$

$$
\min_{x \in X} \quad (F_1(x, y))_{\alpha_n}^L = c_{11}{}_{\alpha_n}^L x + d_{11}{}_{\alpha_n}^L y
$$

$$
\min_{x \in X} \quad (F_1(x, y))_{\alpha_0}^R = c_{11}{}_{\alpha_0}^R x + d_{11}{}_{\alpha_0}^R y
$$

$$\vdots$$

$$
\min_{x \in X} \quad (F_1(x, y))_{\alpha_n}^R = c_{11}{}_{\alpha_n}^R x + d_{11}{}_{\alpha_n}^R y
$$

$$\vdots$$

$$
\min_{x \in X} \quad (F_s(x, y))_{\alpha_0}^L = c_{s1}{}_{\alpha_0}^L x + d_{s1}{}_{\alpha_0}^L y
$$

$$\vdots$$

$$
\min_{x \in X} \quad (F_s(x, y))_{\alpha_n}^L = c_{s1}{}_{\alpha_n}^L x + d_{s1}{}_{\alpha_n}^L y
$$

$$
\min_{x \in X} \quad (F_s(x, y))_{\alpha_0}^R = c_{s1}{}_{\alpha_0}^R x + d_{s1}{}_{\alpha_0}^R y
$$

$$\vdots$$

$$
\min_{x \in X} \quad (F_s(x, y))_{\alpha_n}^R = c_{s1}{}_{\alpha_n}^R x + d_{s1}{}_{\alpha_n}^R y
$$

$$\tag{8.6a}$$

$$\text{s.t.} \quad A_1 {}^L_{\alpha_0} x + B_1 {}^L_{\alpha_0} y \le b_1 {}^L_{\alpha_0},$$

$$\vdots$$

$$A_1 {}^L_{\alpha_n} x + B_1 {}^L_{\alpha_n} y \le b_1 {}^L_{\alpha_n},$$
$$A_1 {}^R_{\alpha_0} x + B_1 {}^R_{\alpha_0} y \le b_1 {}^R_{\alpha_0}, \tag{8.6b}$$

$$\vdots$$

$$A_1 {}^R_{\alpha_n} x + B_1 {}^R_{\alpha_n} y \le b_1 {}^R_{\alpha_n}$$

$$\min_{y \in Y} \quad (f_1(x,y))^L_{\alpha_0} = c_{12} {}^L_{\alpha_0} x + d_{12} {}^L_{\alpha_0} y$$

$$\vdots$$

$$\min_{y \in Y} \quad (f_1(x,y))^L_{\alpha_n} = c_{12} {}^L_{\alpha_n} x + d_{12} {}^L_{\alpha_n} y$$

$$\min_{y \in Y} \quad (f_1(x,y))^R_{\alpha_0} = c_{12} {}^R_{\alpha_0} x + d_{12} {}^R_{\alpha_0} y$$

$$\vdots$$

$$\min_{y \in Y} \quad (f_1(x,y))^R_{\alpha_n} = c_{12} {}^R_{\alpha_n} x + d_{12} {}^R_{\alpha_n} y$$

$$\vdots \tag{8.6c}$$

$$\min_{y \in Y} \quad (f_t(x,y))^L_{\alpha_0} = c_{t2} {}^L_{\alpha_0} x + d_{t2} {}^L_{\alpha_0} y$$

$$\vdots$$

$$\min_{y \in Y} \quad (f_t(x,y))^L_{\alpha_n} = c_{t2} {}^L_{\alpha_n} x + d_{t2} {}^L_{\alpha_n} y$$

$$\min_{y \in Y} \quad (f_t(x,y))^R_{\alpha_0} = c_{t2} {}^R_{\alpha_0} x + d_{t2} {}^R_{\alpha_0} y$$

$$\vdots$$

$$\min_{y \in Y} \quad (f_t(x,y))^R_{\alpha_n} = c_{t2} {}^R_{\alpha_n} x + d_{t2} {}^R_{\alpha_n} y$$

$$\text{s.t.} \quad A_2 {}^L_{\alpha_0} x + B_2 {}^L_{\alpha_0} y \le b_2 {}^L_{\alpha_0},$$

$$\vdots$$

$$A_2 {}^L_{\alpha_n} x + B_2 {}^L_{\alpha_n} y \le b_2 {}^L_{\alpha_n},$$
$$A_2 {}^R_{\alpha_0} x + B_2 {}^R_{\alpha_0} y \le b_2 {}^R_{\alpha_0}, \tag{8.6d}$$

$$\vdots$$

$$A_2 {}^R_{\alpha_n} x + B_2 {}^R_{\alpha_n} y \le b_2 {}^R_{\alpha_n}.$$

**Theorem 8.3** *For $x \in X \subset R^n$, $y \in Y \subset R^m$, if all the fuzzy parameters $\tilde{a}_{ij}$, $\tilde{b}_{ij}$, $\tilde{e}_{ij}$, $\tilde{s}_{ij}$, $\tilde{c}_{ij}$, $\tilde{b}_1$, $\tilde{b}_2$, and $\tilde{d}_{ij}$ have piecewise trapezoidal membership functions (8.3) in the FMO-BLP problem (8.1a)–(8.1d), then $(x^*, y^*)$ is a Pareto optimal solution to the problem (8.2a)–(8.2d) if and only if $(x^*, y^*)$ is a Pareto optimal solution to problem (8.4a)–(8.4d).*

*Proof* Clearly, it follows from the proof of Theorem 8.2 that the feasible regions of problems (8.2a)–(8.2d) and (8.4a)–(8.4d) are the same. Let $(x^*, y^*)$ be a Pareto optimal solution to problem (8.2a)–(8.2d). We prove it is also a Pareto optimal solution to problem (8.4a)–(8.4d) by contradiction, and suppose that there exists a $(\bar{x}, \bar{y}) \in IR_2$ such that, for $\lambda = \alpha, \beta$,

$$(F_1(\bar{x}, \bar{y}))_\lambda^L \le (F_1(x^*, y^*))_\lambda^L,$$
$$(F_1(\bar{x}, \bar{y}))_\lambda^R \le (F_1(x^*, y^*))_\lambda^R,$$

$$\vdots$$

$$(F_s(\bar{x}, \bar{y}))_\lambda^L \le (F_s(x^*, y^*))_\lambda^L,$$
$$(F_s(\bar{x}, \bar{y}))_\lambda^R \le (F_s(x^*, y^*))_\lambda^R.$$

By using Lemma 8.1, for any $\lambda \in [\beta, \alpha]$, we have

$$(F_i(\bar{x}, \bar{y}))_\lambda^L \le (F_i(x^*, y^*))_\lambda^L, (F_i(\bar{x}, \bar{y}))_\lambda^R \le (F_i(x^*, y^*))_\lambda^R, \quad i = 1, \ldots, s,$$

which contradicts the assumption that $(x^*, y^*)$ is a Pareto optimal solution to problem (8.2a)–(8.2d). Thus, $(x^*, y^*)$ is a Pareto optimal solution to problem (8.4a)–(8.4d)

Let $(x^*, y^*)$ be a Pareto optimal solution to problem (8.4a)–(8.4d). If $(x^*, y^*)$ is not a Pareto optimal solution to problem (8.2a)–(8.2d), then there exist $(\bar{x}, \bar{y}) \in IR_1$ such that

$$(F_1(\bar{x}, \bar{y}))_\lambda^L \le (F_1(x^*, y^*))_\lambda^L,$$

$$(F_1(\bar{x}, \bar{y}))_\lambda^R \le (F_1(x^*, y^*))_\lambda^R,$$

$$\vdots$$

$$(F_s(\bar{x}, \bar{y}))_\lambda^L \le (F_s(x^*, y^*))_\lambda^L,$$

$$(F_s(\bar{x}, \bar{y}))_\lambda^R \le (F_s(x^*, y^*))_\lambda^R.$$

Hence, for $\lambda = \alpha$ and $\lambda = \beta$, we have

$$\left(F_j(\bar{x},\bar{y})\right)_\alpha^L \leq \left(F_j(x^*,y^*)\right)_\alpha^L,$$
$$\left(F_j(\bar{x},\bar{y})\right)_\alpha^R \leq \left(F_j(x^*,y^*)\right)_\alpha^R,$$
$$\left(F_j(\bar{x},\bar{y})\right)_\beta^L \leq \left(F_j(x^*,y^*)\right)_\beta^L,$$
$$\left(F_j(\bar{x},\bar{y})\right)_\beta^R \leq \left(F_j(x^*,y^*)\right)_\beta^R, \quad j = 1,2,\ldots,s$$

which contradicts the assumption that $(x^*,y^*)$ is a Pareto optimal solution to problem (8.4a)–(8.4d), and then $(x^*,y^*)$ is a Pareto optimal solution to problem (8.2a)–(8.2d). This completes the proof. □

**Theorem 8.4** *For $x \in X \subset R^n$, $y \in Y \subset R^m$, if all the fuzzy parameters $\tilde{a}_{ij}$, $\tilde{b}_{ij}$, $\tilde{e}_{ij}$, $\tilde{s}_{ij}$, $\tilde{c}_{ij}$, $\tilde{b}_1$, $\tilde{b}_2$, and $\tilde{d}_{ij}$ have piecewise trapezoidal membership functions (8.3) in the FMO-BLP problem (8.1a)–(8.1d), then $(x^*,y^*)$ is a weak Pareto optimal solution to the problem (8.2a)–(8.2d) if and only if $(x^*,y^*)$ is a weak Pareto optimal solution to problem (8.4a)–(8.4d).*

*Proof* The proof is similar to that of Theorem 8.3. □

Similarly, by the result of Corollary 8.1, we have the following results.

**Corollary 8.2** *For $x \in X \subset R^n$, $y \in Y \subset R^m$, if all the fuzzy parameters $\tilde{a}_{ij}$, $\tilde{b}_{ij}$, $\tilde{e}_{ij}$, $\tilde{s}_{ij}$, $\tilde{c}_{ij}$, $\tilde{b}_1$, $\tilde{b}_2$, and $\tilde{d}_{ij}$ have piecewise trapezoidal membership functions (8.5) in the FMO-BLP problem (8.1a)–(8.1d), then $(x^*,y^*)$ is a Pareto optimal solution to problem (8.2a)–(8.2d) if and only if $(x^*,y^*)$ is a Pareto optimal solution to problem (8.4a)–(8.4d).*

**Corollary 8.3** *For $x \in X \subset R^n$, $y \in Y \subset R^m$, if all the fuzzy parameters $\tilde{a}_{ij}$, $\tilde{b}_{ij}$, $\tilde{e}_{ij}$, $\tilde{s}_{ij}$, $\tilde{c}_{ij}$, $\tilde{b}_1$, $\tilde{b}_2$, and $\tilde{d}_{ij}$ have piecewise trapezoidal membership functions (8.5) in the FMO-BLP problem (8.1a)–(8.1d), then $(x^*,y^*)$ is a weak Pareto optimal solution to problem (8.2a)–(8.2d) if and only if $(x^*,y^*)$ is a weak Pareto optimal solution to problem (8.4a)–(8.4d).*

## 8.3 Fuzzy Approximation Kuhn-Tucker Approach

To obtain a solution of the FMO-BLP problem (8.1a)–(8.1d), we only need to solve problem (8.4a)–(8.4d). However, the lower level problem of problem (8.4a)–(8.4d) is a multi-objective optimization. It is not true if we assume that the follower's rational reaction set $P(x)$ is a point-to-point map with respect to $x$. To deal with this undesirable situation, there are at least three approaches, for example, optimistic formulation, pessimistic formulation (Dempe 2002) and selection function approach (Dempe and Starostina 2006). Since the follower has preferences (weighting) of the objectives, here we adopt the third approach, that is the selection function approach, for solving problem (8.4a)–(8.4d). Using the weighting method (Sakawa 1993), we can transform problem (8.4a)–(8.4d) as a linear bi-level programming problem, and the resulting problem can be solved by Kuhn-Tucker approach. This section will introduce a fuzzy approximation Kuhn-Tucker approach to solve the FMO-BLP problem and a case-based example is used to illustrate the proposed approach.

## 8.3.1  Fuzzy Approximation Kuhn-Tucker Approach

A fuzzy approximation Kuhn-Tucker approach is described in Algorithm 8.1 with seven main steps to solve the FMO-BLP problem.

---

**Algorithm 8.1: Fuzzy Approximation Kuhn-Tucker Approach**

---

[Begin]

**Step 1:** Set $l = 1$. Give their weights for fuzzy objectives of every leader and follower and $\sum_{j=1}^{s} w_{j1} = 1$ and $\sum_{j=1}^{t} w_{j2} = 1$, respectively.

**Step 2:** Transfer the problem (8.1) to the problem (8.2) by using Theorem 8.3.

**Step 3:** Let the interval $[0, 1]$ be decomposed into $2^{l-1}$ mean sub-intervals with $(2^{l-1} + 1)$ nodes $\lambda_i (i = 0, 1, \dots, 2^{l-1})$, which are arranged in the order of $0 = \lambda_0 < \lambda_1 < \cdots < \lambda_{2^{l-1}} = 1$ and a range of errors $\varepsilon > 0$.

**Step 4:** Solve $(\text{MOL} - \text{BLP})_2^l$, i.e. problem (8.4) by using the weighting method and Kuhn-Tucker approach when $\beta = 0$ and $\alpha = 1$, and obtain an optimal solution $(x, y)_{2^l}$.

**Step 5:** Solve $(\text{MOL} - \text{BLP})_2^{l+1}$, and obtain an optimal solution $(x, y)_{2^{l+1}}$.

**Step 6:** If $\left\| (x, y)_{2^{l+1}} - (x, y)_{2^l} \right\| < \varepsilon$, then $(x, y)_{2^{l+1}}$ is an approximation solution of the FMO-BLP problem (8.1), otherwise, update $l$ to $2l$ and go back to Step 4.

**Step 7:** Show the approximation solution.

[End]

---

A case-based example will be given to illustrate the proposed approach in the next section

## 8.3.2  A Case-Based Example

We first present an example to explain how to build a FMO-BLP model and then apply the proposed fuzzy approximation Kuhn-Tucker approach to solve the problem defined by the FMO-BLP model.

*Example 8.1* There are two logistics channels in a supply chain system. A forward logistics channel concerns the movement of goods from source to their point of consumption. A backward movement will happen when we need to return goods to suppliers, which is called reverse logistics. In general, the forward logistics brings profits to all operational stages/departments involved; while a reverse logistics usually only brings costs. However, many companies have discovered that cost reductions in inventory carrying costs, transportation and waste disposal can be substantial with an efficient reverse logistics program and related strategies. To effectively manage a reverse logistics chain often involves finding an optimal solution from multiple stages/departments involved in a goods return process. Bi-level or multi-level decision-making approaches are very promising to be applied in supporting this kind of decision-making.

Two main operational stages in a reverse logistics chain are the supplier and the distributor. They both aim to minimize their own cost but have individual constraints for a goods return. A decision about sharing cost for goods return made by the supplier will affect the decision made by the distributor, such as the service quality provided to customers for a product return. Also, the distributor executes its policies after, and in view of, decisions made at the supplier stage. As the two stages in the chain are interrelated in a way that decisions made at one stage affect the performance of others, this can be seen as a bi-level decision problem. The supplier is the leader and the distributor is the follower in the decision process. In almost all of the cases from the real world, supplier and distributor each independently minimizes its cost on a reverse logistic chain without well considering other's benefits. Obviously, these two stages are interrelated in an uncooperative way. In practice, logistics managers often imprecisely know the possible values of related costs. For example, they can only estimate possible inventory carrying cost and transportation cost of a particular set of goods to be returned. This situation brings about a demand on the proposed bi-level decision-making model of reverse logistics management to be able to handle uncertain information.

We define the supplier's objective function $\min_{x \in X} F(x, y)$ as to minimize the cost to the supplier, under the constraints from the requirements of customer service and the environment pollution issue. The distributor, as the follower, attempts to minimize its cost from the reverse logistics $\min_{y \in Y} f(x, y)$ for each policy made by the supplier.

When modeling the bi-level decision problem, the parameters for the objectives and constraints of both the leader and the follower are hard to be set by real numbers as they are described by some uncertain experience data and statistic reports from managers, such as "about $2 K". Therefore a fuzzy-number-based bi-level decision model is created for the reverse logistics decision problem. In order to facilitate an easy understanding of the application for the proposed fuzzy approximation Kuhn-Tucker approach, the decision model is established by simplifying it into the following linear FMO-BLP model.

Consider the following FMO-BLP problem with $x \in R$, $y \in R$, and $X = \{x|x \geq 0\}$, $Y = \{y|y \geq 0\}$,

$$\min_{x \in X} \quad F_1(x,y) = -\tilde{1}x + \tilde{2}y$$

$$\min_{x \in X} \quad F_2(x,y) = \tilde{2}x - \tilde{4}y$$

$$\text{s.t.} \quad -\tilde{1}x + \tilde{3}y \leq \tilde{4},$$

$$\min_{y \in Y} \quad f_1(x,y) = -\tilde{1}x + \tilde{2}y$$

$$\min_{y \in Y} \quad f_2(x,y) = \tilde{2}x - \tilde{1}y$$

$$\text{s.t.} \quad \tilde{1}x - \tilde{1}y \leq \tilde{0},$$

$$\quad -\tilde{1}x - \tilde{1}y \leq \tilde{0},$$

where

$$\mu_{\tilde{1}} = \begin{cases} 0 & t < 0 \\ t^2 & 0 \leq t < 1 \\ 2-t & 1 \leq t < 2 \\ 0 & 2 \leq t \end{cases}, \quad \mu_{\tilde{2}} = \begin{cases} 0 & t < 1 \\ t-1 & 1 \leq t < 2 \\ 3-t & 2 \leq t < 3 \\ 0 & 3 \leq t \end{cases},$$

$$\mu_{\tilde{3}} = \begin{cases} 0 & t < 2 \\ t-2 & 2 \leq t < 3 \\ 4-t & 3 \leq t < 4 \\ 0 & 4 \leq t \end{cases}, \quad \mu_{\tilde{4}} = \begin{cases} 0 & t < 3 \\ t-3 & 3 \leq t < 4 \\ 5-t & 4 \leq t < 5 \\ 0 & 5 \leq t \end{cases},$$

$$\mu_{\tilde{0}} = \begin{cases} 0 & t < -1 \\ t+1 & -1 \leq t < 0 \\ 1-t^2 & 0 \leq t < 1 \\ 0 & 1 \leq t \end{cases}.$$

We now solve this problem by using the proposed fuzzy approximation Kuhn-Tucker approach.

Step 1: We give the weights $(0.5, 0.5)$ for the two fuzzy objectives of the leader and of the follower.

Step 2: This example is first transformed to the following problem:

$$\min_{x \in X} \; (F_1(x,y))_\lambda^L = \left(-\tilde{1}\right)_\lambda^L x + \left(\tilde{2}\right)_\lambda^L y, \quad \lambda \in [0,1]$$

$$\min_{x \in X} \; (F_1(x,y))_\lambda^R = \left(-\tilde{1}\right)_\lambda^R x + \left(\tilde{2}\right)_\lambda^R y, \quad \lambda \in [0,1]$$

$$\min_{x \in X} \; (F_2(x,y))_\lambda^L = \left(\tilde{2}\right)_\lambda^L x + \left(-\tilde{4}\right)_\lambda^L y, \quad \lambda \in [0,1]$$

$$\min_{x \in X} \; (F_2(x,y))_\lambda^R = \left(\tilde{2}\right)_\lambda^R x + \left(-\tilde{4}\right)_\lambda^R y, \quad \lambda \in [0,1]$$

$$\text{s.t.} \quad \left(-\tilde{1}\right)_\lambda^L x + \left(\tilde{3}\right)_\lambda^L y \le \left(\tilde{4}\right)_\lambda^L,$$

$$\left(-\tilde{1}\right)_\lambda^R x + \left(\tilde{3}\right)_\lambda^R y \le \left(\tilde{4}\right)_\lambda^R, \quad \lambda \in [0,1]$$

$$\min_{y \in Y} \; (f_1(x,y))_\lambda^L = \left(\tilde{2}\right)_\lambda^L x + \left(-\tilde{1}\right)_\lambda^L y, \quad \lambda \in [0,1]$$

$$\min_{y \in Y} \; (f_1(x,y))_\lambda^R = \left(\tilde{2}\right)_\lambda^R x + \left(-\tilde{1}\right)_\lambda^R y, \quad \lambda \in [0,1]$$

$$\min_{y \in Y} \; (f_2(x,y))_\lambda^L = \left(-\tilde{1}\right)_\lambda^L x + \left(\tilde{2}\right)_\lambda^L y, \quad \lambda \in [0,1]$$

$$\min_{y \in Y} \; (f_2(x,y))_\lambda^R = \left(-\tilde{1}\right)_\lambda^R x + \left(\tilde{2}\right)_\lambda^R y, \quad \lambda \in [0,1]$$

$$\text{s.t.} \quad \left(\tilde{1}\right)_\lambda^L x + \left(-\tilde{1}\right)_\lambda^L y \le \left(\tilde{0}\right)_\lambda^L,$$

$$\left(\tilde{1}\right)_\lambda^R x + \left(-\tilde{1}\right)_\lambda^R y \le \left(\tilde{0}\right)_\lambda^R,$$

$$\left(-\tilde{1}\right)_\lambda^L x + \left(-\tilde{1}\right)_\lambda^L y \le \left(\tilde{0}\right)_\lambda^L,$$

$$\left(-\tilde{1}\right)_\lambda^R x + \left(-\tilde{1}\right)_\lambda^R y \le \left(\tilde{0}\right)_\lambda^R, \quad \lambda \in [0,1].$$

Step 3: Let the interval $[0,1]$ be decomposed into $2^{l-1}$ even sub-intervals with $(2^{l-1}+1)$ nodes $\lambda_i \, (i = 0, \ldots, 2^{l-1})$ which is arranged in the order of $0 = \lambda_0 < \lambda_1 < \cdots < \lambda_{2^{l-1}} = 1$ and a range of error $\varepsilon = 10^{-6} > 0$.

Step 4: When $l = 1$, we solve the following MOL-BLP problem:

$$\min_{x \in X} \; (F_1(x,y))_1^{L(R)} = -1x + 2y$$

$$\min_{x \in X} \; (F_1(x,y))_0^L = -2x + 1y$$

$$\min_{x \in X} \; (F_1(x,y))_0^R = 0x + 3y$$

$$\min_{x \in X} \; (F_2(x,y))_1^{L(R)} = 2x - 4y$$

$$\min_{x \in X} \; (F_2(x,y))_0^L = 1x - 5y$$

$$\min_{x \in X} \; (F_2(x,y))_0^R = 3x - 3y$$

$$\text{s.t.} \quad -1x + 3y \le 4,$$

$$-2x + 2y \le 3,$$

$$0x + 4y \le 5,$$

$$\min_{y \in Y} \ (f_1(x,y))_1^{L(R)} = 2x - 1y$$

$$\min_{y \in Y} \ (f_1(x,y))_0^{L} = 1x - 2y$$

$$\min_{y \in Y} \ (f_1(x,y))_0^{R} = 3x - 0y$$

$$\min_{y \in Y} \ (f_2(x,y))_1^{L(R)} = -1x + 2y$$

$$\min_{y \in Y} \ (f_2(x,y))_0^{L} = -2x + 1y$$

$$\min_{y \in Y} \ (f_2(x,y))_0^{R} = 0x + 3y$$

$$\text{s.t.} \quad 1x - 1y \leq 0,$$
$$0x - 2y \leq -1,$$
$$2x - 0y \leq 1,$$
$$-1x - 1y \leq 0,$$
$$-2x - 2y \leq -1.$$

Using the weighting method, we transform the above problem as follows:

$$\min_{x \in X} \quad F(x,y) = 1.5x - 3y$$
$$\text{s.t.} \quad -1x + 3y \leq 4,$$
$$-2x + 2y \leq 3,$$
$$0x + 4y \leq 5,$$
$$\min_{y \in Y} \quad f(x,y) = 1.5x + 1.5y$$
$$\text{s.t.} \quad 1x - 1y \leq 0,$$
$$0x - 2y \leq -1,$$
$$2x - 0y \leq 1,$$
$$-1x - 1y \leq 0,$$
$$-2x - 2y \leq -1.$$

Then, using the Kuhn-Tucker approach, we get a solution $(x, y) = (0, 0.5)$ with

$$(F_1(x,y))_1^{L(R)} = 1,$$
$$(F_1(x,y))_0^{L} = 0.5,$$
$$(F_1(x,y))_0^{R} = 1.5,$$
$$(F_2(x,y))_1^{L(R)} = -2,$$
$$(F_2(x,y))_0^{L} = -2.5,$$
$$(F_2(x,y))_0^{R} = -1.5,$$

$$(f_1(x,y))_1^{L(R)} = -0.5,$$
$$(f_1(x,y))_0^{L} = -1,$$
$$(f_1(x,y))_0^{R} = 0,$$
$$(f_2(x,y))_1^{L(R)} = 1,$$
$$(f_2(x,y))_0^{L} = 0.5,$$
$$(f_2(x,y))_0^{R} = 1.5,$$

Step 5: When $l = 2$, we solve the following MOL-BLP problem

$$\min_{x\in X} \quad (F_1(x,y))_1^{L(R)} = -1x + 2y$$

$$\min_{x\in X} \quad (F_1(x,y))_{\frac{1}{2}}^{L} = -\frac{3}{2}x + \frac{3}{2}y$$

$$\min_{x\in X} \quad (F_1(x,y))_0^{L} = -2x + 1y$$

$$\min_{x\in X} \quad (F_1(x,y))_{\frac{1}{2}}^{R} = -\frac{\sqrt{2}}{2}x + \frac{5}{2}y$$

$$\min_{x\in X} \quad (F_1(x,y))_0^{R} = 0x + 3y$$

$$\min_{x\in X} \quad (F_2(x,y))_1^{L(R)} = 2x - 4y$$

$$\min_{x\in X} \quad (F_2(x,y))_{\frac{1}{2}}^{L} = \frac{3}{2}x - \frac{9}{2}y$$

$$\min_{x\in X} \quad (F_2(x,y))_0^{L} = 1x - 5y$$

$$\min_{x\in X} \quad (F_2(x,y))_{\frac{1}{2}}^{R} = \frac{5}{2}x - \frac{7}{2}y$$

$$\min_{x\in X} \quad (F_2(x,y))_0^{R} = 3x - 3y$$

s.t. $-1x + 3y \le 4,$

$$-\frac{3}{2}x + \frac{5}{2}y \le \frac{7}{2},$$

$$-2x + 2y \le 3,$$

$$-\frac{\sqrt{2}}{2}x + \frac{7}{2}y \le \frac{9}{2},$$

$$0x + 4y \le 5,$$

$$\min_{y\in Y} \quad (f_1(x,y))_1^{L(R)} = 2x - 1y$$

$$\min_{y\in Y} \quad (f_1(x,y))_{\frac{1}{2}}^{L} = \frac{3}{2}x - \frac{3}{2}y$$

$$\min_{y\in Y} \quad (f_1(x,y))_0^{L} = 1x - 2y$$

$$\min_{y\in Y} \quad (f_1(x,y))_{\frac{1}{2}}^{R} = \frac{5}{2}x - \frac{\sqrt{2}}{2}y$$

$$\min_{y \in Y} \quad (f_1(x,y))_0^R = 3x - 0y$$

$$\min_{y \in Y} \quad (f_2(x,y))_1^{L(R)} = -1x + 2y$$

$$\min_{y \in Y} \quad (f_2(x,y))_{\frac{1}{2}}^L = -\frac{3}{2}x + \frac{3}{2}y$$

$$\min_{y \in Y} \quad (f_2(x,y))_0^L = -2x + 1y$$

$$\min_{y \in Y} \quad (f_2(x,y))_{\frac{1}{2}}^R = -\frac{\sqrt{2}}{2}x + \frac{5}{2}y$$

$$\min_{y \in Y} \quad (f_2(x,y))_0^R = 0x + 3y$$

$$\text{s.t.} \quad 1x - 1y \le 0,$$
$$\frac{\sqrt{2}}{2}x - \frac{3}{2}y \le -\frac{1}{2},$$
$$0x - 2y \le -1,$$
$$\frac{3}{2}x - \frac{\sqrt{2}}{2}y \le \frac{\sqrt{2}}{2},$$
$$2x - 0y \le 1,$$
$$-\frac{3}{2}x - \frac{3}{2}y \le -\frac{1}{2},$$
$$-1x - 1y \le 0,$$
$$-\frac{\sqrt{2}}{2}x - \frac{\sqrt{2}}{2}y \le \frac{\sqrt{2}}{2},$$
$$-2x - 2y \le -1.$$

We solve the above problem by using the weighting method and the Kuhn-Tucker approach:

$$\min_{x \in X} \quad F(x,y) = \left(3 + \frac{5 - \sqrt{2}}{2}\right)x - 10y$$

$$\text{s.t.} \quad -1x + 3y \le 4,$$
$$-\tfrac{3}{2}x + \tfrac{5}{2}y \le \tfrac{7}{2},$$
$$-2x + 2y \le 3,$$
$$-\tfrac{\sqrt{2}}{2}x + \tfrac{7}{2}y \le \tfrac{9}{2},$$
$$0x + 4y \le 5,$$

$$\min_{y \in Y} \quad f(x,y) = \left(\frac{5 - \sqrt{2}}{2} + 3\right)x + \left(\frac{5 - \sqrt{2}}{2} + 3\right)y$$

$$\text{s.t.} \quad 1x - 1y \le 0,$$
$$\tfrac{\sqrt{2}}{2}x - \tfrac{3}{2}y \le -\tfrac{1}{2},$$
$$0x - 2y \le -1,$$
$$\tfrac{3}{2}x - \tfrac{\sqrt{2}}{2}y \le \tfrac{\sqrt{2}}{2},$$
$$2x - 0y \le 1,$$
$$-\tfrac{3}{2}x - \tfrac{3}{2}y \le -\tfrac{1}{2},$$
$$-1x - 1y \le 0,$$
$$-\tfrac{\sqrt{2}}{2}x - \tfrac{\sqrt{2}}{2}y \le \tfrac{\sqrt{2}}{2},$$
$$-2x - 2y \le -1.$$

The result of the optimal solution occurs at the point $(x, y) = (0, 0.5)$ with

$$(F_1(x, y))_1^{L(R)} = 1,$$
$$(F_1(x, y))_{\frac{1}{2}}^L = 0.75,$$
$$(F_1(x, y))_0^L = 0.5,$$
$$(F_1(x, y))_{\frac{1}{2}}^R = 1.25,$$
$$(F_1(x, y))_0^R = 1.5,$$
$$(F_2(x, y))_1^{L(R)} = -2,$$
$$(F_2(x, y))_{\frac{1}{2}}^L = -2.25,$$
$$(F_2(x, y))_0^L = -2.5,$$
$$(F_2(x, y))_{\frac{1}{2}}^R = -1.75,$$
$$(F_2(x, y))_0^R = -1.5,$$
$$(f_1(x, y))_1^{L(R)} = -0.5,$$
$$(f_1(x, y))_{\frac{1}{2}}^L = -0.75,$$
$$(f_1(x, y))_0^L = -1,$$
$$(f_1(x, y))_{\frac{1}{2}}^R = -\frac{\sqrt{2}}{4},$$
$$(f_1(x, y))_0^R = 0,$$
$$(f_2(x, y))_1^{L(R)} = 1,$$
$$(f_2(x, y))_{\frac{1}{2}}^L = 0.75,$$
$$(f_2(x, y))_0^L = 0.5,$$
$$(f_2(x, y))_{\frac{1}{2}}^R = 1.25,$$
$$(f_2(x, y))_0^R = 1.5.$$

Step 6: Since $\|(x, y)_{2^2} - (x, y)_{2^1}\| = 0 < \varepsilon$, $(x, y) = (0, 0.5)$ is the optimal solution for the example.

Step 7: The solution of the problem is $(x^*, y^*) = (0, 0.5)$ with

$$F_1(x^*, y^*) = 0.5 \times \tilde{2},$$
$$F_2(x^*, y^*) = -0.5 \times \tilde{4},$$
$$f_1(x^*, y^*) = 0.5 \times \tilde{2},$$
$$f_1(x^*, y^*) = -0.5 \times \tilde{1}.$$

This example shows how the fuzzy approximation Kuhn-Tucker approach is used to solve the FMO-BLP problem in a reverse logistic decision making.


## 8.4 Summary

Uncertainty often occurs in a bi-level decision problem. Moreover, multiple objectives may be involved for the leader, or the follower, or both in a bi-level decision problem. To deal with these two problems together and to obtain a compromise solution, this chapter first presents a fuzzy parameter-based multiple objectives bi-level decision model, called the FMO-BLP model, and then presents a fuzzy approximation Kuhn-Tucker approach for solving the FMO-BLP problem. A case-based example is given to illustrate the feasibility of the proposed approach.

# Chapter 9
# Fuzzy Multi-objective Bi-level Goal Programming

In Chap. 8, we presented the definitions, solutions, and algorithms for the *fuzzy multi-objective bi-level programming* (FMO-BLP) problems. This chapter still addresses the fuzzy multi-objective bi-level problem but applies a goal programming approach. We call it *fuzzy multi-objective bi-level goal programming* (FMO-BLGP). This chapter will discuss related definitions, solution concepts, and algorithms for the FMO-BLGP problem and will focus on the linear version of the FMO-BLGP problem. First, a fuzzy ranking method is used to give a mathematical definition for a FMO-BLGP problem, and then, based on a fuzzy vectors distance measure definition, a fuzzy bi-level goal programming (FBLGP) model is proposed. An algorithm for solving the FMO-BLGP problem is also developed.

This chapter is organized as follows: the identification of the FMO-BLGP problem is presented in Sect. 9.1, and a fuzzy bi-level goal decision model and related theorems are developed in Sect. 9.2. Section 9.3 proposes a fuzzy bi-level goal-programming algorithm for solving FMO-BLGP problems. In Sect. 9.4, a numerical example is adopted to illustrate the executing procedure of the algorithm and experiments are carried out. Finally, we discuss and analyze the performance of this algorithm in Sect. 9.5.

## 9.1 Problem Identification

In many real-world bi-level decision applications, the leader or the follower not only have multiple objectives but also have their individual predefined decision targets (called goals) to achieve the objectives through a decision procedure. Therefore, goal programming could be integrated with the FMO-BLP approach to handle the FMO-BLGP problem well.

*Example 9.1* In a company, the CEO is the leader, and the heads of branches of the company are the followers in making an annual budget for the company. The CEO has two objectives: maximizing profit and maximizing marketing occupation with two goals $8M and 80 % of the local market respectively. The branch heads have two objectives: maximizing profit with the goal of "$4M profit" and maximizing

© Springer-Verlag Berlin Heidelberg 2015
G. Zhang et al., *Multi-Level Decision Making*,
Intelligent Systems Reference Library 82, DOI 10.1007/978-3-662-46059-7_9

customer satisfaction with the goal of "increasing customer satisfaction by 10 % compared to last year". To achieve the two sets of goals, we can establish a FMO-BLP model and develop a goal programming-based algorithm.


## 9.2 Solution Concepts

Based on the fuzzy ranking method in Definition 7.19, a FMO-BLP problem is defined as:

For $x \in X \subset R^n$, $y \in Y \subset R^m$, $F: X \times Y \to (F^*(R))^s$, and $f: X \times Y \to (F^*(R))^t$,

$$\min_{x \in X} \quad F(x,y) = \left( \tilde{\alpha}_{11}x + \tilde{\beta}_{11}y, \ldots, \tilde{\alpha}_{s1}x + \tilde{\beta}_{s1}y \right) \tag{9.1a}$$

$$\text{s.t.} \quad \tilde{A}_1 x + \tilde{B}_1 y \leq_\alpha \tilde{b}_1, \tag{9.1b}$$

$$\min_{y \in Y} \quad f(x,y) = \left( \tilde{\alpha}_{12}x + \tilde{\beta}_{12}y, \ldots, \tilde{\alpha}_{t1}x + \tilde{\beta}_{t1}y \right) \tag{9.1c}$$

$$\text{s.t.} \quad \tilde{A}_2 x + \tilde{B}_2 y \leq_\alpha \tilde{b}_2, \tag{9.1d}$$

where $\tilde{\alpha}_{h1}$, $\tilde{\alpha}_{i2} \in (F^*(R))^n$, $\tilde{\beta}_{h1}$, $\tilde{\beta}_{i2} \in (F^*(R))^m$, $\tilde{b}_1 \in (F^*(R))^p$, $\tilde{b}_2 \in (F^*(R))^q$, $\tilde{A}_1 = (\tilde{a}_{ij})_{p \times n}$, $\tilde{B}_1 = (\tilde{b}_{ij})_{p \times m}$, $\tilde{A}_2 = (\tilde{e}_{ij})_{q \times n}$, $\tilde{B}_2 = (\tilde{s}_{ij})_{q \times m}$, $\tilde{a}_{ij}$, $\tilde{b}_{ij}$, $\tilde{e}_{ij}$, $\tilde{s}_{ij} \in F^*(R)$, $h = 1, \ldots, s$, $i = 1, \ldots, t$.

To build a FMO-BLGP model, a distance measure between two fuzzy vectors needs to be developed to measure the distance between a decision and the predefined goal. To do so, a certain number of $\lambda$-cuts is used to approximate a fuzzy number, and a final solution is considered to be reached when solutions under two adjacent $\lambda$-cuts are nearly equal. To help implement this strategy, a $\lambda$-cut based fuzzy vector distance measure is defined below:

**Definition 9.1** Let $\tilde{a} = (\tilde{a}_1, \tilde{a}_2, \ldots, \tilde{a}_n)$, $\tilde{b} = (\tilde{b}_1, \tilde{b}_2, \ldots, \tilde{b}_n)$ be two $n$-dimensional fuzzy vectors, $\phi = \{\alpha \leq \lambda_0 < \lambda_1 < \cdots < \lambda_l \leq 1\}$ be a division of $[\alpha, 1]$, the distance between $\tilde{a}$ and $\tilde{b}$ under $\phi$ is defined as:

$$D(\tilde{a}, \tilde{b}) \triangleq \frac{1}{l+1} \sum_{i=1}^n \sum_{j=0}^l \left\{ \left| a_{i\lambda_j}^L - b_{i\lambda_j}^L \right| + \left| a_{i\lambda_j}^R - b_{i\lambda_j}^R \right| \right\}, \tag{9.2}$$

where $\alpha$ is a predefined satisfactory degree.

In Definition 9.1, a satisfactory degree $\alpha$ is used to give flexibility to compare two fuzzy vectors. It is possible that two fuzzy vectors might not be compared, that is no ranking relation, by using Definition 9.1. For example, when we compare two fuzzy vectors $\tilde{a}$ and $\tilde{b}$, if some of the left $\lambda$-cuts of $\tilde{a}$ are less than those of $\tilde{b}$, while some right $\lambda$-cuts of $\tilde{a}$ are larger than those of $\tilde{b}$, there is no ranking relation between $\tilde{a}$ and $\tilde{b}$. To solve the incomparable problem, we can enhance the aspiration levels of the attributes, that is, we can adjust the satisfactory degree $\alpha$ to a point where all incomparable parts are discarded. It can be understood as a risk taken by a decision maker who neglects all values with the possibility of occurrence smaller than $\alpha$. In such a situation, a solution is supposed to be reached under this aspiration level. So, normally, we take the same $\alpha$ for both objectives and constraints in a bi-level programming problem.

**Lemma 9.1** *For any n-dimensional fuzzy vectors $\tilde{a}$, $\tilde{b}$, $\tilde{c}$, fuzzy distance D defined in (9.1a)–(9.1d) satisfies the following properties:*

1. $D(\tilde{a}, \tilde{b}) = 0$, if $\tilde{a}_i = \tilde{b}_i$, $i = 1, 2, \ldots, n$;
2. $D(\tilde{a}, \tilde{b}) = D(\tilde{b}, \tilde{a})$;
3. $D(\tilde{a}, \tilde{b}) \le D(\tilde{a}, \tilde{c}) + D(\tilde{c}, \tilde{b})$.

Goals set for the objectives of a leader ($\tilde{g}_L$) and a follower ($\tilde{g}_F$) in (9.1a)–(9.1d) are defined as:

$$\tilde{g}_L = (\tilde{g}_{L1}, \tilde{g}_{L2}, \ldots, \tilde{g}_{Ls}), \tag{9.3a}$$

$$\tilde{g}_F = (\tilde{g}_{F1}, \tilde{g}_{F2}, \ldots, \tilde{g}_{Ft}), \tag{9.3b}$$

where $\tilde{g}_{Li}(i = 1, \ldots, s)$ and $\tilde{g}_{Fj}(j = 1, \ldots, t)$ are fuzzy numbers with membership functions of $\mu_{\tilde{g}_{L_i}}$ and $\mu_{\tilde{g}_{F_j}}$ respectively.

Our concern is to make the objectives of both the leader and the follower as near to their goals as possible. Using the distance measure defined in (9.1a)–(9.1d), we transform the FMO-BLGP problem into a FBLGP problem as follows:

For $x \in X \subset R^n$, $y \in Y \subset R^m$, $F: X \times Y \to (F^*(R))^s$, $f: X \times Y \to (F^*(R))^t$,

$$\min_{x \in X} \quad D(F(x, y), \tilde{g}_L) \tag{9.4a}$$

$$\text{s.t.} \quad \tilde{A}_1 x + \tilde{B}_1 y \le_\alpha \tilde{b}_1, \tag{9.4b}$$

$$\min_{y \in Y} \quad D(f(x, y), \tilde{g}_F) \tag{9.4c}$$

$$\text{s.t.} \quad \tilde{A}_2 x + \tilde{B}_2 y \le_\alpha \tilde{b}_2, \tag{9.4d}$$

where $\tilde{A}_1 = (\tilde{a}_{ij})_{p \times n}$, $\tilde{B}_1 = (\tilde{b}_{ij})_{p \times m}$, $\tilde{A}_2 = (\tilde{e}_{ij})_{q \times n}$, $\tilde{B}_2 = (\tilde{s}_{ij})_{q \times m}$, $\tilde{a}_{ij}, \tilde{b}_{ij}, \tilde{e}_{ij}, \tilde{s}_{ij} \in (F^*(R))$, and $\alpha$ is a predefined satisfactory degree.

From Definitions 9.1, we transfer problem (9.4a)–(9.4d) into:

$$\min_{x \in X} \quad \frac{1}{l+1} \sum_{h=1}^{s} \sum_{j=0}^{l} \left\{ \left| \alpha_{h1\lambda_j}^{L} x + \beta_{h1\lambda_j}^{L} y - g_{Lh\lambda_j}^{L} \right| \right. \tag{9.5a}$$

$$\left. + \left| \alpha_{h1\lambda_j}^{R} x + \beta_{h1\lambda_j}^{R} y - g_{Lh\lambda_j}^{R} \right| \right\}$$

$$\text{s.t.} \quad A_{1\lambda_j}^{L} x + B_{1\lambda_j}^{L} y \le b_{1\lambda_j}^{L},$$
$$A_{1\lambda_j}^{R} x + B_{1\lambda_j}^{R} y \le b_{1\lambda_j}^{R}, \quad j = 0, 1, \ldots, l, \tag{9.5b}$$

$$\min_{y \in Y} \quad \frac{1}{l+1} \sum_{i=1}^{t} \sum_{j=0}^{l} \left\{ \left| \alpha_{i2\lambda_j}^{L} x + \beta_{i2\lambda_j}^{L} y - g_{Fi\lambda_j}^{L} \right| \right. \tag{9.5c}$$

$$\left. + \left| \alpha_{i2\lambda_j}^{R} x + \beta_{i2\lambda_j}^{R} y - g_{Fi\lambda_j}^{R} \right| \right\}$$

$$\text{s.t.} \quad A_{2\lambda_j}^{L} x + B_{2\lambda_j}^{L} y \le b_{2\lambda_j}^{L},$$
$$A_{2\lambda_j}^{R} x + B_{2\lambda_j}^{R} y \le b_{2\lambda_j}^{R}, \quad j = 0, 1, \ldots, l \tag{9.5d}$$

where $\phi = \{ \alpha \le \lambda_0 < \lambda_1 < \cdots < \lambda_l \le 1 \}$ is a division of $[\alpha, 1]$.
For a clear understanding of the idea adopted, we define:

$$V_{h1}^{L-} = \frac{1}{2} \left\{ \left| \sum_{j=0}^{l} \alpha_{h1\lambda_j}^{L} x + \sum_{j=0}^{l} \beta_{h1\lambda_j}^{L} y - \sum_{j=0}^{l} g_{Lh\lambda_j}^{L} \right| - \left( \sum_{j=0}^{l} \alpha_{h1\lambda_j}^{L} x + \sum_{j=0}^{l} \beta_{h1\lambda_j}^{L} y \right. \right.$$
$$\left. \left. - \sum_{j=0}^{l} g_{Lh\lambda_j}^{L} \right) \right\},$$

$$V_{h1}^{L+} = \frac{1}{2} \left\{ \left| \sum_{j=0}^{l} \alpha_{h1\lambda_j}^{L} x + \sum_{j=0}^{l} \beta_{h1\lambda_j}^{L} y - \sum_{j=0}^{l} g_{Lh\lambda_j}^{L} \right| + \left( \sum_{j=0}^{l} \alpha_{h1\lambda_j}^{L} x + \sum_{j=0}^{l} \beta_{h1\lambda_j}^{L} y \right. \right.$$
$$\left. \left. - \sum_{j=0}^{l} g_{Lh\lambda_j}^{L} \right) \right\},$$

$$V_{h1}^{R-} = \frac{1}{2} \left\{ \left| \sum_{j=0}^{l} \alpha_{h1\lambda_j}^{R} x + \sum_{j=0}^{l} \beta_{h1\lambda_j}^{R} y - \sum_{j=0}^{l} g_{Lh\lambda_j}^{R} \right| - \left( \sum_{j=0}^{l} \alpha_{h1\lambda_j}^{R} x + \sum_{j=0}^{l} \beta_{h1\lambda_j}^{R} y \right. \right.$$
$$\left. \left. - \sum_{j=0}^{l} g_{Lh\lambda_j}^{R} y \right) \right\},$$

$$V_{h1}^{R+} = \frac{1}{2}\left\{\left|\sum_{j=0}^{l}\alpha_{h1\lambda_j}^R x + \sum_{j=0}^{l}\beta_{h1\lambda_j}^R y - \sum_{j=0}^{l}g_{Lh\lambda_j}^R\right| + \left(\sum_{j=0}^{l}\alpha_{h1\lambda_j}^R x\right.\right.$$

$$\left.\left. + \sum_{j=0}^{l}\beta_{h1\lambda_j}^R y - \sum_{j=0}^{l}g_{Lh\lambda_j}^R\right)\right\},\tag{9.6}$$

$$h = 1,2,\ldots,s,$$

$$V_{i2}^{L-} = \frac{1}{2}\left\{\left|\sum_{j=0}^{l}\alpha_{i2\lambda_j}^L x + \sum_{j=0}^{l}\beta_{i2\lambda_j}^L y - \sum_{j=0}^{l}g_{F_i\lambda_j}^L\right| - \left(\sum_{j=0}^{l}\alpha_{i2\lambda_j}^L x + \sum_{j=0}^{l}\beta_{i2\lambda_j}^L y\right.\right.$$

$$\left.\left. - \sum_{j=0}^{l}g_{F_i\lambda_j}^L\right)\right\},$$

$$V_{i2}^{L+} = \frac{1}{2}\left\{\left|\sum_{j=0}^{l}\alpha_{i2\lambda_j}^L x + \sum_{j=0}^{l}\beta_{i2\lambda_j}^L y - \sum_{j=0}^{l}g_{F_i\lambda_j}^L\right| + \left(\sum_{j=0}^{l}\alpha_{i2\lambda_j}^L x + \sum_{j=0}^{l}\beta_{i2\lambda_j}^L y\right.\right.$$

$$\left.\left. - \sum_{j=0}^{l}g_{F_i\lambda_j}^L\right)\right\},$$

$$V_{i2}^{R-} = \frac{1}{2}\left\{\left|\sum_{j=0}^{l}\alpha_{i2\lambda_j}^R x + \sum_{j=0}^{l}\beta_{i2\lambda_j}^R y - \sum_{j=0}^{l}g_{F_i\lambda_j}^R\right| + \left(\sum_{j=0}^{l}\alpha_{i2\lambda_j}^R x + \sum_{j=0}^{l}\beta_{i2\lambda_j}^R y\right.\right.$$

$$\left.\left. - \sum_{j=0}^{l}g_{F_i\lambda_j}^R y\right)\right\},$$

$$V_{i2}^{R+} = \frac{1}{2}\left\{\left|\sum_{j=0}^{l}\alpha_{i2\lambda_j}^R x + \sum_{j=0}^{l}\beta_{i2\lambda_j}^R y - \sum_{j=0}^{l}g_{F_i\lambda_j}^R\right| + \left(\sum_{j=0}^{l}\alpha_{i2\lambda_j}^R x + \sum_{j=0}^{l}\beta_{i2\lambda_j}^R y\right.\right.$$

$$\left.\left. - \sum_{j=0}^{l}g_{F_i\lambda_j}^R\right)\right\},$$

$$i = 1,2,\ldots,t,$$

where $V_{h1}^{L-}$ and $V_{h1}^{L+}$ are deviational variables representing the under-achievement and over-achievement of the $h$th goal for a leader under the left $\lambda$-cut respectively. $V_{h1}^{R-}$ and $V_{h1}^{R+}$ are deviational variables representing the under-achievement and over-achievement of the $h$th goal for a leader under the right $\lambda$-cut respectively. $V_{i2}^{L-}$, $V_{i2}^{L+}$, $V_{i2}^{R-}$, and $V_{i2}^{R+}$ are for a follower respectively.

For $(v_{11}^{L-}, v_{11}^{L+}, v_{11}^{R-}, v_{11}^{R+}, \ldots, v_{s1}^{L-}, v_{s1}^{L+}, v_{s1}^{R-}, v_{s1}^{R+}) \in R^{4s}, X' \subseteq X \times R^{4s}, (v_{12}^{L-}, v_{12}^{L+}, v_{12}^{R-}, v_{12}^{R+}, \ldots, v_{t2}^{L-}, v_{t2}^{L+}, v_{t2}^{R-}, v_{t2}^{R+}) \in R^{4t}, Y' \subseteq Y \times R^{4t}$, let

$$x = (x_1, \ldots, x_n) \in X,$$

$$x' = \left(x_1, \ldots, x_n, v_{11}^{L-}, v_{11}^{L+}, v_{11}^{R-}, v_{11}^{R+}, \ldots, v_{s1}^{L-}, v_{s1}^{L+}, v_{s1}^{R-}, v_{s1}^{R+}\right) \in X',$$

$$y = (y_1, \ldots, y_m) \in Y,$$

$$y' = \left(y_1, \ldots, y_m, v_{12}^{L-}, v_{12}^{L+}, v_{12}^{R-}, v_{12}^{R+}, \ldots, v_{t2}^{L-}, v_{t2}^{L+}, v_{t2}^{R-}, v_{t2}^{R+}\right) \in Y',$$

and $v_1, v_2 : X' \times Y' \to R$.

Associated with problem (9.5a)–(9.5d), we now consider the following bi-level programming problem:

$$\min_{x' \in X'} \quad v_1 = \sum_{h=1}^{s} (v_{h1}^{L-} + v_{h1}^{L+} + v_{h1}^{R-} + v_{h1}^{R+}) \tag{9.7a}$$

$$\text{s.t.} \quad \sum_{j=0}^{l} \alpha_{h1\lambda_j}^{L} x + \sum_{j=0}^{l} \beta_{h1\lambda_j}^{L} y + v_{h1}^{L-} - v_{h1}^{L+} = \sum_{j=0}^{l} g_{Lh\lambda_j}^{L},$$

$$\sum_{j=0}^{l} \alpha_{h1\lambda_j}^{R} x + \sum_{j=0}^{l} \beta_{h1\lambda_j}^{R} y + v_{h1}^{R-} - v_{h1}^{R+} = \sum_{j=0}^{l} g_{Lh\lambda_j}^{R},$$

$$v_{h1}^{L-}, v_{h1}^{L+}, v_{h1}^{R-}, v_{h1}^{R+} \geq 0,$$

$$v_{h1}^{L-} \cdot v_{h1}^{L+} = 0, v_{h1}^{R-} \cdot v_{h1}^{R+} = 0, \tag{9.7b}$$

$$A_{1\lambda_j}^{L} x + B_{1\lambda_j}^{L} y \leq b_{1\lambda_j}^{L},$$

$$A_{1\lambda_j}^{R} x + B_{1\lambda_j}^{R} y \leq b_{1\lambda_j}^{R},$$

$$h = 1, \ldots, s, \quad j = 0, 1, \ldots, l,$$

$$\min_{y' \in Y'} \quad v_2 = \sum_{i=1}^{t} (v_{i2}^{L-} + v_{i2}^{L+} + v_{i2}^{R-} + v_{i2}^{R+}) \tag{9.7c}$$

$$\text{s.t.} \quad \sum_{j=0}^{l} \alpha_{i2\lambda_j}^{L} x + \sum_{j=0}^{l} \beta_{i2\lambda_j}^{L} y + v_{i2}^{L-} - v_{i2}^{L+} = \sum_{j=0}^{l} g_{Fi\lambda_j}^{L},$$

$$\sum_{j=1}^{l} \alpha_{i2\lambda_j}^{R} x + \sum_{j=0}^{l} \beta_{i2\lambda_j}^{R} y + v_{i2}^{R-} - v_{i2}^{R+} = \sum_{j=0}^{l} g_{Fi\lambda_j}^{R},$$

$$v_{i2}^{L-}, v_{i2}^{L+}, v_{i2}^{R-}, v_{i2}^{R+} \geq 0,$$

$$v_{i2}^{L-} \cdot v_{i2}^{L+} = 0, v_{i2}^{R-} \cdot v_{i2}^{R+} = 0, \tag{9.7d}$$

$$A_{2\lambda_j}^{L} x + B_{2\lambda_j}^{L} y \leq b_{2\lambda_j}^{L},$$

$$A_{2\lambda_j}^{R} x + B_{2\lambda_j}^{R} y \leq b_{2\lambda_j}^{R},$$

$$i = 1, \ldots, t, \quad j = 0, 1, \ldots, l.$$

**Theorem 9.1** *Let* $(x'^*, y'^*) = (x^*, v_{11}^{L-*}, v_{11}^{L+*}, v_{11}^{R-*}, v_{11}^{R+*}, \ldots, v_{s1}^{L-*}, v_{s1}^{L+*}, v_{s1}^{R-*}, v_{s1}^{R+*},$
$y^*, v_{12}^{L-*}, v_{12}^{L+*}, v_{12}^{R-*}, v_{12}^{R+*}, \ldots, v_{t1}^{L-*}, v_{t1}^{L+*}, v_{t1}^{R-*}, v_{t1}^{R+*})$ *be the optimal solution to problem* (9.7a)–(9.7d), *then* $(x^*, y^*)$ *is the optimal solution to problem* (9.5a)–(9.5d).

*Proof* Let the notations associated with problem (9.5a)–(9.5d) are denoted by:

$$S = \left\{ (x,y) | A_{k\lambda_j}^L x + B_{k\lambda_j}^L y \le b_{k\lambda_j}^L, A_{k\lambda_j}^R x + B_{k\lambda_j}^R y \le b_{k\lambda_j}^R, \right.$$
$$\left. j = 0, \ldots, l, \quad k = 1, 2 \right\}, \tag{9.8a}$$

$$S(X) = \left\{ x \in X | \exists y \in Y, A_{k\lambda_j}^L x + B_{k\lambda_j}^L y \le b_{k\lambda_j}^L, A_{k\lambda_j}^R x \right.$$
$$\left. + B_{k\lambda_j}^R y \le b_{k\lambda_j}^R, \quad k = 1, 2, j = 0, \ldots, l \right\}, \tag{9.8b}$$

$$S(x) = \{ y \in Y | (x,y) \in S \}, \tag{9.8c}$$

$$P(x) = \{ y \in Y | y \in \text{argmin } \Psi \} \tag{9.8d}$$

where

$$\Psi = \frac{1}{l+1} \sum_{i=1}^{t} \sum_{j=0}^{l} \left\{ \left| \alpha_{i2\lambda_j}^L x + \beta_{i2\lambda_j}^L - g_{Lh\lambda_j}^L \hat{y} - g_{Fi\lambda_j}^L \right| \right.$$
$$\left. + \left| \alpha_{i2\lambda_j}^R x + \beta_{i2\lambda_j}^R \hat{y} - g_{Fi\lambda_j}^R \right| \right\}$$

$$IR = \{ (x,y) | (x,y) \in S, y \in P(x) \}. \tag{9.8e}$$

Problem (9.5a)–(9.5d) can be written as:

$$\min_{x,y} \quad \frac{1}{l+1} \sum_{h=1}^{s} \sum_{j=0}^{l} \left\{ \left| \alpha_{h1\lambda_j}^L x + \beta_{h1\lambda_j}^L y - g_{Lh\lambda_j}^L \right| \right.$$
$$\left. + \left| \alpha_{h1\lambda_j}^R x + \beta_{h1\lambda_j}^R y - g_{Lh\lambda_j}^R \right| \right\} \tag{9.9a}$$

$$\text{s.t.} \quad (x,y) \in IR. \tag{9.9b}$$

Similarly, we denote those for problem (9.7a)–(9.7d) by:

$$S' = \left\{ (x',y') | A_{(k\lambda_j)}^L x + B_{(k\lambda_j)}^L y \le b_{(k\lambda_j)}^L, A_{(k\lambda_j)}^R x + B_{(k\lambda_j)}^R y \le b_{(k\lambda_j)}^R, \quad k = 1,2, \quad j = 0,1,\ldots,l, \right.$$

$$\sum_{j=0}^{l} \alpha_{h1\lambda_j}^L x + \sum_{j=0}^{l} \beta_{h1\lambda_j}^L y + v_{h1}^{L-} - v_{h1}^{L+} = \sum_{j=0}^{l} g_{Lh\lambda_j}^L,$$

$$\sum_{j=0}^{l} \left(\alpha_{h1\lambda_j}^R\right) x + \sum_{j=0}^{l} \beta_{h1\lambda_j}^R y + v_{h1}^{R-} - v_{h1}^{R+} = \sum_{j=0}^{l} g_{Lh\lambda_j}^R,$$

$$v_{h1}^{L-}, v_{h1}^{L+}, v_{h1}^{R-}, v_{h1}^{R+} \ge 0,$$

$$v_{h1}^{L-} \cdot v_{h1}^{L+} = 0, v_{h1}^{R-} \cdot v_{h1}^{R+} = 0, h = 1,\ldots,s,$$

$$\sum_{j=0}^{l} \alpha_{i2\lambda_j}^L x + \sum_{j=0}^{l} \beta_{i2\lambda_j}^L y + v_{i2}^{L-} - v_{i2}^{L+} = \sum_{j=0}^{l} g_{Fi\lambda_j}^L,$$

$$\sum_{j=0}^{l} \alpha_{i2\lambda_j}^R x + \sum_{j=0}^{l} \beta_{i2\lambda_j}^R y + v_{i2}^{R-} - v_{i2}^{R+} = \sum_{j=0}^{l} g_{Fi\lambda_j}^R,$$

$$v_{i2}^{L-}, v_i 2^{L+}, v_{i2}^{R-}, v_{i2}^{R+} \ge 0, \tag{9.10}$$

$$\left. v_{i2}^{L-} \cdot v_{i2}^{L+} = 0, v_{i2}^{R-} \cdot v_{i2}^{R+} = 0, i = 1,\ldots,t \right\},$$

$$S(X') = \{x' \in X' \exists y' \in Y', (x',y') \in S'\}, \tag{9.11}$$

$$S(x') = \{y' \in Y'(x',y') \in S'\}, \tag{9.12}$$

$$P(x') = \left\{ y' \in Y' | y' \in \text{argmin}\left[ \sum_{i=1}^{t} \left(\hat{v}_{i2}^{L-} + \hat{v}_{i2}^{L+} + \hat{v}_{i2}^{R-} + \hat{v}_{i2}^{R+}\right) \right] \right\}, \tag{9.13}$$

$$IR' = \{(x',y') | (x',y') \in S', y' \in P(x')\}. \tag{9.14}$$

Problem (9.7a)–(9.7d) can be written as

$$\min_{(x',y')} \left\{ \sum_{h=1}^{s} \left(v_{h_1}^{L-} + v_{h1}^{L+} + v_{h1}^{R-} + v_{h1}^{R+}\right) : (x',y') \in IR' \right\} \tag{9.15}$$

As $(x'^*, y'^*)$ is the optimal solution to problem (9.7a)–(9.7d), from (9.15), it can be seen that, for any $(x', y') \in IR'$, we have

$$\sum_{h=1}^{s} \left(v_{h_1}^{L-} + v_{h1}^{L+} + v_{h1}^{R-} + v_{h1}^{R+}\right) \ge \sum_{h=1}^{s} \left(v_{h_1}^{L-*} + v_{h1}^{L+*} + v_{h1}^{R-*} + v_{h1}^{R+*}\right).$$

It follows from the definitions of $v_{h1}^{L-}$ and $v_{h1}^{L+}$ that

$$v_{h1}^{L-} + v_{h1}^{L+} = \left| \sum_{j=0}^{l} \alpha_{h1\lambda_j}^{L} x + \sum_{j=0}^{l} \beta_{h1\lambda_j}^{L} y - \sum_{j=0}^{l} g_{Lh\lambda_j}^{L} \right|,$$

$$v_{h1}^{L-*} + v_{h1}^{L+*} = \left| \sum_{j=0}^{l} \alpha_{h1\lambda_j}^{L} x^* + \sum_{j=0}^{l} \beta_{h1\lambda_j}^{L} y^* - \sum_{j=0}^{l} g_{Lh\lambda_j}^{L} \right|,$$

for $h = 1, \ldots, s$. Similarly, we have

$$v_{h1}^{R-} + v_{h1}^{R+} = \left| \sum_{j=0}^{l} \alpha_{h1\lambda_j}^{R} x + \sum_{j=0}^{l} \beta_{h1\lambda_j}^{R} y - \sum_{j=0}^{l} g_{Lh\lambda_j}^{R} \right|,$$

$$v_{h1}^{R-*} + v_{h1}^{R+*} = \left| \sum_{j=0}^{l} \alpha_{h1\lambda_j}^{R} x^* + \sum_{j=0}^{l} \beta_{h1\lambda_j}^{R} y^* - \sum_{j=0}^{l} g_{Lh\lambda_j}^{R} \right|,$$

for $h = 1, \ldots, s$. So, for any $(x', y') \in IR'$, we can obtain

$$\begin{aligned}
& \left| \sum_{j=0}^{l} \alpha_{h1\lambda_j}^{L} x' + \sum_{j=0}^{l} \beta_{h1\lambda_j}^{L} y' - \sum_{j=0}^{l} g_{Lh\lambda_j}^{L} \right| \\
& + \left| \sum_{j=0}^{l} \alpha_{h1\lambda_j}^{R} x' + \sum_{j=0}^{l} \beta_{h1\lambda_j}^{R} y' - \sum_{j=0}^{l} g_{Lh\lambda_j}^{R} \right| \\
& \geq \left| \sum_{j=0}^{l} \alpha_{h1\lambda_j}^{L} x^* + \sum_{j=0}^{l} \beta_{h1\lambda_j}^{L} y^* - \sum_{j=0}^{l} g_{Lh\lambda_j}^{L} \right| \\
& + \left| \sum_{j=0}^{l} \alpha_{h1\lambda_j}^{R} x^* + \sum_{j=0}^{l} \beta_{h1\lambda_j}^{R} y^* - \sum_{j=0}^{l} g_{Lh\lambda_j}^{R} \right|.
\end{aligned} \tag{9.16}$$

We now prove that the projection of $S'$ onto the $X \times Y$ space, denoted by $S'|_{X \times Y}$ is equal to $S$.

On the one hand, for any $(x, y) \in S'|_{X \times Y}$, from the constraints: $A_{k\lambda_j}^{L} x + B_{k\lambda_j}^{L} y \leq b_{k\lambda_j}^{L}, A_{k\lambda_j}^{R} x + B_{k\lambda_j}^{R} y \leq b_{j\lambda_j}^{R}$, $k = 1, 2, j = 0, \ldots, l$, we have $(x, y) \in S$, so $S'|_{X \times Y} \subseteq S$.

On the other hand, for any $(x, y) \in S$, by (9.6), we can always find $v_{11}^{L-}, v_{11}^{L+}, v_{11}^{R-}, v_{11}^{R+}, \ldots, v_{s1}^{L-}, v_{s1}^{L+}, v_{s1}^{R-}, v_{s1}^{R+}, v_{12}^{L-}, v_{12}^{L+}, v_{12}^{R-}, v_{12}^{R+}, \ldots, v_{t2}^{L-}, v_{t2}^{L+}, v_{t2}^{R-}, v_{t2}^{R+}$, which satisfies the constraints of (9.7b) and (9.7d). Together with the inequalities of $A_{k\lambda_j}^{L} x + B_{k\lambda_j}^{L} y \leq b_{k\lambda_j}^{L}$, and $A_{k\lambda_j}^{R} x + B_{k\lambda_j}^{R} y \leq b_{k\lambda_j}^{R}$, $k = 1, 2$, $j = 0, 1, \ldots, l$, requested by $S$, we have $x, v_{11}^{L-}, v_{11}^{L+}, v_{11}^{R-}, v_{11}^{R+}, \ldots, v_{s1}^{L-}, v_{s1}^{L+}, v_{s1}^{R-}, v_{s1}^{R+}, y, v_{12}^{L-}, v_{12}^{L+}, v_{12}^{R-}, v_{12}^{R+}, \ldots, v_{t2}^{L-}, v_{t2}^{L+}, v_{t2}^{R-}, v_{t2}^{R+} \in S'$, thus $(x, y) \in S'|_{X \times Y}$, $S \subseteq S'|_{X \times Y}$.

So, we can prove that

$$S'|_{X \times Y} = S. \tag{9.17}$$

Similarly, we have

$$S(x)'|_{X \times Y} = S(x), \tag{9.18}$$

$$S(X)'|_{X \times Y} = S(X). \tag{9.19}$$

Also, from $\sum_{j=0}^{l} \alpha_{i2\lambda_j}^{L} x + \sum_{j=0}^{l} \beta_{i2\lambda_j}^{L} y + v_{i2}^{L-} - v_{i2}^{L+} = \sum_{j=0}^{l} g_{Fi\lambda_j}^{L}$, and $v_{i2}^{L-} \cdot v_{i2}^{L+} = 0$, we have

$$v_{i2}^{L-} \mp v_{i2}^{L+} = \left| \sum_{j=0}^{l} \alpha_{i2\lambda_j}^{L} x + \sum_{j=0}^{l} \beta_{i2\lambda_j}^{L} y - \sum_{j=0}^{l} g_{Fi\lambda_j}^{L} \right| \tag{9.20}$$

for $i = 1, \ldots, t$. Similarly, we have

$$v_{i2}^{R-} \mp v_{i2}^{R+} = \left| \sum_{j=0}^{l} \alpha_{i2\lambda_j}^{R} x + \sum_{j=0}^{l} \beta_{i2\lambda_j}^{R} y - \sum_{j=0}^{l} g_{Fi\lambda_j}^{R} \right| \tag{9.21}$$

for $i = 1, \ldots, t$. Thus, we obtain

$$P(x') = \{y' \in Y' | y' \in \arg\min \Psi'\} \tag{9.22}$$

where

$$\Psi' = \sum_{i=1}^{t} \sum_{j=0}^{l} \left\{ \left| \alpha_{i2\lambda_j}^{L} x + \beta_{i2\lambda_j}^{L} - g_{Lh\lambda_j}^{L} \hat{y} - g_{Fi\lambda_j}^{L} \right| \right.$$
$$\left. + \left| \alpha_{i2\lambda_j}^{R} x + \beta_{i2\lambda_j}^{R} \hat{y} - g_{Fi\lambda_j}^{R} \right|, \hat{y} \in S(x') \right\}.$$

From (9.17) and (9.22), we obtain

$$p(x')|_{X \times Y} = p(x). \tag{9.23}$$

From (9.8e), (9.14), (9.17), and (9.23), we obtain

$$IR'|_{X \times Y} = IR, \tag{9.24}$$

which means that, in $X \times Y$ space, the leaders of problem (9.5a)–(9.5d) and (9.7a)–(9.7d) have the same optimizing space.

Thus, from (9.16) and (9.24), for any $(x, y) \in IR$, we have

$$\frac{1}{l+1} \sum_{h=1}^{s} \sum_{j=0}^{l} \left\{ \left| \alpha_{h1\lambda_j}^L x + \beta_{h1\lambda_j}^L y - g_{Lh\lambda_j}^L \right| + \left| \alpha_{h1\lambda_j}^R x + \beta_{h1\lambda_j}^R y - g_{Lh\lambda_j}^R \right| \right\}$$

$$\geq \frac{1}{l+1} \sum_{h=1}^{s} \sum_{j=0}^{l} \left( \left| \alpha_{h1\lambda_j}^L x^* + \beta_{h1\lambda_j}^L y^* - g_{Lh\lambda_j}^L \right| + \left| \alpha_{h1\lambda_j}^R x^* + \beta_{h1\lambda_j}^R y^* - g_{Lh\lambda_j}^R \right| \right).$$

Consequently, $(x^*, y^*)$ is the optimal solution of problem (9.5a)–(9.5d). $\qquad \square$

Adopting the weighting method, (9.7a)–(9.7d) can be further transferred into (9.25a)–(9.25d):

$$\min_{x' \in X'} \quad v_1^- + v_1^+ \tag{9.25a}$$

$$\text{s.t.} \quad \alpha_1 x + \beta_1 y + v_1^- - v_1^+ = \sum_{h=1}^{s} \sum_{j=0}^{l} \left( g_{Lh\lambda_j}^L + g_{Lh\lambda_j}^R \right)$$

$$
\begin{aligned}
& v_1^-, v_1^+ \geq 0, \\
& v_1^- \cdot v_1^+ = 0, \\
& A_{1\lambda_j}^L x + B_{1\lambda_j}^L y \leq b_{1\lambda_j}^L, \\
& A_{1\lambda_j}^R x + B_{1\lambda_j}^R y \leq b_{1\lambda_j}^R, \\
& j = 0, 1, \ldots, l,
\end{aligned}
\tag{9.25b}
$$

$$\min_{y' \in Y'} \quad v_2^- + v_2^+ \tag{9.25c}$$

$$\text{s.t.} \quad \alpha_2 x + \beta_2 y + v_2^- - v_2^+ = \sum_{i=1}^{t} \sum_{j=0}^{l} (g_{Fi\lambda_j}^L + g_{Fi\lambda_j}^R)$$

$$
\begin{aligned}
& v_2^-, v_2^+ \geq 0, \\
& v_2^- \cdot v_2^+ = 0, \\
& A_{2\lambda_j}^L x + B_{2\lambda_j}^L y \leq b_{2\lambda_j}^L, \\
& A_{2\lambda_j}^R x + B_{2\lambda_j}^R y \leq b_{2\lambda_j}^R, \\
& j = 0, 1, \ldots, l,
\end{aligned}
\tag{9.25d}
$$

where $x' = (x_1, \ldots, x_n, v_1^-, v_1^+)$, $y' = (y_1, \ldots, y_m, v_2^-, v_2^+)$, $v_1^- = \sum_{h=1}^{s} \left( v_{L1}^{L-} + v_{L1}^{R-} \right)$, $v_1^+ = \sum_{h=1}^{s} \left( v_{h1}^{L+} + v_{h1}^{R+} \right)$, $v_2^- = \sum_{i=1}^{t} \left( v_{i2}^{L-} + v_{i2}^{R-} \right)$, $v_2^+ = \sum_{i=1}^{t} \left( v_{i2}^{L+} + v_{i2}^{R+} \right)$, $\alpha_1 = \sum_{h=1}^{s} \sum_{j=0}^{l} \left( \alpha_{h1\lambda_j}^L + \alpha_{h1\lambda_j}^R \right)$, $\beta_1 = \sum_{h=1}^{s} \sum_{j=0}^{l} \left( \beta_{h1\lambda_j}^L + \beta_{h1\lambda_j}^R \right)$, $\alpha_2 = \sum_{i=1}^{t} \sum_{j=0}^{l} \left( \alpha_{i2\lambda_j}^L + \alpha_{i2\lambda_j}^R \right)$, $\beta_2 = \sum_{i=1}^{t} \sum_{j=0}^{l} \left( \beta_{i2\lambda_j}^L + \beta_{i2\lambda_j}^R \right)$.

In this formula, $v_1^-$ and $v_1^+$ are deviational variables representing the under-achievement and over-achievement of goals for a leader, and $v_2^-$ and $v_2^+$ are deviational variables representing the under-achievement and over-achievement of goals for a follower respectively.

The non-linear conditions of $v_1^- \cdot v_1^+ = 0$, and $v_2^- \cdot v_2^+ = 0$ need not to be maintained if the Kuhn-Tucker approach together with the simplex algorithm are adopted, since only equivalence at an optimum is wanted. Further explanation can be found from (Charnes and Cooper 1961). Thus, the problem (9.25a)–(9.25d) is further transformed into:

$$\min_{(x,v_1^-,v_1^+)\in X'} \quad v_1 = v_1^- + v_1^+ \tag{9.26a}$$

$$\text{s.t.} \quad \alpha_1 x + \beta_1 y + v_1^- - v_1^+ = \sum_{h=1}^{s}\sum_{j=0}^{l}\left(g_{Lh\lambda_j}^L + g_{Lh\lambda_j}^R\right),$$

$$v_1^-, v_1^+ \geq 0,$$

$$v_1^- \cdot v_1^+ = 0, \tag{9.26b}$$

$$A_{1\lambda_j}^L x + B_{1\lambda_j}^L y \leq b_{1\lambda_j}^L,$$

$$A_{1\lambda_j}^R x + B_{1\lambda_j}^R y \leq b_{1\lambda_j}^R,$$

$$j = 0, 1, \ldots, l,$$

$$\min_{(y,v_2^-,v_2^+)\in Y'} \quad v_2 = v_2^- + v_2^+ \tag{9.26c}$$

$$\text{s.t.} \quad \alpha_2 x + \beta_2 y + v_2^- - v_2^+ = \sum_{h=1}^{s}\sum_{j=0}^{l}\left(g_{Lh\lambda_j}^L + g_{Lh\lambda_j}^R\right),$$

$$v_2^-, v_2^+ \geq 0,$$

$$v_2^- \cdot v_2^+ = 0, \tag{9.26d}$$

$$A_{2\lambda_j}^L x + B_{2\lambda_j}^L y \leq b_{2\lambda_j}^L,$$

$$A_{2\lambda_j}^R x + B_{2\lambda_j}^R y \leq b_{2\lambda_j}^R,$$

$$j = 0, 1, \ldots, l,$$

Problem (9.26a)–(9.26d) is a standard linear bi-level problem that can be solved by the Kuhn-Tucker approach.

Based on the definitions and theorems for the FMO-BLP problem, we will present a solution algorithm for such a problem in the next section.

## 9.3 Fuzzy Bi-level Goal-Programming Algorithm

Based on the analysis above, the fuzzy bi-level goal-programming algorithm is detailed as:

---

**Algorithm 9.1: Fuzzy Bi-level Goal-programming Algorithm**

[Begin]

**Step 1:** (Input) Obtain relevant coefficients which include:

   (1)   Coefficients of (9.1);

   (2)   Coefficients of (9.3);

   (3)   Satisfactory degree$\alpha$;

   (4)   $\varepsilon > 0$.

**Step 2:** (Initialize) Let $k = 1$, which is the counter to record current loop. In (9.5), where $\lambda_j \in [\alpha, 1]$, let $\lambda_0 = \alpha$ and $\lambda_1 = 1$ respectively, then each objective will be transferred into four non-fuzzy objective functions, and each fuzzy constraint is converted into four non-fuzzy constraints.

**Step 3:** (Compute) By introducing auxiliary variables $v_1^-, v_1^+, v_2^-$, and $v_2^+$, we obtain the format of problem (9.26). The solution $(x, v_1^-, v_1^+, y, v_2^-, v_2^+)$ of (9.26) is obtained by the Kuhn-Tucker approach.

**Step 4:** (Compare)

   If $k = 1$,

        then $(x, v_1^-, v_1^+, y, v_2^-, v_2^+)_1 = (x, v_1^-, v_1^+, y, v_2^-, v_2^+)_2$;

        go to Step 5;

   else if $||(x, v_1^-, v_1^+, y, v_2^-, v_2^+)_2 - (x, v_1^-, v_1^+, y, v_2^-, v_2^+)_1|| < \varepsilon$,

        go to Step 7.

**Step 5:** (Split) Suppose that there are $(L + 1)$ nodes $\lambda_j, j = 0, 1, \ldots, L$ in the interval $[\alpha, 1]$, insert $L$ new nodes $\delta_t, (t = 1, 2, \ldots, L)$ in $[\alpha, 1]$ such that $\delta_t = (\lambda_{t-1} + \lambda_t)/2$.

**Step 6:** (Loop) $k = k + 1$.

**Step 7:** (Output) $(x, y)_2$ is obtained as the final solution.

[End]

---

## 9.4 A Numerical Example and Experiments

In this section, we apply the fuzzy bi-level goal-programming algorithm proposed in Sect. 9.3 on a numerical example to illustrate its operation. Experiments are then carried out on some numerical examples with different scales to test the algorithm's performance.

### 9.4.1 A Numerical Example

To illustrate the fuzzy bi-level goal-programming algorithm, we consider the following FMO-BLP problem.

*Example 9.2* **Step 1:** (Input the relevant coefficients).

1. Coefficients of (9.1a)–(9.1d).

   Suppose that the problem has one leader and one follower with two objectives $F_1$ and $F_2$ for the leader and $f_1$ and $f_2$ for the follower respectively. This FMO-BLP problem is as follows:

$$\max_{x \in X} \quad F_1(x,y) = \tilde{6}x + \tilde{3}y$$

$$\max_{x \in X} \quad F_2(x,y) = -\tilde{3}x + \tilde{6}y$$

$$\text{s.t.} \quad -\tilde{1}x + \tilde{3}y \leq \widetilde{21},$$

$$\min_{y \in Y} \quad f_1(x,y) = \tilde{4}x + \tilde{3}y$$

$$\min_{y \in Y} \quad f_2(x,y) = \tilde{3}x + \tilde{1}y$$

$$\text{s.t.} \quad -\tilde{1}x - \tilde{3}y \leq \widetilde{27},$$

where $x \in R, y \in R$, and $X = \{x | x \geq 0\}$, $Y = \{y | y \geq 0\}$.
The membership functions for this FMO-BLP problem are as follows:

$$\mu_{\tilde{6}}(x) = \begin{cases} 0, & x < 5 \\ \frac{x^2 - 25}{11}, & 5 \leq x < 6 \\ 1, & x = 6 \\ \frac{64 - x^2}{28}, & 6 < x \leq 8 \\ 0, & x > 8 \end{cases} \quad , \quad \mu_{\tilde{3}}(x) = \begin{cases} 0 & x < 2 \\ \frac{x^2 - 4}{5} & 2 \leq x < 3 \\ 1 & x = 3, \\ \frac{25 - x^2}{16} & 3 < x \leq 5 \\ 0 & x > 5 \end{cases}$$

$$\mu_{-\tilde{3}}(x) = \begin{cases} 0, & x < -5 \\ \frac{16-x^2}{7}, & -4 \le x < -3 \\ 1, & x = -3 \\ \frac{25-x^2}{16}, & -3 < x \le -1 \\ 0, & x > -1 \end{cases}, \quad \mu_{\tilde{4}}(x) = \begin{cases} 0, & x < 3 \\ \frac{x^2-9}{7}, & 3 \le x < 4 \\ 1, & x = 4, \\ \frac{36-x^2}{20}, & 4 < x \le 6 \\ 0, & x > 6 \end{cases}$$

$$\mu_{\tilde{1}}(x) = \begin{cases} 0, & x < 0.5 \\ \frac{x^2-0.25}{0.75}, & 0.5 \le x < 1 \\ 1, & x = 1 \\ \frac{4-x^2}{3}, & 1 < x \le 2 \\ 0, & x > 2 \end{cases}, \quad \mu_{-\tilde{1}}(x) = \begin{cases} 0, & x < -2 \\ \frac{4-x^2}{3}, & -2 \le x < -1 \\ 1, & x = -1 \\ \frac{x^2-0.25}{0.75}, & -1 < x \le -0.5 \\ 0, & x > -0.5 \end{cases},$$

$$\mu_{\widetilde{21}}(x) = \begin{cases} 0, & x < 19 \\ \frac{x^2-361}{80}, & 19 \le x < 21 \\ 1, & x = 21 \\ \frac{625-x^2}{184}, & 21 < x \le 25 \\ 0, & x > 25 \end{cases}, \quad \mu_{\widetilde{27}}(x) = \begin{cases} 0, & x < 25 \\ \frac{x^2-625}{104}, & 25 \le x < 27 \\ 1, & x = 27 \\ \frac{961-x^2}{232}, & 27 < x \le 31 \\ 0, & x > 31 \end{cases}.$$

2. Suppose the membership functions of the fuzzy goals set for the leader are:

$$\mu_{\widetilde{g_{L1}}}(x) = \begin{cases} 0, & x < 15 \\ \frac{x^2-225}{175}, & 15 \le x < 20 \\ 1, & x = 20 \\ \frac{900-x^2}{500}, & 20 < x \le 30 \\ 0, & x > 30 \end{cases}, \quad \mu_{\widetilde{g_{L2}}}(x) = \begin{cases} 0, & x < 4 \\ \frac{x^2-16}{48}, & 4 \le x < 8 \\ 1, & x = 8 \\ \frac{225-x^2}{161}, & 8 < x \le 15 \\ 0, & x > 15 \end{cases}.$$

The membership functions of the fuzzy goals set for the follower are:

$$\mu_{\widetilde{g_{F1}}}(x) = \begin{cases} 0, & x < 10 \\ \frac{x^2-100}{225}, & 10 \le x < 15 \\ 1, & x = 15 \\ \frac{400-x^2}{175}, & 15 < x \le 20 \\ 0, & x > 20 \end{cases}, \quad \mu_{\widetilde{g_{F2}}}(x) = \begin{cases} 0, & x < 7 \\ \frac{x^2-49}{32}, & 7 \le x < 9 \\ 1, & x = 9 \\ \frac{121-x^2}{40}, & 9 < x \le 11 \\ 0, & x > 11 \end{cases}.$$

3. Satisfactory degree: $\alpha = 0.2$.
4. $\varepsilon = 0.15$.

**Step 2:** (Initialize) Let $k = 1$. Associated with this example, we have

$$\max_{x \in X} \quad \left| \sqrt{11\lambda + 25}x + \sqrt{5\lambda + 4}y - \sqrt{175\lambda + 225} \right|$$

$$+ \left| \sqrt{64 - 2811\lambda}x - \sqrt{25 - 16\lambda}y - \sqrt{900 - 500\lambda} \right|$$

$$\max_{x \in X} \quad \left| -\sqrt{16 - 7\lambda}x + \sqrt{11\lambda + 25}y - \sqrt{48\lambda + 16} \right|$$

$$+ \left| -\sqrt{8\lambda + 1}x + \sqrt{64 - 28\lambda}y - \sqrt{225 - 161\lambda} \right|$$

$$\text{s.t.} \quad -\sqrt{4 - 2\lambda}x + \sqrt{5\lambda + 4}y \le \sqrt{80\lambda + 361},$$

$$-\sqrt{-0.75\lambda + 0.25}x + \sqrt{25 - 16\lambda}y \le \sqrt{625 - 184\lambda},$$

$$\min_{y \in Y} \quad \left| \sqrt{7\lambda + 9}x + \sqrt{5\lambda + 4}y + \sqrt{225\lambda + 100} \right| + \left| \sqrt{36 - 20\lambda}x \right.$$

$$\left. -\sqrt{25 - 16\lambda}y - \sqrt{400 - 175\lambda} \right|$$

$$\min_{y \in Y} \quad -\left| \sqrt{5\lambda + 4}x + \sqrt{0.75\lambda + 0.25}y - \sqrt{32\lambda + 49} \right| + \left| \right.$$

$$\left. -\sqrt{25 - 16\lambda}x + \sqrt{4 - 3\lambda}y - \sqrt{121 - 40\lambda} \right|$$

$$\text{s.t.} \quad \sqrt{0.75\lambda + 0.25}x + \sqrt{5\lambda + 4}y \le \sqrt{104\lambda + 625},$$

$$\sqrt{4 - 3\lambda}x + \sqrt{25 - 16\lambda}y \le \sqrt{901 - 232\lambda},$$

where $\lambda \in [0.2, 1]$.

Referring to the algorithm, only $\lambda_0 = 0.2$ and $\lambda_1 = 1$ are considered initially. Thus four non-fuzzy objective functions and four non-fuzzy constraints for the leader and the follower are generated respectively:

$$\max_{x \in X} \quad \frac{1}{4} \left\{ \left| \sqrt{27.2}x + \sqrt{5}y - \sqrt{260} \right| + |6x + 3y - 20| \right.$$

$$+ \left| \sqrt{58.4}x + \sqrt{21.8}y - 20\sqrt{2} \right| + |6x + 3y - 20|$$

$$+ \left| -\sqrt{14.6}x + \sqrt{27.2}y - \sqrt{25.6} \right| + |-3x + 6y - 8| + |$$

$$\left. -\sqrt{2.6}x + \sqrt{58.4}y - \sqrt{192.8} \right| + |-3x + 6y - 8| \right\}$$

$$\text{s.t.} \quad -\sqrt{3.4}x + \sqrt{5}y \le \sqrt{377},$$

$$-x + 3y \le 21,$$

$$-\sqrt{0.4}x + \sqrt{5}y \le \sqrt{645.8},$$

$$-x + 3y \le 21,$$

$$\min_{y \in Y} \quad \frac{1}{4} \{|3x + 2y - 12.04| + |4x + 3y - 19.1| + |6x - 5y - 7.4|$$

$$+ |4x - 3y - 10.63| + |-2x + 0.5y - 18.3|$$

$$+ |-3x + y - 15| + |-5x + 2y - 9| + |-3x + y - 9|\}$$

$$\text{s.t.} \quad \sqrt{0.4}x + \sqrt{5}y \le \sqrt{645.8},$$

$$x + 3y \le 27,$$

$$\sqrt{3.4}x + \sqrt{21.8}y \le \sqrt{914.6},$$

$$x + 3y \le 27.$$

**Step 3:** (Compute) By introducing auxiliary variables $v_1^-$, $v_1^+$, $v_2^-$, and $v_2^+$, we have

$$\min_{x,v_1^-,v_1^+} \quad v_1^- + v_1^+$$

$$\text{s.t.} \quad 3.083x + 20.076y + v_1^- - v_1^+ = 54.73,$$
$$-1.8x + 2.2y \le 19.4,$$
$$-x + 3y \le 21,$$
$$-0.6x + 4.7y \le 24.3,$$
$$-x + 3y \le 21$$

$$\min_{y,v_2^-,v_2^+} \quad v_2^- + v_2^+$$

$$\text{s.t.} \quad 16.498x + 8.205y + v_2^- - v_2^+ = 51.337,$$
$$0.6x + 2.2y \le 25.4,$$
$$x + 3y \le 7,$$
$$1.8x + 4.7y \le 30.2,$$
$$x + 3y \le 27.$$

Using the Kuhn-Tucker approach, the current solution is (1.901, 0, 0, 2.434, 0, 0).
**Step 4:** (Compare) Because $k = 1$, go to Step 5.
**Step 5:** (Split) By inserting a new node $\lambda_1 = (0.2 + 1)/2$, there are a total three nodes of $\lambda_0 = 0.2$, $\lambda_1 = 0.6$ and $\lambda_2 = 1$. Then a total of six non-fuzzy objective functions for the leader and follower, together with six non-fuzzy constraints for the leader and follower respectively, are generated.
**Step 6:** (Loop) $k = 1 + 1 = 2$, go to Step 3, and a current solution of (2.011, 0, 0, 2.356, 0, 0) is obtained. As $|2.011 - 1.901| + |2.356 - 2.434| = 0.188 > \varepsilon = 0.15$, the algorithm continues until the solution of (1.957, 0, 0, 2.388, 0, 0) is obtained. The computing results are listed in Table 9.1.
**Step 7:** (Output) As $|1.957 - 1.872| + |2.388 - 2.2446| = 0.14 < \varepsilon = 0.15$, $(x^*, y^*) = (1.957, 2.388)$ is the final solution of this example. The objectives for the leader and follower under $(x^*, y^*) = (1.957, 2.388)$ are:

$$\begin{cases} F_1(x^*, y^*) = F_1(1.957, 2.388) = \tilde{6} \cdot 1.957 + \tilde{3} \cdot 2.388, \\ F_2(x^*, y^*) = F_2(1.957, 2.388) = -\tilde{3} \cdot 1.957 + \tilde{6} \cdot 2.388, \\ f_1(x^*, y^*) = f_1(1.957, 2.388) = \tilde{4} \cdot 1.957 + \tilde{3} \cdot 2.388, \\ f_2(x^*, y^*) = f_2(1.957, 2.388) = \tilde{3} \cdot 1.957 + \tilde{1} \cdot 2.388. \end{cases}$$

**Table 9.1** Summary of the running solution

| $k$ | $x$ | $y$ | $v_{1\lambda}^+$ | $v_{1\lambda}^-$ | $v_{2\lambda}^+$ | $v_{2\lambda}^-$ |
|---|---|---|---|---|---|---|
| 1 | 1.901 | 2.434 | 0 | 0 | 0 | 0 |
| 2 | 2.011 | 2.356 | 0 | 0 | 0 | 0 |
| 3 | 1.872 | 2.466 | 0 | 0 | 0 | 0 |
| 4 | 1.957 | 2.388 | 0 | 0 | 0 | 0 |

## 9.4.2 Experiments and Evaluation

The fuzzy bi-level goal-programming algorithm was implemented by Visual Basic 6.0, and run on a desktop computer with CPU Pentium 4 2.8 GHz, RAM 1G, Windows XP. To test the performance of the proposed algorithm, the following experiments are carried out.

1. To test the efficiency of the proposed algorithm, we employ ten numerical examples and enlarge the problem scales by changing the numbers of decision variables, objective functions and constraints for both leaders and followers from two to ten simultaneously. For each of these examples, the final solution has been obtained within 5 s.
2. To test the performance of the fuzzy distance measure in Definition 9.1, we adjust the satisfactory degree values from 0 to 0.5 on the ten numerical examples again. At the same time, we change some of the fuzzy coefficients in the constraints by moving the points whose membership values equal 0 by 10 % from the left and right respectively. Experiments reveal that, when a satisfactory degree is set as 0, the average solution will change by about 6 % if some of the constraint coefficients are moved as discussed above. When we increase satisfactory degrees, the average solution change decreases. For the point in which satisfactory degrees are equal to 0.5, the average solution change is 0.

From Experiment (1), we can see that the proposed algorithm is quite efficient. The reason is the fact that final solutions can be reached by solving corresponding linear bi-level programming problems, which can be handled by the Kuhn-Tucker approach.

From Experiment (2), we can see that if we change some coefficients of fuzzy numbers within a small range, solutions will be less sensitive to this change under a higher satisfactory degree. The reason is that, when the satisfactory degree is set to 0, every $\lambda$-cut of fuzzy coefficients from 0 to 1 will be considered. Thus, the decision maker can certainly be influenced by minor information.

For a decision-making process involved with fuzzy parameters, decision makers may sometimes make small adjustment on the uncertain information about the preference or circumstances. If the change occurs to the minor information, that is with smaller satisfactory degrees, there should normally be no significant change to the final solution. For example, when estimating future profit, the manufacturer may adjust the possibility of five thousand dollars profit from 2 to 3 %, while the possibility of one hundred thousand dollars profit remains 100 %. In such a situation, there should be no outstanding change for his or her final decision on the device investment. Therefore, to increase the satisfactory degrees is an acceptable strategy for a feasible solution.

From the above analysis, the advantages and disadvantages of the proposed fuzzy bi-level goal-programming algorithm are as follows:

1. This algorithm is quite efficient, as it adopts strategies to transform a non-linear bi-level decision problem into a linear decision problem.
2. When pursuing optimality, the negative effect from conflicting objectives can be avoided and a leader can finally reach his or her satisfactory solution by setting goals for the objectives.
3. The information of the original fuzzy numbers is considered adequately by using a certain number of $\lambda$-cuts to approximate the final precise solution.
4. In some situations, this algorithm might suffer from expensive calculation, as the size of $\lambda$-cuts will increase exponentially with respect to iteration counts.

## 9.5  Summary

In a bi-level decision model, the leader and/or the follower may have more than one objective to achieve. This kind of bi-level decision problem is studied by goal programming in this chapter. Meanwhile, we take into consideration the situation where coefficients to formulate a bi-level decision model are not precisely known to us. A fuzzy set method is applied to handle these coefficients, and a fuzzy bi-level goal-programming algorithm is proposed to solve the FMO-BLP problems. A numerical example is then adopted to explain this algorithm. Experiments reveal that the algorithm is quite effective and efficient in solving the FMO-BLP problems.

# Part IV
# Rule-set-based Bi-level Decision Making

# Chapter 10
# Rule-Set-Based Bi-level Decision Making

As discussed in previous chapters, bi-level decision-making problems are normally modeled by bi-level programming. Because many uncertain factors are involved in a bi-level decision-making, it is sometimes very difficult to formulate the objective functions and constraints of decision makers. When a bi-level decision problem cannot be formulated by normal bi-level programming, we can consider using rules to express the objective functions and constraints of a decision problem.

This chapter presents a *rule-set-based bi-level decision* (RSBLD) concept, approach and solution. It first introduces related theories to prove the feasibility of modeling a bi-level decision problem by a rule-set. It then proposes a *rule-set-based bi-level decision modeling approach* (i.e. Algorithm 10.1) for modeling a bi-level decision problem which contains rule generation, rule reduction and other steps. For solving a rule-set-based bi-level decision problem, this chapter also presents a *rule-set-based bi-level decision solution algorithm* (i.e., Algorithm 10.2) and a *transformation-based solution algorithm* (i.e., Algorithm 10.3) through developing *attribute-importance-degree* (AID) based rule trees (Zheng and Wang 2004). A case-based example is used to illustrate the functions and effectiveness of the proposed RSBLD modeling approach and the solution algorithms.

This chapter is organized as follows. Section 10.1 identifies the non-programming bi-level decision problem. Section 10.2 introduces the concepts and notions of information tables and rule-set, which are given as preliminaries in this chapter. Section 10.3 presents a RSBLD model. A RSBLD modeling approach is presented in Sect. 10.4. In Sect. 10.5, two RSBLD solution algorithms are provided to solve a bi-level decision problem which is modeled by rule-sets. A case-based example is shown in Sect. 10.6. Section 10.7 gives experiment results, and the summary is presented in Sect. 10.8.

## 10.1 Problem Identification

In general, there are two main uncertainties in modeling a bi-level decision problem. One is that the parameter values in the objective functions and constraints of the leader and the followers may be indefinite or inaccurate. Fuzzy optimization

© Springer-Verlag Berlin Heidelberg 2015
G. Zhang et al., *Multi-Level Decision Making*,
Intelligent Systems Reference Library 82, DOI 10.1007/978-3-662-46059-7_10

approaches can handle this issue, as discussed in previous chapters. Another type of uncertainty involves the form of the objective functions and constraints. That is, how to determine the relationships among the proposed decision variables and formulate the functions for a real decision problem. The challenge can be handled by a rule-set-based approach. Below, we give an example by way of explanation.

*Example 10.1* A factory's human resource management system is distributed over two levels. The upper level is the factory executive committee and the lower level is the workshop management committee. When deciding whether a person can be recruited for a particular position, the factory executive committee principally considers the following two factors; the "age" and "education level (edulevel)" of the person. The workshop management committee is largely concerned with two other factors: the "seniority" and "health" of the person. Suppose the condition attributes in ascending order according to the importance degree are "age", "edulevel", "seniority", "health".

It is clearly difficult to express the worker selection conditions of the two committees as linear or non-linear functions, but the two committees have all necessary information about the workers. We can build rules using the information to form the selection conditions and objectives of this decision (selection) problem.

## 10.2 Information Tables and Rule-Sets

For the convenience of describing proposed models and algorithms, we will first introduce some basic notions of information tables, formulas, rules, decision rule set functions and rule trees. In addition, we will give some related definitions and theorems which will be used in subsequent sections.

### 10.2.1 Information Tables

To present the definition of a rule, we first describe information table and decision table.

An information table is a knowledge-expressing system which is an important tool for representing and processing knowledge in machine learning, data mining and many other fields. It provides a convenient way to describe a finite set of objects called the universe of discourse by a finite set of attributes (Pawlak 1991).

**Definition 10.1** (*Information table*, Pawlak 1991) An information table can be formulated as a tuple:

$$S = (U, At, L, \{V_a \mid a \in At\}, \{I_a \mid a \in At\}),$$

where $U$ is a finite non-empty set of objects, $At$ is a finite non-empty set of attributes, $L$ is a language defined using attributes in $At$, $V_a$ is a non-empty set of values for $a \in At$, $I_a : U \to V_a$ is an information function. Each information function $I_a$ is a total function that maps an object of $U$ to exactly one value in $V_a$.

An information table can represent all the available information and knowledge about a situation. Here the objects are only perceived, observed, or measured by using a finite number of properties. We can easily extend the information function $I_a$ to some subsets of attributes. For a subset $A \subseteq At$, the value of an object $x$ over $A$ is denoted by $I_A(x)$.

A decision table is a special case of information tables. A decision table is commonly viewed as a functional description, which maps inputs (conditions) to outputs (actions) without necessarily specifying the manner in which the mapping is to be implemented (Lew and Tamanaha 1976). The formal definition of a decision table is given as follows.

**Definition 10.2** (*Decision table*, Pawlak 1991) A decision table is an information table for which the attributes in $A$ are further classified into disjoint sets of condition attributes $C$ and decision attributes $D$, i.e., $At = C \cup D$, $C \cap D = \emptyset$.

A decision table can be seen as a special and important knowledge expression system. It shows that, when some conditions are satisfied, decisions (actions, operations, or controls) can be made. Decision attributes in a decision table can be unique or not. In the latter case, the decision table can be converted to one with a unique decision attribute. Therefore, in this chapter, we suppose that there is only one decision attribute in a decision table.

Based on the two definitions above, we introduce the following definitions.

### 10.2.2 Formulas and Rules

**Definition 10.3** (*Formulas*, Yao and Yao 2002) In the language $L$ of an information table, an atomic formula is given by $a = v$, where $a \in At$ and $v \in V_a$. If $\phi$ and $\psi$ are formulas, then so as $\neg\phi$, $\phi \wedge \psi$, and $\phi \vee \psi$.

The semantics of the language $L$ can be defined in the Tarski style through the notions of a model and satisfiability. The model is an information table $S$, which provides interpretation for the symbols and formulas of $L$.

**Definition 10.4** (*Satisfiability of formula*, Yao and Yao 2002) The satisfiability of a formula $\phi$ by an object $x$, written $x \models_S \phi$ or in short $x \models \phi$ if $S$ is defined by the following conditions:

1. $x \models a = v$ if $I_a(x) = v$,
2. $x \models \neg\phi$ if $x \not\models \phi$,

3. $x \models \phi \wedge \psi$ if $x \models \phi$ and $x \models \psi$.
4. $x \models \phi \vee \psi$ if $x \models \phi$ or $x \models \psi$.

If $\phi$ is a formula, the set

$$m_s(\phi) = \{x \in U \mid x \models \phi\}$$

is called the meaning of the formula $\phi$ in $S$. If $S$ is understood, we simply write $m(\phi)$. The meaning of a formula $\phi$ is therefore the set of all objects having the property expressed by the formula $\phi$. In other words, $\phi$ can be viewed as the description of the set of objects $m(\phi)$. Thus, a connection between the formulas of $L$ and subsets of $U$ is established.

To illustrate the idea, consider an information table given by Table 10.1 (Quinlan 1983). The following expressions are some of the formulas of the language $L$:

$$\mathbf{height} = \text{tall}, \quad \mathbf{hair} = \mathbf{dark},$$
$$\mathbf{height} = \text{tall} \wedge \mathbf{hair} = \text{dark},$$
$$\mathbf{height} = \text{tall} \vee \mathbf{hair} = \text{dark}.$$

The meanings of the formulas are given by:

$$m(\mathbf{height} = \text{tall}) = \{o_3, o_4, o_5, o_6, o_7\},$$
$$m(\mathbf{hair} = \text{dark}) = \{o_4, o_5, o_7\},$$
$$m(\mathbf{height} = \text{tall} \wedge \mathbf{hair} = \text{dark}) = \{o_4, o_5, o_7\},$$
$$m(\mathbf{height} = \text{tall} \vee \mathbf{hair} = \text{dark}) = \{o_3, o_4, o_5, o_6, o_7\}.$$

From Definition 10.1, we know that an information table records the attribute values of a set of objects and can be an object database. The aim of knowledge acquisition is to discover useful and regular knowledge from the object databases. Usually, the knowledge is expressed by rules which can be formulated as follows (Pawlak 1991; Yao and Yao 2002).

**Table 10.1** An information table

| Object | Height | Hair | Eyes | Class |
|--------|--------|-------|-------|-------|
| $o_1$ | Short | Blond | Blue | + |
| $o_2$ | Short | Blond | Brown | − |
| $o_3$ | Tall | Dark | Blue | + |
| $o_4$ | Tall | Dark | Blue | − |
| $o_5$ | Tall | Dark | Blue | − |
| $o_6$ | Tall | Blond | Blue | + |
| $o_7$ | Tall | Dark | Brown | − |
| $o_8$ | Short | Blond | Brown | − |

**Definition 10.5** (*Rules*) Let $S = (U, At, L, \{V_a \,|\, a \in At\}, \{I_a \,|\, a \in At\})$ be an information table, then a rule $r$ is a formula with the form

$$\phi \Rightarrow \psi,$$

where $\phi$ and $\psi$ are formulas of information table $S$ for any $x \in U$,

$$x \vDash \phi \Rightarrow \psi \text{ iff } x \vDash \neg \phi \vee \psi.$$

**Definition 10.6** (*Decision Rules*) Let $S = (U, C \cup D, L, \{V_a \,|\, a \in At\}, \{I_a \,|\, a \in C \cup D\})$ be a decision table, where $C$ is the set of condition attributes and $D$ is the set of decision attributes. A decision rule $dr$ is a rule with the form $\phi \Rightarrow \psi$, where $\phi, \psi$ are both conjunctions of atomic formulas, for any atomic formula $c = v$ in $\phi$, $c \in C$, and for any atomic formula $d = v$ in $\psi$, $d \in D$.

It is evident that each object in a decision table can be expressed by a decision rule.

**Definition 10.7** An object $x$ is said to be consistent with a decision rule $dr : \phi \Rightarrow \psi$, iff $x \vDash \phi$ and $x \vDash \psi$; $x$ is said to be conflict with $dr$, iff $x \vDash \phi$ and $x \vDash \neg \phi$.

Based on these definitions, we introduce decision rule set functions.

## 10.2.3 Decision Rule Set Function

To present a bi-level decision model based on obtaining a decision rule set, we first need to define decision rule set functions. A decision rule set function can be defined as a mapping from $n$ objects to a decision. Given a decision table $S = (U, At, L, \{V_a \,|\, a \in At\}, \{I_a \,|\, a \in At\})$, where $At = C \cup D$ and $D = \{d\}$. Suppose $x$ and $y$ are two variables, where $x \in X$ and $X = V_{a1} \times \cdots \times V_{am}$, $y \in Y$ and $Y = V_d$. $V_{ai}$ is the set of attribute $a_i$'s values, $a_i \in C$, $i = 1$ to $m$, $m$ is the number of condition attributes. $RS$ is a decision rule set generated from $S$.

**Definition 10.8** (*Decision rule set function*) A decision rule set function $rs$ from $X$ to $Y$ is a subset of the Cartesian product $X \times Y$, such that for each $x$ in $X$, there is a unique $y$ in $Y$ generated with $RS$ such that the ordered pair $(x, y)$ is in $rs$, where $RS$ is called a decision rule set, $x$ is called a condition variable, $y$ is called a decision variable, $X$ is the definitional domain, and $Y$ is the value domain.

Calculating the value of a decision rule set function is to make decisions for objects with a decision rule set. To present the method of calculating the value of a decision rule set function, we introduce a definition below about matching objects to decision rules.

**Definition 10.9** (*Matching an object to a decision rule*) An object $o$ is said to match a decision rule $\phi \Rightarrow \psi$, if $o \vDash \phi$. Given a decision rule set $RS$, all decision rules in $RS$ that are matched by object $o$ are denoted as $MR^o_{RS}$.

With the definition, a brief method of calculating the result of a decision rule set function is described as follows:

Step 1:  Calculate $MR_{RS}^o$;
Step 2:  Select a decision rule $dr$ from $MR_{RS}^o$, where $dr : \wedge\{(a, v_a)\} \Rightarrow (d, v_d)$;
Step 3:  Set the decision value of object $o$ to be $v_d$, i.e. $rs(o) = v_d$.

In Step 2, how to select a decision rule from $MR_{RS}^o$ is the key task of the process. For example, there is a decision rule set $RS$:

1. $(a, 1) \wedge (b, 2) \Rightarrow (d, 2)$,
2. $(a, 2) \wedge (b, 3) \Rightarrow (d, 1)$,
3. $(b, 4) \Rightarrow (d, 2)$,
4. $(b, 3) \wedge (c, 2) \Rightarrow (d, 3)$,

and an undecided object:

$$o = (a, 2) \wedge (b, 3) \wedge (c, 2).$$

With Step 1, $R_{RS}^o = \{(a, 2) \wedge (b, 3) \Rightarrow (d, 1); (b, 3) \wedge (c, 2) \Rightarrow (d, 3)\}$.

With Step 2, if the final rule is selected as $(a, 2) \wedge (b, 3) \Rightarrow (d, 1)$, then with Step 3, $rs(o) = 1$; if the final rule is selected as $(b, 3) \wedge (c, 2) \Rightarrow (d, 3)$, then with Step 3, $rs(o) = 3$.

From the above example, we know that there may be more than one rule in $MR_{RS}^o$. In this case, when the decision values of these rules are different, the result can be controlled according to above method, which is called uncertainty of a decision rule set function. The method of selecting the final rule from $MR_{RS}^o$ is thus very important, and is called the uncertainty solution method. In this chapter, we use the AID-based rule tree to deal with the problem, and more details about the method will be discussed in Sects. 10.4 and 10.5.

### 10.2.4 Rule Trees

A rule tree is a compact and efficient structure for expressing a rule set. We first introduce the definition of rule trees (Zheng and Wang 2004) as follows.

**Definition 10.10** (*Rule tree*)

1. A rule tree is composed of one root node, multiple leaf nodes and middle nodes;
2. The root node represents the whole rule set;
3. Each path from the root node to a leaf node represents a rule;
4. Each middle node represents an attribute testing. Each possible value of an attribute in a rule set is represented by a branch. Each branch generates a new child node. If an attribute is reduced in some rules, then a special branch is needed to represent it and the value of the attribute in this rule is assumed to be "*", which is different from any possible values of the attribute.

**Fig. 10.1** An example of a
rule tree

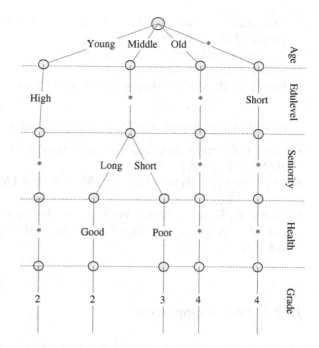

When two nodes are connected with a branch, we call the upper node a start
node and the lower node an end node.

Figure 10.1 gives an example of a rule tree, where "Age", "Educational level
(Edulevel)", "Seniority", and "Health" are the names of its conditional attributes,
and "Grade" is the name of its decision attribute. The values of these attributes are
noted beside branches.

We define the number of nodes between a branch and the root node as the level
of the branch (including the root node) in the path. For each rule tree, we have two
assumptions as follows:

**Assumption 10.1** The branches at the same level represent the possible values of
the same attribute.

Here, an attribute is expressed by the level of a rule tree.

**Assumption 10.2** If a rule tree expresses a decision rule set, the branches at the
bottom level represent the possible values of the decision attribute.

Based on Definition 10.10 and the two assumptions, we can improve the rule
tree structure with two constraints.

**Definition 10.11** (*Attribute importance degree (AID) based rule tree*) An AID-
based rule tree is a rule tree which satisfies the following two additional conditions:

1. The conditional attribute expressed at the upper level is more important than that expressed at any lower level;
2. Among the branches with the same start node, the value represented by the left branch is more important (or better) than the value represented by any right branch. Each possible value is more important (or better) than the value "*".

In the rule tree illustrated in Fig. 10.1, if we suppose

- $ID(a)$ is the importance degree of attribute $a$, and $ID(\text{Age}) > ID(\text{Edulevel}) > ID(\text{Seniority}) > ID(\text{Health})$;
- (Age, Young) is better than (Age, Middle), and (Age, Middle) is better than (Age, Old);
- (Seniority, Long) is better than (Seniority, Short), and (Health, Good) is better than (Health, Poor), then the rule tree illustrated by Fig. 10.1 is an AID-based rule tree.

### 10.2.5 Rules Comparison

There are many reasons for uncertainty in a decision rule set function. First, the causality between two events cannot be measured easily, because there are many factors involved in the events and many different relations among these factors. When generating a decision rule set, there are some strict constraints and some less-important elements that are ignored in uncertainty analysis. Second, some empirical knowledge, especially perceptual knowledge, cannot be expressed precisely and accurately. Third, the decision rule language can induce uncertainty issues. In addition, knowledge stored in a rule base or database for a decision problem is often finite and incomplete, which is also a reason for uncertainty. More obviously, different learned objects, incomplete learning algorithms and some learning processes can cause uncertainty in a decision rule set function as well.

The uncertainty can be eliminated through a process of selection. We can select a rule correctly only when related information is known. In other words, we are said to be informed only when we can select rules accurately and definitely. In this chapter, we present a rule tree-based model to deal with the uncertainty. After the ordering of importance degrees and the possible values of attributes, a rule tree (Definition 10.10) is improved to become an AID-based rule tree (Definition 10.11). It can be proved that the following theorem holds.

**Theorem 10.1** *If we suppose the isomorphic trees to be the same, then there is a one-to-one correspondence between a rule set and an AID-based rule tree.*

From the definitions of rule-set and AID-based rule trees, it is clear that the theorem holds.

Compared to a decision rule set, an AID-based rule tree has the following advantages:

1. An AID-based rule tree is a more concise structure, especially when the scale of a rule set is huge;
2. An AID-based rule tree is a more structured model which provides a number of useful properties such as confliction and repetition in an original decision rule set;
3. It can speed up the searching and matching process on the decision rules;
4. The rules in an AID-based rule tree are ordered, which provides a way to solve uncertainty problems in decision rule set functions.

**Definition 10.12** (*Comparison of rules*) Suppose the condition attributes are ordered by their importance degrees as $a_1, \ldots, a_p$. Rule $dr_1: \wedge \{(a_i, v_{a1i})\} \Rightarrow (d_1, v_{d1})$ is said to be better than rule $dr_2: \wedge \{(a_i, v_{a2i})\} \Rightarrow (d_2, v_{d2})$, if $v_{a1k}$ is better than $v_{a2k}$ or the value of $a_k$ is deleted from rule $dr_2$, where $k \in \{1, \ldots, p\}$, and for each $j < k$, $v_{a1j} = v_{a2j}$.

For example, we have two rules:
$dr_1$: (Age, Middle) $\wedge$ (Working Seniority, Long) $\Rightarrow$ 2,
$dr_2$: (Age, Middle) $\wedge$ (Working Seniority, Short) $\Rightarrow$ 3, and the value "Long" of the attribute "Working Seniority" is better than the value "Short". With Definition 10.12, we know $dr_1$ is better than $dr_2$.

**Theorem 10.2** *In an AID-based rule tree, the rule expressed by the left branch is better than the rule expressed by the right branch.*

This is evident from Definitions 10.11 and 10.12.

**Theorem 10.3** *After transformation to an AID-based rule tree, the rules in a rule set are totally in order, that is, every two rules can be compared.*

It is evident that the theorem holds from Definition 10.11 and Theorem 10.2.

*Example 10.2* We can order the rules expressed by the rule tree shown in Fig. 10.1 as follows:

1. (Age, Young) $\wedge$ (Edulevel, High) $\Rightarrow$ 2,
2. (Age, Middle) $\wedge$ (Working Seniority, Long) $\Rightarrow$ 2,
3. (Age, Middle) $\wedge$ (Working Seniority, Short) $\Rightarrow$ 3,
4. (Age, Old) $\Rightarrow$ 4,
5. (Edulevel, Short) $\Rightarrow$ 4,

where rule $i$ is better than rule $i + 1$, $i = 1, 2, 3, 4$.

## 10.3  Rule-Set-Based Bi-level Decision Model

This section will present a RSBLD model, which uses a rule set rather than mathematic functions to express a bi-level decision problem. We will first discuss the representation of objectives and constraints in a RSBLD model, and then present the formulation of the model. Lastly, we will describe a modeling approach for solving RSBLD problems.

### 10.3.1  Objectives

As we have discussed before, a decision table can be used to lay out all possible situations in tabular form where a decision may be encountered, and to specify which action to take in each of these situations. Decision tables are commonly thought to be restricted in applicability to procedures involving sequencing of tests, nested-IFs, or CASE statements. In fact, a decision table can implement any computable function. It is observed that any Turing Machine program can be "emulated" by a decision table by letting each Turing Machine instruction of the form (input, state) + (output, tape movement, state) be represented by a decision rule (or in a decision table) where (input, state) are conditions and (output, tape movement, state) are actions. From a more practical point of view, it can also be shown that all computer program flowcharts can be emulated by decision tables (Lew and Tamanaha 1976).

In principle, therefore, after emulating all possible situations in a decision domain, all objective functions can be transformed to decision tables, called *objective decision tables*. That is, the objectives of the leader and the follower in a bi-level decision problem can be transformed into a set of decision tables, where decision variables are represented by the objects in these decision tables.

Decision rule sets are general knowledge generated from decision tables and they have stronger knowledge-expressing ability than decision tables because they overcome the following disadvantages of decision tables:

1. For complex situations, decision tables may become extremely large;
2. The objects in decision tables lack adaptability. They are hard to adapt to new situations and one object can only record a single situation.

Thus, in the proposed model, we use a decision rule set to represent the objectives of the leader and the follower in a bi-level decision problem, whereas decision tables can be viewed as special cases of a decision rule set.

## 10.3.2 Constraints

Constraints (constraint conditions) can be seen as the description of the search space in a decision problem. Here, we use a rule set to represent constraints. Similar to the discussion in Sect. 10.3.1, after emulating all possible situations in the constraint field, the constraints can be formulated to an information table. When the information tables are too big to be processed, they can be transformed to a rule set using the methods provided by Agrawal et al. (1993) and Agrawal and Srikant (1994).

A rule set can be viewed as knowledge generated from information tables, but it has stronger knowledge-expressing ability and better adaptability than information tables. An information table can be viewed as a special case of rule sets. By using rule sets, we give the following definition about constraint conditions.

**Definition 10.13** (*Constraint Condition*) Suppose $x$ is a decision variable and $RS$ is a rule set, then a constraint condition $cf(x, RS)$ is defined as

$$cf(x, RS) = \begin{cases} True, & \text{if for } \forall r \in RS,\ x \in m(r) \\ False, & \text{else.} \end{cases} \tag{10.1}$$

The meaning of the constraint condition $cf(x, RS)$ is whether variable $x$ belongs to the region constrained by $RS$.

## 10.3.3 Rule-Set-Based Bi-level Decision Model

We can describe the objectives and constraints of the leader and the follower by rule sets, called a rule-set-based bi-level decision model, as follows.

**Definition 10.14** (Rule-set-based bi-level decision model)

$$\begin{aligned} &\min_{x} \quad f_L(x, y) \\ &\text{s.t.} \quad cf(x, G_L) = True \\ &\quad\quad \min_{y} \quad f_F(x, y) \\ &\quad\quad \text{s.t.} \quad cf(y, G_F) = True \end{aligned} \tag{10.2}$$

where $x$ and $y$ are decision variables (vectors) of the leader and the follower respectively. $f_L$ and $f_F$ are the objective decision rule set functions of the leader and the follower respectively. $cf$ is the constraint condition. $F_L$ and $G_L$ are the objective decision rule set and constraint rule set respectively of the leader, and $F_L$ and $G_L$ are the objective decision rule set and constraint rule set respectively of the follower.

In this model, we suppose that the decision rule set in the objectives can cover the objects in the constraints. That is, each object in the constraints can be matched by one decision rule at least in the objective decision rule set. The assumption is not too restricted, because when the objective decision tables used to generate objective decision rule set are huge and the objects in them are uniformly distributed, the resulting decision rule set usually covers most of the objects to be decided. In other cases, where some objects in the constraint fields cannot be matched by decision rules in the objective decision rule set, additional methods should be introduced, such as similarity matching, fuzzy marching, etc. In this chapter, we will discuss the models and decision methods based on the above assumption.

## 10.4  Rule-Set-Based Bi-level Decision Modeling Approach

In the following, we propose an approach for modeling a bi-level decision problem by rule sets.

---

**Algorithm 10.1: Rule-set-based Bi-level Decision Modeling Approach**

**Input:** A bi-level decision problem with its objectives and constraints of the leader and the follower;
**Output:** A rule-set-based bi-level decision model;

[Begin]
**Step 1:** Transform the bi-level decision problem with rule-set (information tables are as special cases);
**Step 2:** Pre-process $F_L$, such as delete reduplicate rules from the rule-set sets, eliminate noise, etc.;
**Step 3:** If $F_L$ needs to be reduced, then use a reduction algorithm to reduce $F_L$;
**Step 4:** Pre-process $G_L$, such as delete reduplicate rules from the rule-set sets, eliminate noise, etc;
**Step 5:** If $G_L$ needs to be reduced, then use a reduction algorithm to reduce $G_L$;
**Step 6:** Pre-process $F_F$, such as delete reduplicate rules from the rule-set sets, eliminate noise, etc.;
**Step 7:** If $F_F$ needs to be reduced, then use a reduction algorithm to reduce $F_F$;
**Step 8:** Pre-process $G_F$, such as delete reduplicate rules from the rule-set sets, eliminate noise, etc.;
**Step 9:** If $G_F$ needs to be reduced, then use a reduction algorithm to reduce $G_F$;
[End]

---

Figure 10.2 is the flow charts of the proposed rule-set-based bi-level decision problem modeling approach.

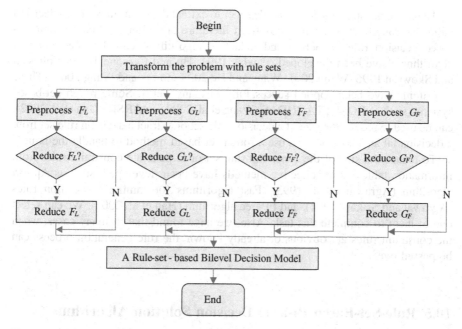

**Fig. 10.2** Flow chart of the RSBLD modeling approach for modeling rule- set-based bi-level decision problems (Algorithm 10.1)

We provide explanations for the approach. Step 1 is the key step of the modeling process. Decision makers can complete the step by laying out all possible situations that is, transforming the decision problem to information tables. When the decision makers have the general knowledge (rules) for the problem, they can directly transform the problem to related rule sets. Therefore, the key in the step is decision makers' knowledge about the problem and their ability to write the knowledge into rules.

In Steps 2, 4 and 6, each of the four rule sets is pre-processed. As incompleteness, noise and inconsistency are common characteristics of a huge real dataset we need to use related techniques to eliminate these problems before using the data to make a bi-level decision (Han and Kamber 2001).

In Steps 5, 7, and 9 of this algorithm, related rule sets are reduced by applying a reduction algorithm. This is for at least one of the following three reasons:

1. When modeling a real-world bi-level decision problem, the rule set in the model is often on a large scale, which is not convenient to process, and cannot be easily interpreted and understood by decision makers.
2. The rules in the rule set lack adaptability. In this case, the rule set cannot adapt to new situations well, so it is unable or has poor ability to support decision making.
3. The rule sets in the model are original data sets, so the patterns in the data sets need to be extracted, and the results are more general rules.

To reduce the size of a decision rule set or to extract decision rules from a decision table, the rough set-based methods from literature are efficient. Many rough set-based decision rule extraction and reduction algorithms, called value reduction algorithms, have been developed (Pawlak 1991; Hu and Cercone 1995; Mollestad and Skowron 1996; Wang 2001; Wang and He 2003; Zheng and Wang 2004). These algorithms have been applied successfully in many fields. Some rough set-based systems, such as ROSETTA, RIDAS (Wang et al. 2002), and RSES (Jan et al. 2002), can be used to reduce the size of a decision rule set or extract a decision rule set from a decision table. Therefore, we use a rough set-based method to handle the issue.

To reduce the size of the constraint rule set or to generate constraint rules from information tables, some effective methods have been developed, such as Apriori algorithm (Agrawal et al. 1993), Fast algorithms for mining association rules (Agrawal and Srikant 1994), and FP tree algorithm (Han et al. 2000). We can select one of them to complete this task. Also, we have found that in many cases, when the constraint rules are obvious or already known, the rule generation process can be passed over.

## 10.5  Rule-Set-Based Bi-level Decision Solution Algorithms

In this section, we present two algorithms to solve rule-set-based bi-level decision models, in which a key technique is that an AID-based rule tree is used to express a rule set.

### 10.5.1  Concepts and Properties

Based on Bard (1998), we have the following definition.

**Definition 10.15**

(a)  Constraint region of a bi-level decision problem:

$$S = \{(x, y) \mid cf(x, G_L) = True, cf(y, G_F) = True\} \qquad (10.3)$$

(b)  Feasible set for the follower for each fixed $x$:

$$S(x) = \{y \mid (x, y) \in S\} \qquad (10.4)$$

(c)  Projection of S onto the leader's decision space:

$$S(x) = \{x \mid \exists y, (x, y) \in S\} \qquad (10.5)$$

(d)  Follower's rational reaction set for $x \in S(X)$:

$$P(x) = \left\{ y \mid y \in \arg\min_{y'} \left\{ f_F\left(x, y'\right) \mid y' \in S(x) \right\} \right\} \qquad (10.6)$$

(e)  Inducible region:

$$IR = \{(x, y) \mid (x, y) \in S, y \in P(x)\} \qquad (10.7)$$

To ensure that (10.2) is well posed it is common to assume that $S$ is nonempty and compact, and that for all decisions taken by the leader, the follower has some room to respond, i.e. $P(x) \neq \varnothing$. The rational reaction set $P(x)$ defines the response while the inducible region $IR$ represents the set over which the leader may optimize its objective. Thus in terms of the above notation, the bi-level decision problem can be written as

$$\min\{f_L(x, y) \mid (x, y) \in IR\} \qquad (10.8)$$

From the features of the bi-level decision, it is clear that once the leader selects an $x$, the first term in the follower's objective function becomes a constant and can be removed from the problem. In this case, we replace $f_F(x, y)$ with $f_F(y)$.

To begin, let $\left(x_{[1]}, y_{[1]}\right), \left(x_{[2]}, y_{[2]}\right), \ldots, \left(x_{[N]}, y_{[N]}\right)$ denote the $N$ ordered feasible solutions to the rule-set-based one level one objective problem

$$\min_{x}\{f_L(x, y) \mid (x, y) \in S\} \qquad (10.9)$$

such that

$$f_L\left(x_{[i]}, y_{[i]}\right) \leq f_L\left(x_{[i+1]}, y_{[i+1]}\right) \qquad (10.10)$$

and $i = 1, \ldots, N - 1$. Thus, to solve (10.9) is equivalent to finding the global optimum $\left(x_{[k]}, y_{[k]}\right)$.

## 10.5.2  Rule-Based-Based Solution Algorithm

We present an algorithm for solving a rule-set-based bi-level decision problem. The main picture of the algorithm is to repeatedly solve two rule-set-based one-level decision problems. One is for the leader in all of the variables $x$ and a subset of the variables $y$ associated with an optimal basis to the follower's problem, and the other is for the follower with all the variables $x$ fixed. The leader first makes his decision, and the decision will influence the objective rule set function and the constraint rule set function of the follower. In a systematic way, the algorithm explores the optimal solution of the follower's problem for $x$ fixed and then returns to the leader's problem with the corresponding variables $y$. If the variables $y$ are not an optimal solution of the leader's decision problem, then the leader modifies his decision and engages in the procedures repeatedly.

---

**Algorithm 10.2: Rule-set-based Solution Algorithm**

**Input:** The objective decision rule set $F_L = \{dr_{L1}, \ldots, dr_{Lp}\}$ and the constraint rule set $G_L$ of the leader, the objective decision rule set $F_F = \{dr_{F1}, \ldots, dr_{Fq}\}$ and the constraint rule set $G_F$ of the follower;
**Output:** The final decision of the leader and the follower $(x_{[i]}, y_{[i]})$.

---

[Begin]
**Step 1:** Construct the objective rule tree $FT_L$ of the leader by $F_L$;
    **Step 1.1:** Arrange the condition attributes in ascending order according to the importance degrees. Let the attributes be the discernible attributes of levels from the top to the bottom of the tree;
    **Step 1.2:** Initialize $FT_L$ to an empty AID-based rule tree;
    **Step 1.3:** For each rule R of the decision rule set $F_L$;
        **Step 1.3.1:** Let $CN \leftarrow$ root node of the rule tree $FT_L$;
        **Step 1.3.2:** For $i = 1$ to m; /*m is the number of levels in the rule tree */
    If there is a branch of CN representing the ith discernible attribute value of rule R, then
        let $CN \leftarrow$ node $I$;            /*node $I$ is the node generated by the branch*/
        else {Create a branch of CN to represent the ith distensibility attribute value;
    According to the value order of the ith discernible attribute, put the created branch to the right place;
        Let CN←node $J$. /*node $J$ is the end node of the branch*/
**Step 2:** Construct the objective rule tree $FT_F$ of the follower by $F_F$;
    The detail of Step 2 is similar to that of Step 1. It is necessary to replace $FT_L$ with $FT_F$ and replace $F_L$ with $F_F$ in the sub-steps of Step 1.
**Step 3:** Solve problem (10.9) to obtain its optimal solution $(x_{[i]}, y_{[i]})$, and initialize $i = 1$;
    **Step 3.1:** Initialize $FT_L'$ to an empty AID-based rule tree, where $FT_L'$ represents the objective rule tree of the follower pruned by the constraint rule set;
    **Step 3.2:** Use the constraint rule tree sets $G_L$ and $G_F$ to prune $FT_L$;
    For each rule dr in $G_L$ and $G_F$,
    Add the rules in $FT_L$ that are consistent with dr to $FT_L'$;
    Delete the rules in $FT_L$ and $FT_L'$ that are conflict with $dr$;

**Step 3.3:** Search for the rules with the minimal decision value in $FT'_L$ and the result rule set is $RS = \{dr_1, ..., dr_m\}$, where $dr_1$ to $dr_m$ are the rules from left to right in $FT'_L$ ;

**Step 3.4:** Let $dr$ be the first rule in $RS$;

**Step 3.5:** $RS = RS \setminus \{dr\}$;
$OS = \{o|o$ is the objects consistent with dr and the constraint rule sets$\}$;

**Step 3.6:** Order the objects in $OS$ so that the ith object in $OS$ is better than the $(i+1)$th object in $OS$ according to Definition 10.12;

**Step 3.7:** The solution of the problem (10.9) is the first object (Definition 10.12) in $OS$.

**Step 4:** Let $FT'_F$ be the objective rule tree of the follower pruned by the constraint RULE SET of the leader and the follower, and $W = \psi$;

**Step 5:** Prune the rules from $FT'_F$, which are not consistent with rule
$$True \Rightarrow x_{[i]}$$
and suppose the result is a rule tree $FT''_F$;

**Step 6:** Solve the follower's rule-set-based decision problem:

$$\min_y\{f_F(x_{[i]}, y): y \in P(x_{[i]})\} \tag{10.11}$$

and get the optimal solution $(x_{[i]}, y)$;

**Step 6.1:** Search for the rules with the minimal decision value in $FT''_F$ and the result rule set is $RS' = \{dr'_1, ..., dr'_m\}$;

**Step 6.2:** Let $dr'$ be the first rule in $RS'$;

**Step 6.3:** $RS' = RS' \setminus \{dr'\}$; $OS' = \{o' \mid o'$ is the objects consistent with $dr'$ and the constraint rule set$\}$;

**Step 6.4:** Order the objects in $OS'$ so that the ith object in $OS'$ is better than the $(l+1)$th object in $OS'$ according to Definition 10.12;

**Step 6.5:** The solution of the follower's problem is the first object (Definition 10.12) in $OS'$.

**Step 7:** If $y = y_{[i]}$, then {
The optimal solution set is obtained, which is $(x_{[i]}, y_{[i]})$; End;}
Else {
Go to Step 8;
};

**Step 8:** Select another solution for the follower;

**Step 8.1:** $OS' = OS' \setminus \{(x_{[i]}, y)\}$;

**Step 8.2:** If OS' is null, then {
If $RS'$ is null, then {

Go to Step 9;}

else {

Let $dr'$ be the first rule in $RS$;

$$RS' = RS' \setminus \{dr'\};$$

$OS' = \{o' \mid o';$ is the objects consistent with $dr'$ and the constraint rule-set$\}$;

**Step 8.3:** Order the objects in $OS'$ and the next solution of the follower's problem is the first object $(x_{[i]}, y)$ (Definition 10.12) in $OS'$.

**Step 8.4:** Go to Step 7.

**Step 9:** Select another solution for the leader;

**Step 9.1:** $OS = OS \setminus \{(x_{[i]}, y_{[i]})\}$;

**Step 9.2:** $W = W \cup \{(x_{[i]}, y_{[i]})\}$

**Step 9.3:** If $OS$ is null, then {

If RS is null, then {

There is no optimal solution for the problem;

End;}

Else {

Let dr be the first rule in RS;

$RS = RS \setminus \{dr\}$;

$OS = \{o \mid o$ is the objects consistent with dr and the constraint rule set;}

Let $(x_{[i+1]}, y_{[i+1]})$ be the first object in $OS$;

$i = i + 1$;

Go to Step 5;}.

**[End]**

---

Figure 10.3 shows the flow charts of Algorithm 10.2. We will use an example to illustrate the algorithm.


### *10.5.3 Transformation-Based Solution Algorithm*

This section presents an alternative algorithm for solving a rule-set-based bi-level decision problem. The main idea of this algorithm is to transform two level rule-sets to one level first, then to solve this one-level decision. Before developing the algorithm, a transformation theorem will be proposed to show the solution equivalence for the two problems before and after transformation.

First, we give a definition below.

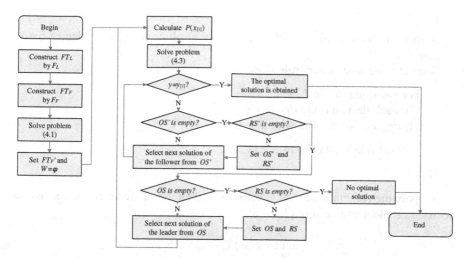

**Fig. 10.3** Flow charts of the approach for solving rule-set based bi-level decision model

**Definition 10.16** (*Combination rule of two decision rules*) Suppose $dr_1$: $\phi_1 \Rightarrow$ $(d_1, v_1)$ and $dr_2$: $\phi_2 \Rightarrow (d_2, v_2)$ are two decision rules and they are not in conflict, then their combination rule is denoted as $dr_1 \cap dr_2$ with the form

$$\phi_1 \wedge \phi_2 \Rightarrow (d, (v_1, v_2)),$$

where $d_1, d_2$, and $d$ are the decision attributes of $dr_1, dr_2$ and $dr$ respectively, $v_1, v_2$ and $(v_1, v_2)$ are the decision values of $dr_1$ and $dr_2$ and $dr$ respectively.

Here, $v_1, v_2$ are called the leader decision and the follower decision of $dr$ respectively. For example, suppose

$$dr_1: (\text{Age, Young}) \Rightarrow 2,$$
$$dr_2: (\text{Working Seniority, Long}) \Rightarrow 2,$$

then the combination of the two rules is

$$dddr: (\text{Age, Young}) \wedge (\text{Working Seniority, Long}) \Rightarrow (d, (2, 2)).$$

Suppose the objective rule sets are expressed by AID-based rule trees, then the transformation process can be presented as follows.

[Begin]
**Step 1**: (Initialization): Let $CT$ be an empty attribute importance degree-based rule tree;
**Step 2**: (Construct a new rule tree):

> For each rule $dr_L$ in $FT_L$
> For each decision rule $dr_F$ in $FT_F$
> {If $dr_L$ are not conflict with $dr_F$, then
>
> > Add rule $dr_L \cap dr_F$ to $CT$;}

[End]

Suppose the combined rule set is noted as $F$, then the single level rule-sets based decision problem can be formulated as:

$$
\begin{aligned}
\min_{x,y} \quad & f(x,y) \\
\text{s.t.} \quad & cf(x, G_L) = \text{True}, \\
& cf(y, G_F) = \text{True},
\end{aligned}
\tag{10.12}
$$

where $x$ and $y$ are variables of the leader and the follower respectively; $f$ is the objective decision rule-set function; $cf$ is the constraint function; $F$, $G_L$, $G_F$ are the objective decision rule set, leader's constraint rule set and follower's constraint rule set respectively.

With the following theorem, we can prove the solution equivalence of the original problems and the transformed problem.

**Theorem 10.4** *The RSBLD model presented in* Eq. 10.2 *has an optimal solution $(x, y)$, iff $(x, y)$ is an optimal solution of its corresponding single level decision model presented in* Eq. 10.12.

*Proof* Suppose $x$ and $y$ are variables of the leader and the follower respectively, $f_L$ and $f_F$ are the objective rule-set functions of the leader and the follower respectively in Eq. 10.2, and $f$ is the objective rule set function in Eq. 10.12. $F_L$ and $F_F$ are the objective rule sets of the leader and the follower in the RSBLD model, and $F$ is the objective rule set in the single level decision model.

If the optimal solution of the RSBLD model presented in Eq. 10.2 is $(x, y)$, and

$$
f_L(x,y) = v_L \quad \text{and} \quad f_F(x,y) = v_F.
$$

Suppose the final matching rules (Sect. 10.2) of $(x, y)$ in rule sets $F_L$ and $F_F$ are $dr_L$ and $dr_F$ respectively. Then, from the process of transformation, we know the rule $dr_L \cap dr_F$ belongs to the combined rule set $F$.

Because $(x, y)$ is the optimal solution of the RSBLD model, $dr_L$ and $dr_F$ must be the best rules, having the minimal decision values in $F_L$ and $F_F$ respectively. Thus, $dr = dr_L \cap dr_F$ must be the best rules matched by $(x, y)$ in $F$. Besides, because $(x, y)$ is the best object satisfying both $dr_L$ and $dr_F$, $(x, y)$ is the best object satisfying $dr$. Thus, $(x, y)$ is the optimal solution of the single level decision model presented in Eq. 10.12.

The sufficient condition of the theorem is proved.

If the optimal solution of the single level decision model presented in Eq. 10.12 is $(x, y)$, and

$$f(x, y) = (d, (v_{Ld}, v_{Fd})).$$

Suppose the final matching rule of $(x, y)$ in rule set $F$ is $dr$, then from the process of transformation, there must be two decision rules $dr_L$ in $F_L$ and $dr_F$ in $F_F$ that $dr = dr_L \cap dr_F$. If there is more than one rule pair $dr_L$ and $dr_F$ satisfying that $dr = dr_L \cap dr_F$, then select the best one among them.

Because $(x, y)$ is the optimal solution of the single level decision model, $dr$ must be the best rules having the minimal decision value in $F$. Thus, $dr_L$ and $dr_F$ must be the best rules matched by $(x, y)$ in $F_L$ and $F_F$ respectively. Besides, because $(x, y)$ is the best object satisfying $dr$, $(x, y)$ is the best object satisfying both $dr_L$ and $dr_F$ both, so $(x, y)$ is the optimal solution of the bi-level decision model.

Thus, the necessary condition of the theorem is proved.                                    □

From Theorem 10.1, it is known that the optimal solution of the RSBLD problem presented in Eq. 10.2 and its transformation problem shown in Eq. 10.12 are equivalent. Therefore, any RSBLD problem can be transformed into a single level decision problem. Furthermore, a solution is achieved by solving the single level decision problem. Note that although the original bi-level decision problem and the transformed one level problem have the same optimal solution, they are not equivalent. Some information of the leader and the follower unrelated to the acquiring of the optimal solution is reduced during the transformation process, because the aim of transformation is only to generate a model which can be easily solved but has the same optimal solution as the original bi-level decision model.

Based on the transformation theorem proposed, we will develop a transformation-based solution algorithm for RSBLD problems in the following sections.

**Algorithm 10.3: Rule-set-based Transformation Solution Algorithm**

**Input:** The objective decision rule set $F_L = \{dr_{L1}, ..., dr_{Lp}\}$ and the constraint rule set $G_L$ of the leader, the objective decision rule set $F_F = \{dr_{F1}, ..., dr_{Fq}\}$ and the constraint rule set $G_F$ of the follower;
**Output:** An optimal solution of the RSBLD problem ($ob$).

[Begin]
**Step 1:** Construct the objective rule tree $FT_L$ of the leader by $F_L$;
   **Step 1.1:** Arrange the condition attributes in ascending order according to the importance degrees. Let the attributes be the discernible attributes of levels from the top to the bottom of the tree;
   **Step 1.2:** Initialize $FT_L$ to an empty AID-based rule tree;
   **Step 1.3:** For each rule $R$ of the decision rule set $F_L$ {
      **Step 1.3.1:** Let $CN$ ←root node of the rule tree $FT_L$;
      **Step 1.3.2:** For $i = 1$ to $m$ /*$m$ is the number of levels in the rule tree*/
   If there is a branch of $CN$ representing the ith discernible attribute value of rule $R$, then
      let $CN$ ←node $I$;                  /*node $I$ is the node generated by the branch*/
      else {Create a branch of CN to represent the $i$th discernible attribute value;
      According to the value order of the $i$th discernible attribute, put the created branch to the right place;
      Let $CN$←node $J$. /*node $J$ is the end node of the branch*/
**Step 2:** Construct the objective rule tree $FT_F$ of the follower by $F_F$;
   The detail of Step 2 is similar to that of Step 1. It is necessary to replace $FT_L$ with $FT_F$ and to replace $F_L$ with $F_F$ in the sub-steps of Step 1.
**Step 3:** Transform the bi-level decision problem to a single level one, and the resultant objective rule tree is $CT$;
**Step 4:** Use the constraint rule set of both the leader and the follower to prune $CT$;
   **Step 4.1:** Generate an empty new AID-based rule tree $CT'$;
   **Step 4.2:** For each rule $dr$ in $G_L$ and $G_F$;
   Add the rules in $CT$ to $CT'$ that are consistent with $dr$ to $FT_L'$;
   Delete the rules in $CT$ and $CT'$ that conflict with $dr$;
   **Step 4.3:** Let $CT = CT'$;
**Step 5:** Search for the leftmost rule $dr$ in $CT$ whose leader decision and follower decision are both minimal;
**Step 6:** If $dr$ does not exist, then there is no optimal solution for the problem and go to end;
**Step 7:** $OB = \{ob | ob \models dr$ and for $r, G_L, G_F, ob \models r\}$;
**Step 8:** If there is more than one object in $OB$, then according to Definition 10.12, select the best or most important object $ob$; else $ob$=the object in $OB$;
**Step 9:** $ob$ is the optimal solution of the RSBLD problem;
[End]

**Fig. 10.4**  Flow chart of Algorithm 10.3

The flow chart of the algorithm is illustrated in Fig. 10.4. By this algorithm, we can obtain a solution for a rule-set-based bi-level decision problem through solving the transformed single level problem.

## 10.6  A Case Study

In this section, we use the RSBLD approach and algorithms introduced to handle the recruit decision problem given in Example 10.1.

### 10.6.1  Problem Modeling

Now, we use Algorithm 10.1 to transform the recruit problem to a RSBLD model:

Step 1 transforms the recruit problem with rule sets (information tables as special cases).

Tables 10.2 and 10.3 represent the objective rule set of the leader, and the follower respectively.

Equations 10.12 and 10.13 give the constraint rule sets of the leader and the follower respectively.

The constraint rule set of the leader is

$$G_L = \{True \Rightarrow (\text{Age, Young}) \vee (\text{Age, Middle})\}. \tag{10.12}$$

The constraint rule set of the follower is

$$G_F = \{True \Rightarrow (\text{Seniority, Long}) \vee (\text{Seniority, Middle})\}. \tag{10.13}$$

Because the scale of the data is very small, the pre-process steps (Steps 2, 4, 6 and 8) are passed over. Besides, the constraint rule set of the leader and the follower are brief enough, so the reduction steps of $G_L$ and $G_F$ (Step 5 and Step 9) can be ignored.

Step 3 and Step 7 reduce the objective rule set of the leader and the follower.

After reducing the decision tables based on the rough set theory given in Sect. 10.2.5, we can obtain the reduced objective rule set of the leader and the follower as shown in Eqs. 10.14 and 10.15. Here, we use the decision matrices-based value reduction algorithm (Ziarko et al. 1996) in the RIDAS system (Wang et al. 2002).

**Table 10.2** Objective rule set of the leader

| Age | Edulevel | Seniority | Health | Grade |
|-----|----------|-----------|--------|-------|
| Young | High | Middle | Good | 2 |
| Middle | High | Long | Middle | 2 |
| Young | Short | Short | Poor | 4 |
| Young | Middle | Middle | Middle | 2 |
| Middle | Middle | Short | Middle | 3 |
| Middle | Middle | Long | Middle | 2 |
| Old | High | Long | Middle | 3 |
| Young | Short | Middle | Poor | 2 |
| Middle | Short | Short | Middle | 4 |
| Old | Short | Middle | Poor | 4 |
| Middle | Short | Long | Good | 3 |
| Middle | Short | Long | Middle | 2 |
| Old | High | Middle | Poor | 3 |
| Old | High | Long | Good | 2 |
| Old | Short | Long | Good | 4 |
| Young | High | Long | Good | 4 |
| Young | Short | Long | Middle | 3 |

**Table 10.3** Objective rule set of the follower

| Age | Edulevel | Seniority | Health | Grade |
|-----|----------|-----------|--------|-------|
| Young | High | Long | Good | 2 |
| Old | Short | Short | Good | 4 |
| Young | High | Short | Good | 2 |
| Old | High | Long | Middle | 3 |
| Young | Short | Long | Middle | 4 |
| Middle | High | Middle | Poor | 3 |
| Middle | Short | Short | Poor | 4 |
| Old | Short | Short | Poor | 4 |
| Old | High | Long | Good | 2 |
| Young | Short | Long | Good | 2 |
| Young | Short | Middle | Middle | 3 |
| Middle | Short | Middle | Good | 3 |
| Old | High | Long | Good | 2 |
| Middle | High | Long | Good | 2 |
| Middle | High | Short | Poor | 4 |

The reduced objective rule set of the leader:

$F_L = \{$

  (Age, Young) $\wedge$ (Seniority, Middle) $\Rightarrow$ (Grade, 2),

  (Age, Middle) $\wedge$ (Edulevel, High) $\Rightarrow$ (Grade, 2),

  (Edulevel, Short) $\wedge$ (Seniority, Short) $\Rightarrow$ (Grade, 4),

  (Edulevel, Middle) $\wedge$ (Seniority, Short) $\Rightarrow$ (Grade, 3),

  (Edulevel, Middle) $\wedge$ (Seniority, Long) $\Rightarrow$ (Grade, 2),

  (Age, Old) $\wedge$ (Health, Middle) $\Rightarrow$ (Grade, 3),

  (Age, Old) $\wedge$ (Edulevel, Short) $\Rightarrow$ (Grade, 4),          (10.14)

  (Age, Middle) $\wedge$ (Health, Good) $\Rightarrow$ (Grade, 3),

  (Age, Middle) $\wedge$ (Seniority, Long) $\wedge$ (Health, Middle) $\Rightarrow$ (Grade, 2),

  (Age, Old) $\wedge$ (Edulevel, High) $\wedge$ (Health, Good) $\Rightarrow$ (Grade, 2),

  (Edulevel, High) $\wedge$ (Health, Poor) $\Rightarrow$ (Grade, 3),

  (Age, Young) $\wedge$ (Edulevel, High) $\wedge$ (Seniority, Long) $\Rightarrow$ (Grade, 4),

  (Age, Young) $\wedge$ (Edulevel, Short) $\wedge$ (Seniority, Long) $\Rightarrow$ (Grade, 3)

  $\}$.

The reduced objective rule set of the follower:

  $F_F = \{$

    (Edulevel, High) $\wedge$ (Health, Good) $\Rightarrow$ (Grade, 2)

    (Edulevel, Short) $\wedge$ (Seniority, Short) $\Rightarrow$ (Grade, 4)

    (Age, Old) $\wedge$ (Health, Middle) $\Rightarrow$ (Grade, 3)

    (Age, Young) $\wedge$ (Seniority, Long) $\wedge$ (Health, Middle) $\Rightarrow$ (Grade, 4)

    (Seniority, Middle) $\Rightarrow$ (Grade, 3)

    (Seniority, Long) $\wedge$ (Health, Good) $\Rightarrow$ (Grade, 2)

    (Seniority, Short) $\wedge$ (Health, Poor) $\Rightarrow$ (Grade, 4)

    $\}$.

$$(10.15)$$

With the above steps, we get the RSBLD model of the decision problem, as follows:

$$\min_{x} \quad f_L(x, y)$$

$$\text{s.t.} \quad cf(x, G_L) = True,$$

$$\min_{y} \quad f_F(x, y)$$

$$\text{s.t.} \quad cf(y, G_F) = True$$

where $f_L, f_F$ are the corresponding decision rule set functions of $F_L, F_F$ respectively.

## 10.6.2 Solution

Now, we use Algorithm 10.2 to solve the bi-level decision problem given in Sect. 10.6.1. We suppose that the four condition attributes are ordered by attribute importance degrees as "age", "edulevel", "seniority", "health".

Step 1: Construct the objective rule tree $FT_L$ of the leader by $F_L$ and the result is as shown in Fig. 10.4;

Step 2: Construct the objective rule tree $FT_F$ of the follower by $F_F$ and the result is as shown in Fig. 10.5;

Step 3: Solve problem (10.9), and initialize $i = 1$.

Step 3.1: Let $FT'_L$ be the objective rule tree of the follower pruned by the constraint rule sets, and initialize $FT'_L$ to an empty AID-based rule tree (Fig. 10.6);

Step 3.2: Use the constraint rule tree $GT_L$ to prune $FT_L$ and the result is $FT'_L$ as Fig. 10.7;

Step 3.3: Search for the rules with the minimal decision value in $FT'_L$ and the result rule set is

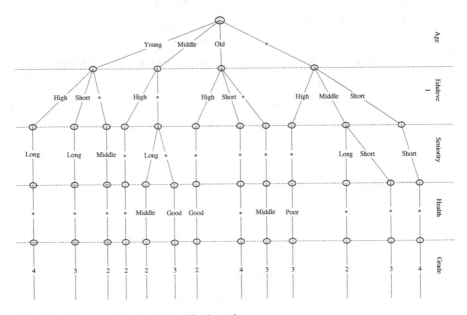

**Fig. 10.5** Rule tree of the leader's objective rule set

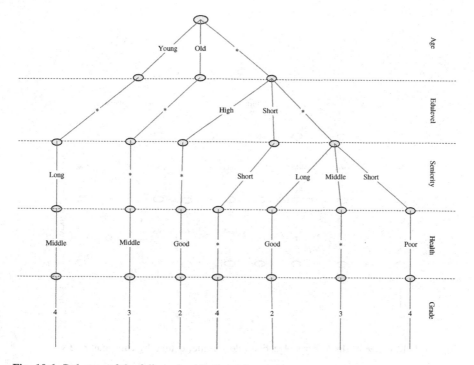

**Fig. 10.6** Rule tree of the follower's objective rule set

$$RS = \{$$
(Age, Young) ∧ (Seniority, Middle) ⇒ (Grade, 2);
(Age, Middle) ∧ (Edulevel, High) ⇒ (Grade, 2)
(Age, Middle) ∧ (Seniority, Long) ∧ (Health, Middle) ⇒ (Grade, 2)
(Edulevel, Middle) ∧ (Seniority, Long) ⇒ (Grade, 2)
$$\};$$

Step 3.4:

$$dr: \text{(Age, Young)} \wedge \text{(Seniority, Middle)} \Rightarrow \text{(Grade, 2)}$$

Steps 3.5–3.6:

$$RS = \{$$
(Age, Middle) ∧ (Edulevel, High) ⇒ (Grade, 2)
(Age, Middle) ∧ (Seniority, Long) ∧ (Health, Middle) ⇒ (Grade, 2)
(Edulevel, Middle) ∧ (Seniority, Long) ⇒ (Grade, 2)
$$\};$$

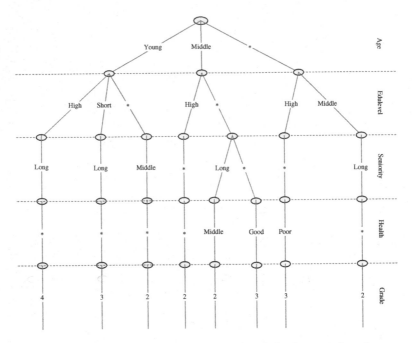

**Fig. 10.7** Rule tree of the leader's objective rule set after cutting by constraint rule set

$OS = \{$
(Age, Young) $\wedge$ (Edulevel, High) $\wedge$ (Seniority, Middle) $\wedge$ (Health, Poor),
(Age, Young) $\wedge$ (Edulevel, Middle) $\wedge$ (Seniority, Middle) $\wedge$ (Health, Good),
(Age, Young) $\wedge$ (Edulevel, Middle) $\wedge$ (Seniority, Middle) $\wedge$ (Health, Middle),
(Age, Young) $\wedge$ (Edulevel, Middle) $\wedge$ (Seniority, Middle) $\wedge$ (Health, Poor),
(Age, Young) $\wedge$ (Edulevel, Poor) $\wedge$ (Seniority, Middle) $\wedge$ (Health, Good),
(Age, Young) $\wedge$ (Edulevel, Middle) $\wedge$ (Seniority, Middle) $\wedge$ (Health, Middle),
(Age, Young) $\wedge$ (Edulevel, Middle) $\wedge$ (Seniority, Middle) $\wedge$ (Health, Poor)
$\}$

Step 3.7:  The solution of problem (4.1) is the first object in $OS$, that is

$$o = (\text{Age, Young}) \wedge (\text{Edulevel, High}) \wedge (\text{Seniority, Middle})$$
$$\wedge (\text{Health, Good})$$

Step 4:  Let $FT'_F$ be the objective rule tree of the follower pruned by the constraint rule set and $FT'_F$ is as Fig. 10.8. $W = \psi$;

Step 5:  $x_{[1]} = (\text{Age, Young}) \wedge (\text{Edulevel, High})$. Prune the rules from $FT'_F$, which are not consistent with

**Fig. 10.8** The Rule tree of
the follower's objective rule
set after cutting by constraint
rule set

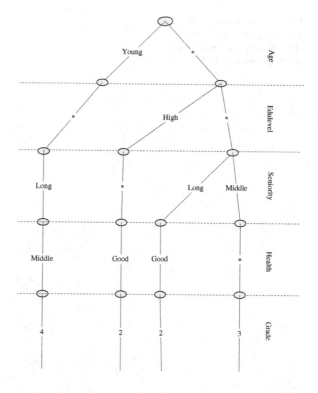

$$True \Rightarrow x_{[1]}$$

and suppose the result is the rule tree $FT''_F$ as shown in Fig. 10.9;

Step 6:  Solve the follower's rule-set-based decision problem below.

$$\min_{y}\{f_F(x_{[i]}, y): y \in P(x_{[i]})\}$$

Step 6.1:  Search for the rules with the minimal decision value in $F''_T$ and
the result rule set is

$$RS' = \{$$
$$\qquad (\text{Edulevel, High}) \wedge (\text{Health, Good})(\text{Grade, 2})$$
$$\qquad (\text{Seniority, Long}) \wedge (\text{Health, Good})(\text{Grade, 2})$$
$$\};$$

**Fig. 10.9** Rule tree of the follower's objective rule set after pruning the rules which are not consistent with $x_{[i]}$

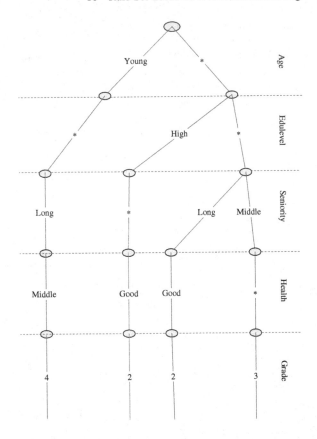

Step 6.2:

$$dr' : (\text{Edulevel}, \text{High})) \wedge (\text{Health}, \text{Good})(\text{Grade}, 2)$$

Steps 6.3–6.4:

$$RS' = \{(\text{Seniority, Long}) \wedge (\text{Health, Good})(\text{Grade}, 2)\};$$

$OS' =$
  (Age, Young) $\wedge$ (Edulevel, High) $\wedge$ (Seniority, Long) $\wedge$ (Health, Good),
  (Age, Young) $\wedge$ (Edulevel, High) $\wedge$ (Seniority, Middle) $\wedge$ (Health, Good),
  (Age, Middle) $\wedge$ (Edulevel, High) $\wedge$ (Seniority, Long) $\wedge$ (Health, Good),
  (Age, Middle) $\wedge$ (Edulevel, High) $\wedge$ (Seniority, Middle) $\wedge$ (Health, Good)
  }

Step 6.5:   The solution of the follower's problem is

$$\text{(Age, Young)} \land \text{(Edulevel, High)} \land \text{(Seniority, Long)}$$
$$\land \text{(Health, Good)}.$$

Step 7:   Because $y \neq y_{[i]}$, Go to Step 8;

Step 8:

Step 8.1–8.2:

$$OS' = \{$$

(Age, Young) $\land$ (Edulevel, High) $\land$ (Seniority, Middle) $\land$ (Health, Good),

(Age, Middle) $\land$ (Edulevel, High) $\land$ (Seniority, Long) $\land$ (Health, Good),

(Age, Middle) $\land$ (Edulevel, High) $\land$ (Seniority, Middle) $\land$ (Health, Good)

$$\},$$

Step 8.3:   The next solution of the follower's problem is

$$o' = \text{(Age, Young)} \land \text{(Edulevel, High)} \land \text{(Seniority, Middle)}$$
$$\land \text{(Health, Good)}$$

Step 8.4:   Go to Step 7;

Step 7:   Because $y = y_{[i]}$, the optimal solution for the bi-level decision problem is
(Age, Young) $\land$ (Edulevel, High) $\land$ (Seniority, Middle) $\land$ (Health, Good)
[End]

The solution with the variables of both the factory executive committee and the workshop management committee will be used in the factory's decision in recruiting new workers. It will maximize the ability to satisfy the objectives of decision making at the two levels.

Now, we use the Algorithm 10.3 to solve the RSBLD problem. We suppose that the four condition attributes are ordered as "age", "edulevel", "seniority", and "health".

Step 1:   Construct the objective rule tree $FT_L$ of the leader by $F_L$, and the result is illustrated by Fig. 10.10;

Step 2:   Construct the objective rule tree $FT_F$ of the follower by $F_F$, and the result is illustrated by Fig. 10.11;

Step 3:   Transform the RSBLD problem to a single level one, and the resulting objective rule tree $CT$ is illustrated by Fig. 10.12;

Step 4:   Use the constraint rule set of both the leader and follower to prune $CT$, and the result is illustrated by Fig. 10.13;

Step 5:   Search for the leftmost rule $dr$ in $CT$ whose leader decision and follower decision are both minimal, and the result is

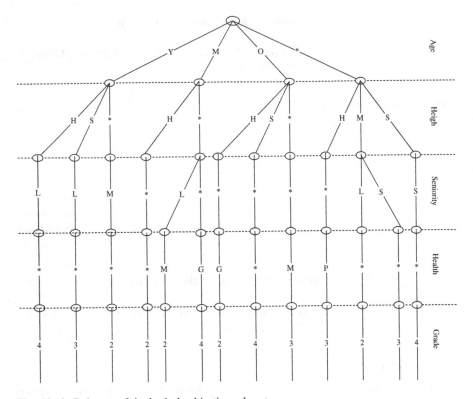

**Fig. 10.10**  Rule tree of the leader's objective rule set

dr:(Age, Young) ( Edulevel, High) (Seniority, Middle) (Health, Good) (d, (2, 2));

Step 6:   $OB = \{ob \mid ob$ is the object satisfying:

(Age, Young) ( Edulevel, High) ( (Seniority, Middle) (Health, Good)};

Step 7:   $ob =$ (Age, Young) (Edulevel, High) (Seniority, Middle) (Health, Good);

Step 8:   $ob$ is the final solution of the RSBLD problem.

In Figs. 10.10, 10.11, 10.12 and 10.13, these attribute values are represented by its first letter.

**Fig. 10.11** Rule tree of the follower's objective rule set

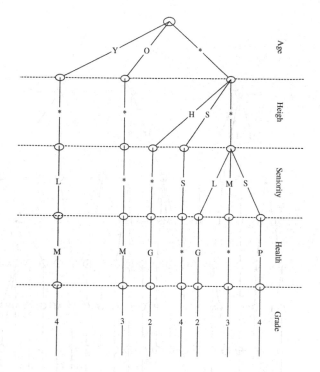

## 10.7 Experiments and Analysis

To test the effectiveness of the proposed RSBLD problem modeling algorithm (Algorithm 10.1) and solution algorithms (Algorithms 10.2 and 10.3), we implemented these algorithms in Matlab 6.5. We then used classical data sets from the UCI database to test them in a set of experiments. The UCI database (http://www.ics.uci.edu/~mlearn/MLRepository.html) consists of many data sets that can be used by the decision systems and machine learning communities for the empirical analysis of algorithms.

For each data set we chose, we first selected half of the data set as the original objective rule set of the leader and the other half as the original objective rule set of the follower. We assumed no constraints, meaning that all objects consistent with the objective rule set were in the constraint region. We also supposed that the first half of the condition attributes were for the leader and the remainder for the follower. The importance degrees of the condition attributes are in descending order from the first condition attribute to the last condition attribute. The two experiments are processed on a computer with 2.33 GHz CPU and 2G memory space. Here, we describe only these two experiments, as follows.

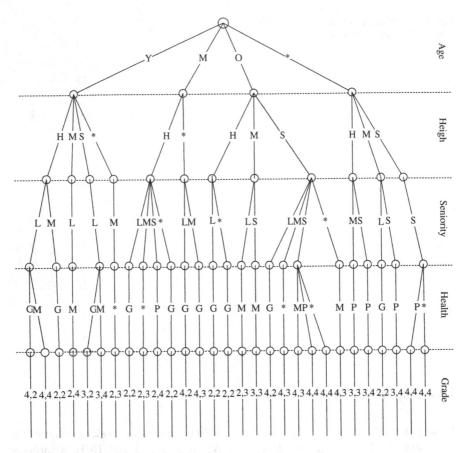

**Fig. 10.12** Transformation result of the objective rule trees

**Experiment 10.1** Testing of Algorithm 10.1 with the data sets in the UCI database.

Step 1: Randomly choose 50 % of the objects from the data set to be the original objective decision rule set of the leader, and the remaining 50 % of the objects to be the original objective decision rule set of the follower;

Step 2: Apply Algorithm 10.1 to construct a rule-set-based bi-level decision model by using the chosen rule sets. Here, we use the decision matrices-based value reduction algorithm (Ziarko et al. 1996) in the RIDAS system (Wang et al. 2002) to reduce the size of the original rule sets

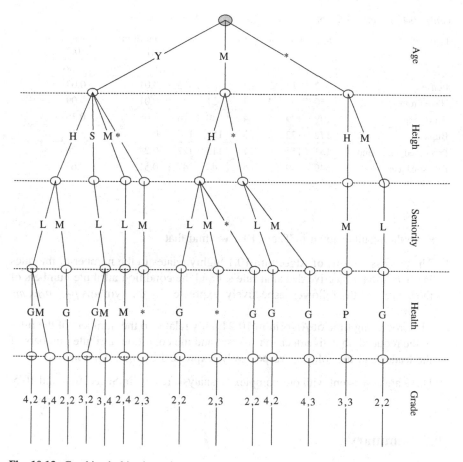

**Fig. 10.13**  Combined objective rule trees after pruning by the constraint rules

**Experiment 10.2**  Testing of Algorithm 10.2 with the data sets in the UCI database. Following Steps 1 and 2 in Experiment 10.1, we have

Step 3:  Apply Algorithm 10.2 to get a solution from the generated rule-set-based bi-level decision model in Experiment 10.1.

The complexity of the two algorithms (Algorithms 10.1 and 10.2) is also tested by conducting these two experiments. As shown in Table 10.4, $p_{OL}$ and $p_{OF}$ are the numbers of objects in the original decision rules of the leader and the follower respectively; $m_L$ and mF are the condition attribute numbers of the leader and the follower respectively; $n_{OL}$ and $n_{OF}$ are the numbers of the rules in the reduced objective decision rule set of the leader and the follower respectively; $t_1$ and $t_2$ are the processing times of Algorithms 10.1 and 10.2 respectively.

**Table 10.4** Testing results of Algorithms 10.1 and 10.2

| Data sets | $p_{OL}$ | $p_{OF}$ | $m_L$ | $m_F$ | $n_{OL}$ | $n_{OF}$ | Algorithm 10.1 | Algorithm 10.2 |
|---|---|---|---|---|---|---|---|---|
| | | | | | | | $t_1(s)$ | $t_2(s)$ |
| Lenses | 12 | 12 | 2 | 3 | 6 | 3 | <0.01 | 0.03 |
| Hayes-roth | 50 | 50 | 2 | 3 | 21 | 24 | <0.01 | 0.09 |
| Auto-mpg | 199 | 199 | 4 | 4 | 80 | 76 | 0.08 | 0.39 |
| Bupa | 172 | 172 | 3 | 3 | 159 | 126 | 0.06 | 3.10 |
| Processed_cleveland | 151 | 151 | 6 | 7 | 115 | 127 | 0.28 | 5.20 |
| Breast-cancer-wisconsin | 349 | 349 | 5 | 5 | 47 | 47 | 0.51 | 0.63 |

From the results shown in Table 10.4 we find that

1. The processing time of Algorithm 10.1 highly relates to the number of the rules in the original objective decision rule set and the condition attribute numbers of the leader and the follower respectively, expressed by the symbols $p_{OL}$, $p_{OF}$, $m_L$ and $m_F$.
2. The processing time of Algorithm 10.2 highly relates to the numbers of the rules in the reduced objective decision rule set and the condition attribute numbers of the leader and the follower respectively, expressed by $n_{OL}$, $n_{OF}$, $m_L$ and $m_F$.

These are consistent with our complexity analysis results in Sects. 10.4 and 10.5.

## 10.8  Summary

In the traditional bi-level decision-making models discussed in previous chapters, objectives and constraints are expressed by linear or nonlinear functions, and bi-level programming or genetic approaches can be effectively used to obtain solutions. However, some real-world bi-level decision problems cannot be easily formulated as linear or non-linear programs. This chapter uses rule sets to handle the issue. It presents how to use rule sets to model non-programming bi-level decision problems, and also develops two algorithms to solve rule-set-based bi-level decision problems.

# Part V
# Multi-level Decision Support Systems and Applications

# Chapter 11
# Fuzzy Bi-level and Tri-level Decision Support Systems

This chapter presents two multi-level decision support systems that implement related algorithms developed in previous chapters to support decision making in practice.

The first is a *fuzzy bi-level decision support system* (called FBLDSS), which incorporates a set of solution algorithms developed in previous chapters for solving bi-level decision problems. This system can handle fuzzy bi-level decision problems, fuzzy multi-objective bi-level decision problems and fuzzy multi-objective bi-level multi-follower decision problems in both linear and non-linear forms. Users (decision makers) are allowed to adjust their subjective preferences to achieve balance between different objectives during the solution process through interacting with the FBLDSS.

The second is a *tri-level decision support system* (called TLDSS) that implements the *decision entity relationship diagram* (DERD) approach and related tri-level optimization algorithms developed in Chap. 6. It has the ability to generate a tri-level decision model based on related parameters entered by users and to obtain a solution for a linear tri-level decision model.

In the following sections, we will present the two decision support system tools and related examples.

## 11.1 A Fuzzy Bi-level Decision Support System

This section introduces a *fuzzy bi-level decision support system* (FBLDSS), which can handle fuzzy bi-level decision problems with different features (linear or non-linear; fuzzy or non-fuzzy; single objective or multi-objective; single follower or multi-follower). This section first presents the configuration and structure of this system, and then illustrates linear and nonlinear bi-level decision processes through two typical examples.

© Springer-Verlag Berlin Heidelberg 2015
G. Zhang et al., *Multi-Level Decision Making*,
Intelligent Systems Reference Library 82, DOI 10.1007/978-3-662-46059-7_11

**Fig. 11.1** The main interface of the FBLDSS

## 11.1.1 System Configuration and Interfaces

The FBLDSS is developed in a Windows environment. We adopt the object-oriented approach to implement it using MS Visual Basic 6.0. The Windows-based interface allows users exploit the capabilities of the system in a friendly way.

In the main interface, as shown in Fig. 11.1, the menu includes the items *File*, *Method*, *Model*, *Result*, and *Help*, all of which perform many kinds of decision support activities. Figure 11.2 shows the overall structure of the menu, from which we can create a bi-level decision model (linear or non-linear) by clicking the item of *New-a linear bi-level model* or *New-a general bi-level model*; clear the current model configuration by clicking *Reset current model item*; trigger the fuzzy approximation branch-and-bound algorithm, fuzzy approximation *K*th-Best algorithm, fuzzy bi-level PSO algorithm or fuzzy bi-level goal-programming algorithm by clicking the items *Approximation B_B*, *Approximation kth-best*, *PSO* or *Approximation goal*; and display the solution for the current bi-level decision model (linear or non-linear) by clicking *Linear optimization result* or *General bi-level optimization result* respectively.

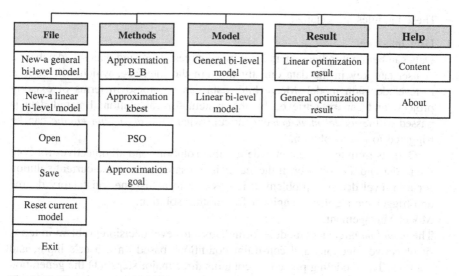

**Fig. 11.2** The menu structure of the FBLDSS

## 11.1.2 System Structure

As a specific type of DSS, this FBLDSS provides computerized assistance to the decision makers in a decentralized organization to gather knowledge about a bi-level decision problem and control the decision-making process to achieve a better-informed outcome.

The structure of the FBLDSS is depicted in Fig. 11.3. Five modules are involved in this architecture, i.e. *user interface*, *model management*, *algorithm engine*, *updating system*, and *visualization*. Data is collected through the user interface, and formatted as a corresponding bi-level decision model by the *model management* module. The core calculations are carried out in the *algorithm engine* of this model, and the solution is outputted through the *visualization* module to the end user by the *user interface*.

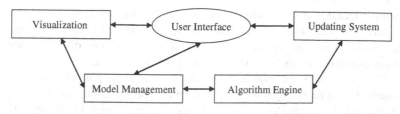

**Fig. 11.3** The system structure of the FBLDSS

These modules are detailed below:

1. User Interface

   Data can be input to the system from two sources: users and data sets/databases. A user can key in the data directly through the interface, which will build a bi-level decision model. Meanwhile, data can be stored in terms of existing projects and can be retrieved by the system for calculation. Entered data is passed to the *model management* module where the *algorithm engine* can be triggered to run a solution.

   Outputs from the system include feasible solutions and all objectives for both the leader and the follower. If the leader is not satisfied with the current solution for a bi-level decision problem, it is possible to adjust the satisfactory degree and trigger the algorithm engineer for another solution.

2. Model Management

   The *model management* module formulates a bi-level decision problem in terms of objective functions and constraint conditions based on a user's input, and controls. The modeling procedure contains three major steps: (1) the generation of objectives, (2) the generation of constraints and (3) the elicitation of a certain satisfactory degree for a bi-level decision problem. These functions are data driven, each requiring a set of coefficients.

3. Algorithm Engine

   The *algorithm engine* module performs the core calculation task of this system including defuzzification, comparison of fuzzy sets, and bi-level model optimization.

   This system provides three fuzzy bi-level decision algorithms: the fuzzy approximation branch-and-bound algorithm, the fuzzy approximation $K$th-Best algorithm, and the fuzzy bi-level PSO algorithm. The first two algorithms can solve fuzzy linear bi-level decision problems. A non-linear bi-level decision problem can only be solved by using the fuzzy bi-level PSO algorithm.

   For multi-objective bi-level decision problems, an extra option of the fuzzy bi-level goal-programming algorithm is provided in this system for decision makers to apply.

   Once the satisfactory degree is adjusted in the *user interface* module, the *algorithm engine* will be triggered to run on the selected bi-level decision model retrieved from the *model management* module.

4. Updating System

   In this FBLDSS, update routines are provided from two aspects, i.e. users can modify an established bi-level decision model retrieved from the model base, and they can also adjust the satisfactory degree to obtain solutions under different aspiration levels for a fuzzy bi-level decision model.

   To define a fuzzy bi-level decision model, a set of crisp numbers is employed to describe the fuzzy objective functions and constraint conditions by fixing four points and the format of a fuzzy number. The capacity to incorporate new information into the established crisp numbers may sometimes introduce conflicts between constraints, or cause an existing feasible solution to be invalid. However, the satisfactory-degree-adjustable mechanism can reduce the possibility of the non-solution situation.

5. Visualization

Since fuzzy numbers are used to interpret the coefficients of objective functions and constraints, the way fuzzy sets are described in modeling a fuzzy bi-level decision problem becomes crucial. The fuzzy membership functions of the coefficients and also the final objective values for both leader and follower are described by graphs.

To solve a fuzzy or non-fuzzy bi-level decision problem, we need to search the equilibrium between the leader and the follower, both of whom achieve optimality, while the leader takes priority within the constraints. To describe this equilibrium, the *visualization* module will complete the presentation and interpretation functions. The output solutions will help the user to identify the preferred equilibrium, and will give insights into how this can be achieved.

To implement the functions of intuitive fuzzy number input and pellucid fuzzy objective output, we describe a fuzzy number by the information of the formats of the left and right functions, the two points where the membership value equals zero, and the other two points where the membership value equals one. Here, the formats of membership functions can be selected as linear, quadratic, cubic, exponential, and logarithmic. Figure 11.16 shows a window to identify a fuzzy number, and Fig. 11.19 shows a window to display a fuzzy objective function.

## *11.1.3  Linear Bi-level Decision Support Process*

The whole decision support procedure for linear bi-level decision problems by the FBLDSS involves three phases, i.e. (1) problem identification, (2) preference elicitation, and (3) solution searching/generation.

The relationship between these phases is illustrated in Fig. 11.4. We use a numerical example to illustrate the linear bi-level decision support process by using this FBLDSS.

**Fig. 11.4** Optimization
process by the FBLDSS

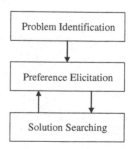

*Example 11.1* Suppose that this is a *fuzzy multi-objective bi-level multi-follower* (FMO-BLMF) decision problem, and let $x = (x_1, x_2, x_3)^T \in R^3$ be the leader's decision variables, $F^1$, $F^2$ and $F^3$ be the leader's objectives.

There are two followers involved in this example. Let $y_1$ and $y_2$ be the decision variables for the first and second follower respectively while $z = (z_1, z_2)^T \in R^2$ be the shared decision variables between the two followers, $f_1^1, f_1^2$ be the first follower's objectives and $f_2^1, f_2^2, f_2^3$ be the second follower's objectives, $X = \{x \geq 0\}, Y_1 = \{y_1 \geq 0\}, Y_2 = \{y_2 \geq 0\}, Z = \{z \geq 0\}$.

Its mathematic model is written as follows:

$$
\begin{aligned}
\max_{x \in X} \quad & F^1(x, y_1, y_2, z) = (\tilde{9}, \tilde{8}, \tilde{6})(x_1, x_2, x_3)^T + (\tilde{4}, \tilde{0}, \tilde{6}, \tilde{8})(y_1, y_2, z_1, z_2)^T \\
\max_{x \in X} \quad & F^2(x, y_1, y_2, z) = (\tilde{1}, \tilde{2}, \tilde{3})(x_1, x_2, x_3)^T + (\tilde{4}, \tilde{0}, \tilde{6}, \tilde{8})(y_1, y_2, z_1, z_2)^T \\
\max_{x \in X} \quad & F^3(x, y_1, y_2, z) = (\tilde{3}, \tilde{4}, \tilde{0})(x_1, x_2, x_3)^T + (\tilde{6}, \tilde{8}, \tilde{1}, \tilde{2})(y_1, y_2, z_1, z_2)^T \\
\text{s.t.} \quad & (\tilde{4}, \tilde{0}, \tilde{6})(x_1, x_2, x_3)^T + (\tilde{8}, \tilde{1}, \tilde{2}, \tilde{3})(y_1, y_2, z_1, z_2)^T \leq \tilde{9}, \\
& (\tilde{4}, \tilde{0}, \tilde{6})(x_1, x_2, x_3)^T + (\tilde{8}, \tilde{3}, \tilde{4}, \tilde{0})(y_1, y_2, z_1, z_2)^T \leq \widetilde{14}, \\
\max_{y \in Y_1, z \in Z} \quad & f_1^1(x, y_1, z) = (\tilde{3}, \tilde{4}, \tilde{0})(x_1, x_2, x_3)^T + (\tilde{6}, \tilde{4}, \tilde{0})(y_1, z_1, z_2)^T \\
\max_{y \in Y_1, z \in Z} \quad & f_1^2(x, y_1, z) = (\tilde{6}, \tilde{8}, \tilde{1})(x_1, x_2, x_3)^T + (\tilde{2}, \tilde{4}, \tilde{0})(y_1, z_1, z_2)^T \\
\text{s.t.} \quad & (\tilde{6}, \tilde{8}, \tilde{1})(x_1, x_2, x_3)^T + (\tilde{8}, \tilde{4}, \tilde{0})(y_1, z_1, z_2)^T \leq \widetilde{16}, \\
& (\tilde{6}, \tilde{3}, \tilde{4})(x_1, x_2, x_3)^T + (\tilde{0}, \tilde{3}, \tilde{4})(y_1, z_1, z_2)^T \leq \tilde{8}, \\
\max_{y \in Y_2, z \in Z} \quad & f_2^1(x, y_2, z) = (\tilde{6}, \tilde{8}, \tilde{8})(x_1, x_2, x_3)^T + (\tilde{3}, \tilde{4}, \tilde{0})(y_2, z_1, z_2)^T \\
\max_{y \in Y_2, z \in Z} \quad & f_2^2(x, y_2, z) = (\tilde{6}, \tilde{8}, \tilde{3})(x_1, x_2, x_3)^T + (\tilde{0}, \tilde{6}, \tilde{8})(y_2, z_1, z_2)^T \\
\max_{y \in Y_2, z \in Z} \quad & f_2^2(x, y_2, z) = (\tilde{1}, \tilde{4}, \tilde{0})(x_1, x_2, x_3)^T + (\tilde{8}, \tilde{3}, \tilde{4})(y_2, z_1, z_2)^T \\
\text{s.t.} \quad & (\tilde{0}, \tilde{6}, \tilde{8})(x_1, x_2, x_3)^T + (\tilde{4}, \tilde{0}, \tilde{6})(y_2, z_1, z_2)^T \leq \tilde{8}, \\
& (\tilde{8}, \tilde{1}, \tilde{2})(x_1, x_2, x_3)^T + (\tilde{4}, \tilde{0}, \tilde{6})(y_2, z_1, z_2)^T \leq \tilde{9}.
\end{aligned}
$$

The membership functions for this FBLMF decision problem are displayed as follows:

$$\mu_{\tilde{0}}(t) = \begin{cases} 0 & t < -1 \\ t+1 & -1 \leq t < 0 \\ 1-t^2 & 0 \leq t < 1 \\ 0 & 1 \leq t, \end{cases} \qquad \mu_{\tilde{1}}(t) = \begin{cases} 0 & t < 0 \\ t^2 & 0 \leq t < 1 \\ 2-t & 1 \leq t < 2 \\ 0 & 2 \leq t, \end{cases}$$

$$\mu_{\tilde{2}}(t) = \begin{cases} 0 & t < 1 \\ t-1 & 1 \leq t < 2 \\ 3-t & 2 \leq t < 3 \\ 0 & 3 \leq t, \end{cases} \qquad \mu_{\tilde{3}}(t) = \begin{cases} 0 & t < 2 \\ t-2 & 2 \leq t < 3 \\ 4-t & 3 \leq t < 4 \\ 0 & 4 \leq t, \end{cases}$$

$$\mu_{\tilde{4}}(t) = \begin{cases} 0 & t < 3 \\ t-3 & 3 \leq t < 4 \\ 5-t & 4 \leq t < 5 \\ 0 & 5 \leq t, \end{cases} \qquad \mu_{\tilde{6}}(t) = \begin{cases} 0 & t < 5 \\ t-5 & 5 \leq t < 6 \\ 7-t & 6 \leq t < 7 \\ 0 & 7 \leq t, \end{cases}$$

$$\mu_{\tilde{8}}(t) = \begin{cases} 0 & t < 7 \\ t-7 & 7 \leq t < 8 \\ 9-t & 8 \leq t < 9 \\ 0 & 9 \leq t, \end{cases} \qquad \mu_{\tilde{9}}(t) = \begin{cases} 0 & t < 8 \\ t-8 & 8 \leq t < 9 \\ 10-t & 9 \leq t < 10 \\ 0 & 10 \leq t, \end{cases}$$

$$\mu_{\widetilde{14}}(t) = \begin{cases} 0 & t < 13 \\ t-13 & 13 \leq t < 14 \\ 15-t & 14 \leq t < 15 \\ 0 & 15 \leq t, \end{cases} \qquad \mu_{\widetilde{16}}(t) = \begin{cases} 0 & t < 15 \\ t-15 & 15 \leq t < 16 \\ \frac{289}{33}-\frac{t^2}{33} & 16 \leq t < 17 \\ 0 & 17 \leq t. \end{cases}$$

Now we describe the solution process of the example by means of the FBLDSS.

The first phase is to set up a framework for the bi-level decision problem in which decisions will be made. In this framework, we define what decision makers wish to achieve as objectives, and any limitations and conditions as constraints. In specification, the following data need to be input to set up the framework.

1. The number of followers and the number of decision variables, objectives and constraints for the leader: In this example, there are two followers, three decision variables, three objectives, and two constraints for the leader, as shown in Fig. 11.5.
2. The variables, objectives and constraints for the leader: In this example, the leader controls decision variables $(x_1, x_2, x_3)$, as shown in Fig. 11.6, having three objectives named L_Obj(1), L_Obj(2), and L_Obj(3), as shown in Fig. 11.7, and having two constraints named L_Con(1) and L_Con(2), as shown in Fig. 11.8.
3. The variables, objectives and constraints for the followers:
The first follower (Follower 1) controls decision variables $(y_1, z_1, z_2)$, as shown in Fig. 11.9, having two objectives named F1_Obj(1) and F1_Obj(2), as shown in Fig. 11.10, and having two constraints named F1_Con(1) and F1_Con(2), as shown in Fig. 11.11.
The second follower (Follower 2) controls decision variables $(y_1, z_1, z_2)$, having two objectives named F2_Obj(1) and F2_Obj(2), and having two constraints named F2_Con(1) and F2_Con(2), as shown in Figs. 11.12, 11.13 and 11.14.

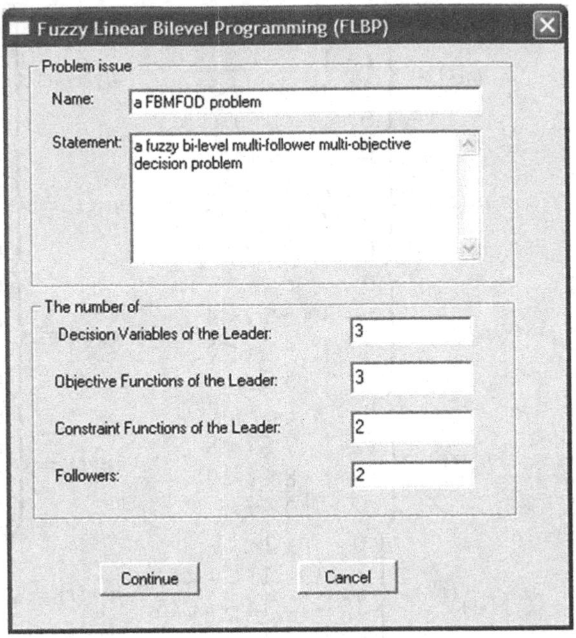

**Fig. 11.5** Interface for
variable input-linear-1

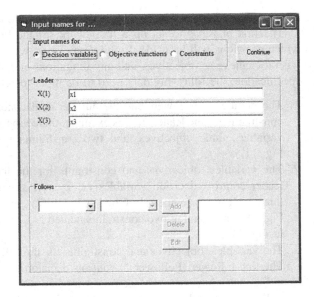

**Fig. 11.6** Interface for
variable input-linear-2

4. The max/min choice for individual objectives: In this example, each objective is
   to be searched for a maximum value, as shown in Fig. 11.15.
5. The fuzzy parameters occurring in the objectives and constraints: In this
   example, the fuzzy coefficient for $x_1$ in the first objective of the leader is entered

**Fig. 11.7** Interface for
variable input-linear-3

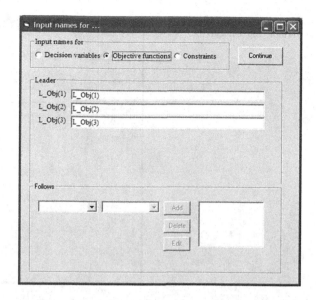

**Fig. 11.8** Interface for
variable input-linear-4

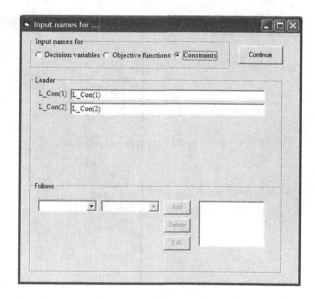

in Fig. 11.16. The other fuzzy parameters should similarly be entered one by one.

The second phase is to elicit the preferences of the decision makers. Once multiple options exist, the decision makers at both levels are allowed to rank these options by assigning specific weights (refer to Figs. 11.17 and 11.18). In this

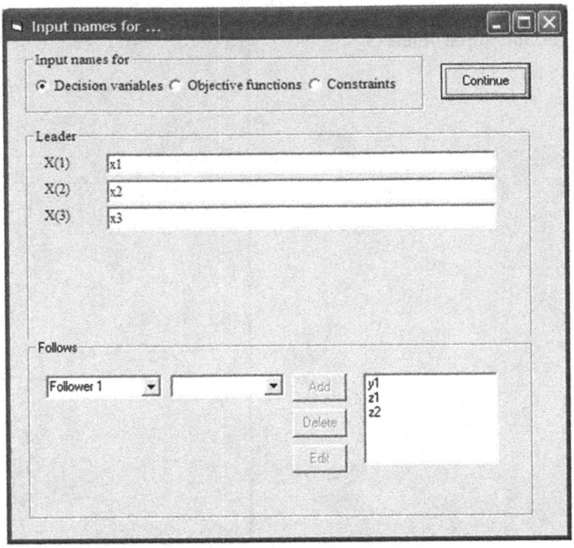

**Fig. 11.9** Interface for
variable input-linear-5

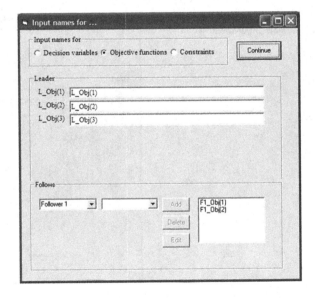

**Fig. 11.10** Interface for
variable input-linear-6

example, all the decision makers give equivalent weights to each of their objectives,
as shown in Fig. 11.17.

The final phase is to search/obtain a solution. Once a bi-level decision model is
built up, the window, as shown in Fig. 11.17, can be activated to reach a solution. In
this window, the fuzzy approximation branch-and-bound algorithm, fuzzy
approximation $K$th-best algorithm, and fuzzy bi-level PSO algorithm are all

**Fig. 11.11** Interface for
variable input-linear-7

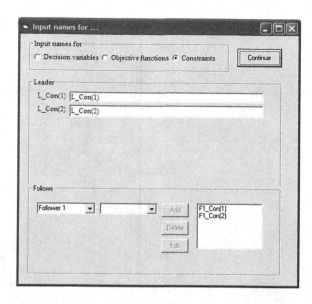

**Fig. 11.12** Interface for
variable input-linear-8

available to be selected as the approach for the linear bi-level problem by clicking
the option button of *B_B*, *K_Best* or *PSO*.

In Example 11.1, the problem is a FMO-BLMF decision problem and the
decision maker chooses the fuzzy bi-level goal-programming algorithm to solve it;
the window, as shown in Fig. 11.18, can be activated to reach a solution.

In the windows shown as Figs. 11.17 and 11.18, a user can set a specific
satisfactory degree for all the membership functions of the fuzzy numbers by

**Fig. 11.13** Interface for variable input-linear-9

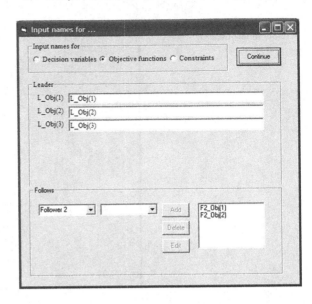

**Fig. 11.14** Interface for variable input-linear-10

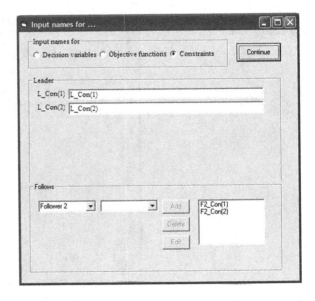

adjusting the slider or keying in a value in the text box above the slider. Thus, a solution under a specific satisfactory degree chosen by the user can be obtained. In this example, the leader selected 0.1 for the satisfactory degree, as shown in Fig. 11.17.

By clicking the *Run* button in the windows shown as Figs. 11.17 and 11.18, the optimization algorithm selected will run, the solutions will be shown, and the corresponding fuzzy objective can be sequentially displayed in the window shown

**Fig. 11.15** Interface for variable input-linear-11

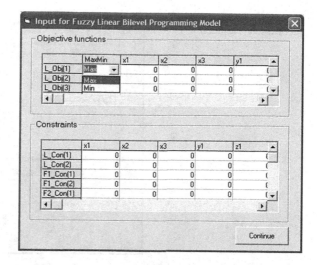

**Fig. 11.16** The input window for the membership function of a fuzzy number

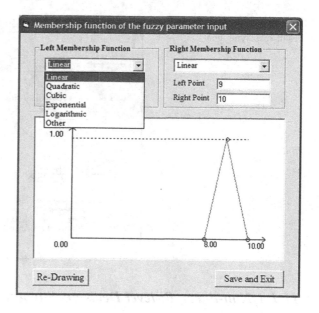

as Fig. 11.19. In this example, under the satisfactory degree 0.1, the final solution $\left(x_1^*, x_2^*, x_3^*, y_1^*, y_2^*, z_1^*, z_2^*\right) = (0, 0.4513, 0, 0.508, 0, 1.7072, 0)$ has been reached, as shown in Fig. 11.17. Under this solution, the first objective value of the leader is shown in Fig. 11.19 by clicking the text box under "L_Obj(1)" in Fig. 11.17. The other objective values can be obtained in the same way.

If the leader is not satisfied with the current solution, the weights for each objective can be changed or the satisfactory degree can be adjusted to obtain another solution.

**Fig. 11.17** Interface for result displaying-linear

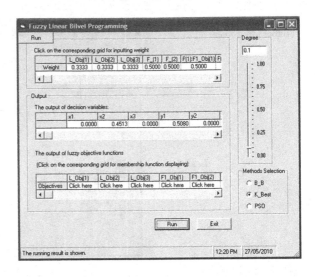

**Fig. 11.18** Interface for result displaying-goal

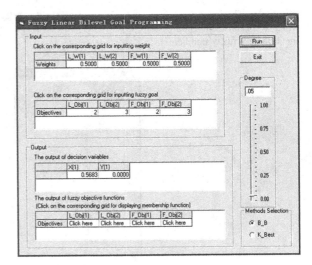

## 11.1.4  Non-linear Bi-level Decision Support Process

We use a non-linear bi-level decision problem to illustrate the bi-level decision support process by means of this FBLDSS.

*Example 11.2* Suppose that a bi-level decision problem has one leader and one follower and each decision entity has one decision variable, one objective function and one constraint, which is described as follows:

**Fig. 11.19** The window for displaying a fuzzy objective

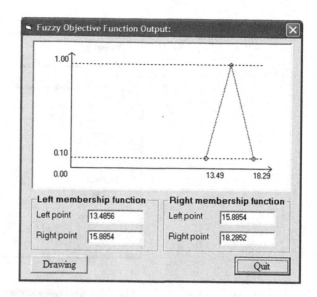

$$\max_{x \in X} \quad F(x,y) = 20{,}800 - \frac{10{,}816}{xy} - 30xy$$

s.t.   $x > 0,$

$$\max_{y \in Y} \quad f(x,y) = 20{,}800 - \frac{8{,}652.8(x-1)}{xy} - 1{,}000y$$

s.t.   $y > 0.$

To solve the non-linear bi-level decision problem, we need to build a structure in which the decision will be made. In this decision structure, we define the decision makers' objectives and constraints. As non-linear formulas have very flexible and unexpected forms, this system gives tremendous flexibility to users by providing text boxes to input objective functions and constraint conditions. In specification, the following data need to be input to set up a non-linear bi-level decision model.

1. The number of leaders and the number of followers: For this problem, there is one leader and one follower, as shown in Fig. 11.20.
2. The variables, objectives and constraints for the leaders and the followers: In this example, the leader has one decision variable $x$, and the follower has one decision variable $y$, as shown in Fig. 11.21. The objectives named $F$ and $f$ respectively for the leader and the follower need to be entered in the window shown in Fig. 11.22. The constraints named $G$ and $g$ respectively for the leader and the follower need to be entered in the window shown in Fig. 11.23.
3. The fuzzy parameters (if there are fuzzy parameters involved in the problem) occur in objectives and constraints: These should be entered as shown in Fig. 11.24, and their membership functions can be entered one by one in Fig. 11.16.

**Fig. 11.20** Interface for
variable input-nonlinear-1

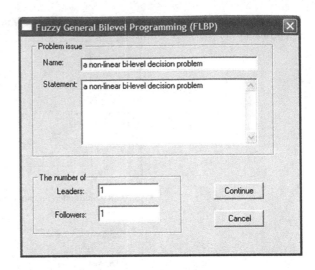

**Fig. 11.21** Interface for
variable input-nonlinear-2

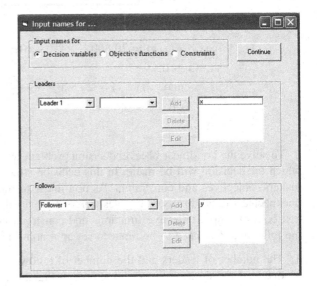

4. The max/min choices for individual objective: In this example, both the leader
   and the follower are looking for the maximum objectives, as shown in
   Fig. 11.24.
5. The ≥, ≤, or = choices for individual constraints: as shown in Fig. 11.24.
6. The function formats for the objectives and constraints of both leaders and
   followers: as shown in Fig. 11.24.

To facilitate formula inputting, a list box, as shown in Fig. 11.24, can be activated to show variable names entered in Fig. 11.21. Thus users can refer to the
variable names to reduce the possibility of mistyping.

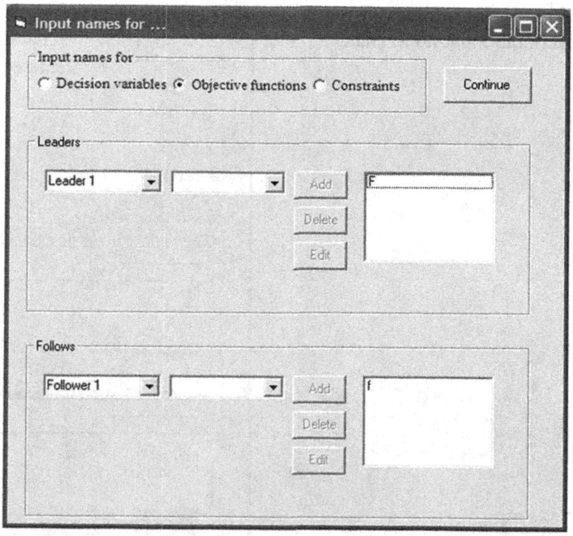

**Fig. 11.22** Interface for
variable input-nonlinear-3

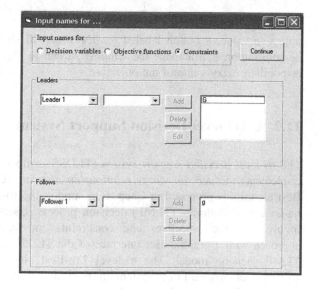

**Fig. 11.23** Interface for
variable input-nonlinear-4

Once a non-linear bi-level decision model is established, a window as shown in
Fig. 11.25 will be shown to display the solutions and objective values for both
leaders and followers. In this example, the solution of $(x, y) = (7.1889, 2.6455)$ has
been reached. Under this solution, the objectives of the leader and follower are able
to achieve 19,660.7357 and 15,338.7115 respectively. If there are fuzzy parameters
involved in this model, the objective values will be fuzzy numbers. In such a
situation, the window shown in Fig. 11.26 will display the solutions and fuzzy
objective values for both leaders and followers. Once a user clicks the text box of an

**Fig. 11.24** Interface for
variable input-nonlinear-5

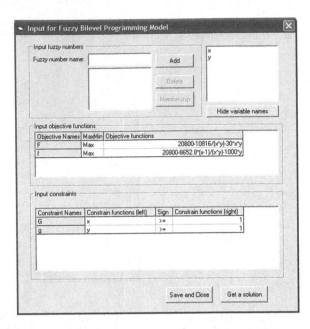

objective value, a button labeled *membership* will show up. Clicking the *membership* button will activate the window shown in Fig. 11.27 to display the corresponding fuzzy value of the objective.

## 11.2  A Tri-level Decision Support System

The *tri-level decision support system* (TLDSS) software presented in this section implements the *decision entity relationship diagram* (DERD) approach and *K*th-Best algorithms proposed in Chap. 6 to support the tri-level decision process and *tri-level multi-follower* (TLMF) decision process. Users can input all coefficients involved in the objectives and constraints conveniently through the DERD approach with a graphic user interface of the TLDSS, and can build a tri-level or TLMF decision model. The tri-level *K*th-Best algorithm and TLMF *K*th-Best algorithm embedded in the system are then invoked to solve the tri-level decision model and TLMF decision model respectively.

### 11.2.1  System Configuration and Tri-level Decision Support Process

The TLDSS is developed with Java language and its decision information is presented in mathematics formula and stored in an XML file in the TLDSS.

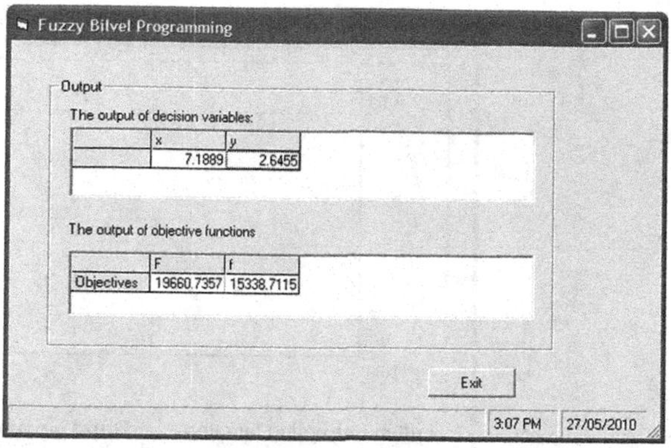

**Fig. 11.25** Interface for output-nonlinear

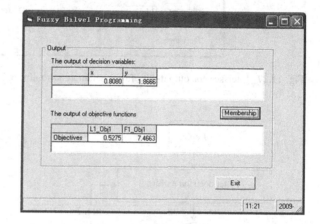

**Fig. 11.26** Interface for output-with-fuzzy-parameters-nonlinear-1

The system flow chart is shown in Fig. 11.28.

Figure 11.28 describes the decision support process of the TLDSS. First, users enter the decision entity structure and names within all three levels, and then determine the relationships between these decision entities. Second, users enter the decision variables and coefficients of objective functions and constraint conditions. A DERD model and a programming model will then be formed. The model established can be saved to a XML file and a user can subsequently load it from the XML file easily. Also, the system can use the corresponding Kth-Best algorithm to obtain an optimal solution to the tri-level or TLMF decision model that is established.

The detailed TLDSS working process is described in Table 11.1.

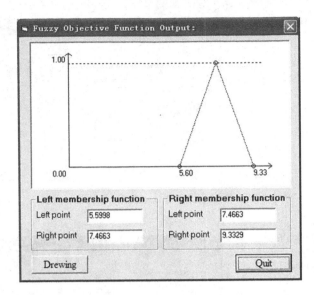

**Fig. 11.27**  Interface for output-with-fuzzy-parameters-nonlinear-2

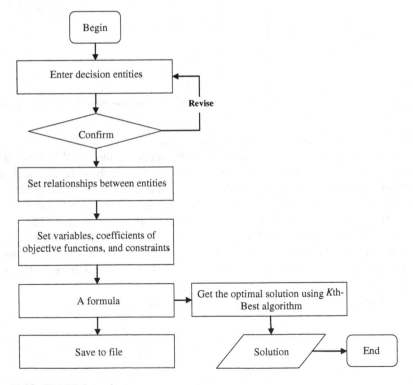

**Fig. 11.28**  TLDSS flow chart

**Table 11.1** The tri-level decision support process

| Step | Process |
|------|---------|
| Step 1 | A user adds child nodes to the root node by pressing the *Add Node* button |
| Step 2 | After all the nodes are added into the tree structure diagram, the user needs to confirm the diagram. A message dialog will pop up to alert the user to add the relationship between the decision entities and related variables |
| Step 3 | The user adds the number of variables, objective functions and constraints |
| Step 4 | The user chooses the corresponding *K*th-Best algorithm to solve the decision problem |
| Step 5 | The user obtains an optimal solution and the related results are displayed |

## 11.2.2 Detailed Operational Process and System Interface

In this section, we will present the detailed operational process and related interfaces of this TLDSS in support of tri-level and TLMF decision-making.

Figure 11.29 shows the initial interface of the TLDSS. At the top of the window, there is a menu bar which has the buttons *File, Models, Results* and *Help*. In the menu *File*, users can save and load decision problems. In the menu *Models*, users can reset decision problems. The *Results* menu is used to show the formula and to obtain an optimal solution and related results. The left-hand side of the window is used to show decision problems in a tree structure. At the bottom, there are four buttons for editing decision problems. The right-hand side is for editing the relationships, node names, variables, objective functions and constraint conditions. There is also an option for users to choose the corresponding *K*th-Best algorithm to solve the model that is established. In addition, there is a blank area at the bottom right-hand side which is used to show the information of each formula, such as node name, variables, objective function, and constraint conditions.

We then introduce the system interface for the modeling process. Users first enter the decision entities for three levels. As shown in the tree structure of Fig. 11.30, each node is a decision entity. From left to the right, the levels from to bottom are shown in hierarchical form. Users can use the buttons to edit the node in the editing panel shown in Fig. 11.31, and they can also edit the relationships between followers, the relationships between the leader and followers, the node name, the variable number, and the objective functions and constraints.

According to the tri-level decision model, three groups of values need to be input to the TLDSS: the number of variables, the coefficients of the objective functions, and the coefficients of the constraint conditions.

When the *Edit Variable Number* button is pressed, a window as shown in Fig. 11.32 appears. The number of decision variables can be entered when one of the decision entities is clicked.

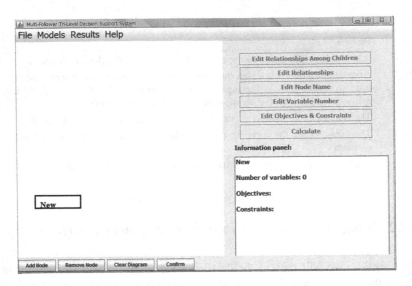

**Fig. 11.29** Initial window of TLDSS

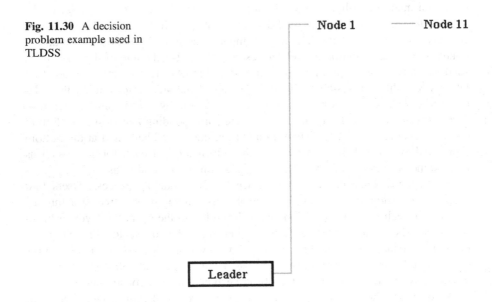

**Fig. 11.30** A decision problem example used in TLDSS

To set the coefficients of the objective functions and constraints, the *Edit Objectives and Constraints* button is selected. The window shown in Fig. 11.33 appears. In this system, the number of objective functions is not predefined and users can add as many functions as they wish.

When the *Add new* button is pressed, another window is displayed, as shown in Fig. 11.34, in which the Variable column lists the variable names involved in the objective function. The Coefficient column shows the coefficients to be entered

**Fig. 11.31** Editing panel in
TLDSS

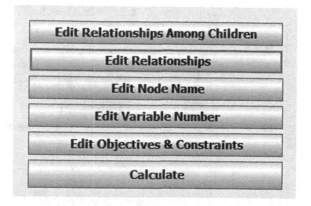

**Fig. 11.32** Edit variable
number

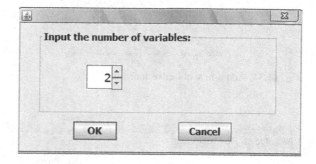

**Fig. 11.33** Objective
functions and constraints

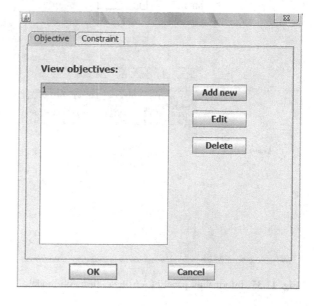

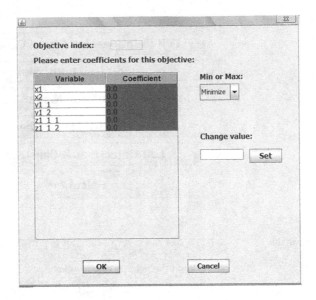

**Fig. 11.34** Add a new objective function

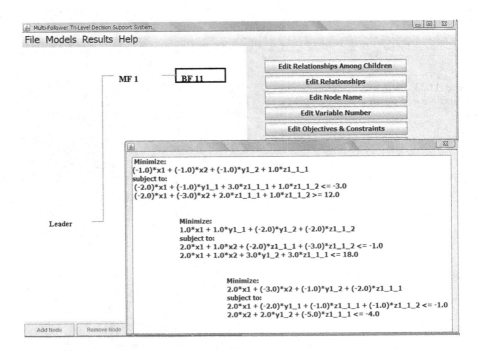

**Fig. 11.35** Information panel of a tri-level decision model

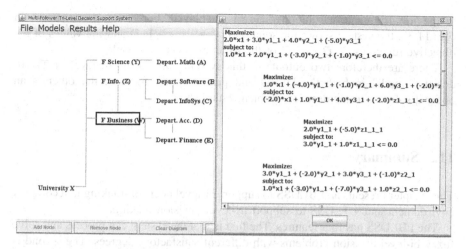

**Fig. 11.36** Information panel of a TLMF decision model

when the default value is 0.0. Users can set the Min or Max according to the requirement of a decision problem. They can also change the values of the coefficients using the button *Set*. The values will be changed and shown in the coefficient column.

At the bottom right-hand side, there is an information panel which shows the information of each node selected. Finally, the decision problem is set up and a tri-level or TLMF decision model is generated, as shown in Figs. 11.35 and 11.36. The information includes the node name, the number of variables, the objective functions, and the constraint conditions.

Lastly, there is a *Calculate* button and a *Run* button to calculate the decision problem shown on the left to obtain the result. When users press one of the buttons

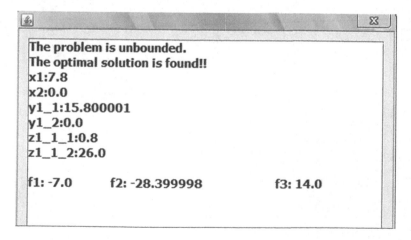

**Fig. 11.37** The optimal solution dialog

in the *Results* menu, a solution dialog is displayed. For example, as shown in Fig. 11.37, the system shows the solution ($x1$, $y1\_1$, $z1\_1\_1$) and the values of the objective functions $f1$, $f2$, and $f3$ of the tri-level decision model.

There are therefore two outputs in this system. One is the tri-level or TLMF decision model in both the DERD and programming formats; the other is an optimal solution to the decision problem established.


## 11.3  Summary

This chapter presents an FBLDSS to support bi-level decision-making in a complex environment and a TLDSS to support tri-level decision-making.

The FBLDSS has four features. The first is that the system can process and solve fuzzy bi-level decision problems with different satisfactory degrees. The second is that the system can deal with any form of membership function of fuzzy parameters in a fuzzy bi-level decision model. The third is that the system involves several methods to suit much wider bi-level decision circumstances under various uncertain environments. The last feature is that this system can recognize and deal with a large number of arbitrary function inputs, which gives great flexibility to decision makers to describe their particular bi-level decision problems.

The TLDSS has two main features. First, it adopts the DERD approach and the programming method to generate a tri-level decision model and a TLMF decision model using a graphic interface. In particular, it is able to describe various relationships between followers in a TLMF decision model. Second, it provides the corresponding solution algorithms to achieve an optimal solution to the proposed tri-level decision model or TLMF decision model.

# Chapter 12
# Bi-level Programming for Competitive Strategic Bidding Optimization in Electricity Markets

We focus on the application of bi-level programming in electricity markets (power market) in this chapter. Competitive strategic bidding optimization of electric power plants (companies) is becoming one of the key issues in electricity markets. This chapter presents a strategic bidding optimization technique developed by applying the bi-level programming. By analyzing the strategic bidding behavior of power plants, we understand that this bidding problem includes several power plants and only one market operator respectively known as multiple leaders and single follower. The problem can be considered as a bi-level multi-leader optimization problem which is introduced in Chap. 5. We therefore build a *bi-level multi-leader* (BLML) decision (programming) model for this bidding problem in day-ahead electricity markets. In the BLML decision model, each power plant is allowed to choose its biddings to maximize its individual profit, and the market operator can find its minimum purchase electricity fare that is determined by the output power of each unit and the uniform marginal prices.

In this chapter, we first give the background of this bi-level programming application in Sect. 12.1. Section 12.2 conducts bidding strategy analysis in competitive electricity markets that is used for modeling. Section 12.3 presents a BLML competitive electricity markets model. A real data-based case study is shown in Sect. 12.4 to illustrate and test the bi-level programming model for competitive strategic bidding optimization in electricity markets. Experimental results on a strategic bidding problem for a day-ahead electricity market have demonstrated the validity of the proposed decision model. Section 12.5 summaries this application.

## 12.1 Background

Throughout the world, electric power industries are undergoing enormous restructuring from nationalized monopolies to individual organizations in a competitive market (Huang and Pai 2002) with the support of digital eco-systems. Because of the significance of electricity energy to national economies and society (Guerrero et al. 2008), electricity markets must be operated under extensive

© Springer-Verlag Berlin Heidelberg 2015
G. Zhang et al., *Multi-Level Decision Making*,
Intelligent Systems Reference Library 82, DOI 10.1007/978-3-662-46059-7_12

conditions of absolute security and stabilization. The research on electricity markets has attracted many researchers, owners and managers from electricity entities. The competitive mechanism of day-ahead markets is a very important research issue in electricity market studies, which can be described as follows: each power plant submits a set of hourly (half-hourly) generation prices and the available capacities for the following day. According to this data and an hourly (half-hourly) load forecast, a market operator allocates the generation output for each unit.

As no determinate operation model for electricity markets exists, the marketing procedure of electric power industries varies from country to country. Generally speaking, there are three kinds of running models in electricity markets: the power pool model, wholesale competitive model, and retail competitive model. These models adopt three kinds of electric power trading methods: long term contract, day-ahead market, and facility service. Among them, the day-ahead market is the most competitive and active imposing great influence on profits for each participant in the market. Specifically, each power plant submits a set of generation prices and other related data, based on which the market operator makes a generating plan for the following day. To optimize this procedure, many models and algorithms have been proposed.

This chapter applies the bi-level optimization approach for dealing with strategic bidding optimization in electricity markets. We propose both a strategic bidding model for power plants and a generation output dispatch model for a market operator in a day-ahead electricity market. Since there are several power plants considered as leaders; and there is only one market operator as the follower, this decision problem is a bi-level multi-leader optimization model which was introduced in Chap. 5. Based on these two models, a specific BLML decision model, which includes ramp rate constraints for competitive electricity markets, is proposed. A real data-based case study on competitive strategic bidding problem in an electricity market is then presented.

## 12.2  Bidding Strategy Analysis in Competitive Electricity Markets

In an auction-based day-ahead electricity market, each power plant will try to maximize its own profit by strategic bidding. Normally, each power plant submits a set of hourly (half-hourly) generation prices and available capacities for the following day. Based on this data and an hourly-load (half-hourly-load) forecast, a market operator will allocate generation output. In this section, under the analysis of bidding strategy optimization problems, we build a competitive strategic bidding model for power plants and a generation output dispatch model for a market operator in a day-ahead electricity market.

## 12.2.1  Strategic Pricing Model for Power Plants

In the upper level, each power plant is concerned with how to choose a bidding strategy, which includes generation price and available capacity. Many bidding functions have been proposed. For a power system, the generation cost function generally adopts a quadratic function of the generation output, i.e. the generation cost function can be represented as:

$$C_j(P_j) = a_j P_j^2 + b_j P_j + c_j \qquad (12.1)$$

where $P_j$ is the generation output of generator $j$, and $a_j, b_j, c_j$ are coefficients of the generation cost function of generator $j$.

The marginal cost of generator $j$ is calculated by:

$$\lambda_j = 2a_j P_j + b_j \qquad (12.2)$$

It is a linear function of its generation output $P_j$. The rule in a goods market may expect each power plant to bid according to its own generation cost. Therefore, we adopt this linear bid function. Suppose that the bidding for the $j$th unit at time $t$ is:

$$R_{tj} = \alpha_{tj} + \beta_{tj} P_{tj} \qquad (12.3)$$

where $t \in T$ is the time interval, $T$ is time interval number, $j$ represents the unit number, $P_{tj}$ is the generation output of unit $j$ at time $t$, and $\alpha_{tj}$ and $\beta_{tj}$ are the bidding coefficients of unit $j$ at time $t$.

According to the justice principle of *the same quality, the same network, and the same price*, we adopt a uniform marginal price (UMP) as the market clearing price. Once the energy market is cleared, each unit will be paid according to its generation output and UMP. The payoff of the $i$th power plant is:

$$F_i = \sum_{t=1}^{T} \left( \sum_{j \in G_i} UMP_t P_{tj} - \sum_{j \in G_i} (a_j P_{tj}^2 + b_{tj} P_{tj} + c_{tj}) \right) \qquad (12.4)$$

where $G_i$ is the suffix set of the units belonging to the $i$th power plant. Each power plant wishes to maximize its own profit $F_i$. In fact, $F_i$ is the function of $P_{tj}$ and $UMP_t$, and $UMP_t$ is the function of all units' bidding $\alpha_{tj}$, $\beta_{tj}$ and output power $P_{tji}$, which will impact on each other.

Therefore, we establish a *strategic pricing model* for power plants as follows:

$$\max_{\alpha_{tj}, \beta_{tj} \in G_i} F_i = F_i(\alpha_{t1}, \beta_{t1}, \ldots, \alpha_{tN}, \beta_{tN}, P_{t1}, \ldots, P_{tN})$$

$$= \sum_{t=1}^{T} \left( UMP_t P_{ti} - \sum_{j \in G_i} \left( a_j P_{tj}^2 + b_{tj} P_{tj} + c_{tj} \right) \right) \qquad (12.5)$$

$$i = 1, \ldots, L$$

where $L$ is the number of power plants, $P_{ti} = \sum_{j \in G_i} P_{tj}$, $t = 1, \ldots, T$.

The profit calculated for each power plant will consider both $P_{ti}$ and $UMP_t$, which can be computed by a market operator, according to the market clearing model.

## 12.2.2 Generation Output Dispatch Model for Market Operator

A market operator actually represents the consumer electricity purchase from power plants, under the conditions of security and stabilization. The objective of a market operator is to minimize the total purchase fare, while encouraging power plants to use a bid price as low as possible. It is reasonable that the lower the price, the more the output. Thus, the function value of a market operator's objective will be calculated according to the bidding price. Most previous strategic bidding models do not include ramp rate constraints, without which the solution for generating dispatch may not be a truly optimal one. We should consider the ramp rate constraints in the real world when modeling a generating dispatch. However, if a model includes ramp rate as a constraint, the number of decision variables involved in the problem will increase dramatically, which requires a more powerful solution algorithm. Based on the analysis above, we build a market operator's *generation output dispatch model* as follows:

$$
\min_{P_{tj}} \quad f = f(\alpha_{t1}, \beta_{t1}, \ldots, \alpha_{tN}, \beta_{tN}, P_{t1}, \ldots, P_{tN}) = \sum_{t=1}^{T} \sum_{j=1}^{N} R_{tj} P_{tj}
$$

$$
\text{s.t.} \quad \sum_{j=1}^{N} P_{tj} = P_{tD} \tag{12.6}
$$

$$
P_{jmin} \leq P_{tj} \leq P_{jmax}
$$

$$
-D_j \leq P_{tj} - P_{t-1,j} \leq U_j, \quad t = 1, 2, \ldots, T
$$

where $t \in T$ is the time interval, $T$ is the time interval number, $j$ represents the unit number, $P_{tj}$ is the generation output of unit $j$ at the time $t$, and $\alpha_{tj}$ and $\beta_{tj}$ are the bidding co-efficients of unit $j$ at the time $t$, $P_{tD}$ is the load demand at the time $t$, $P_{jmin}$ is the minimum output power of the $j$th unit, $P_{jmax}$ is the maximum output power of the $j$th unit, $D_j$ is the maximum downwards ramp rate of the $j$th unit, and $U_j$ is the maximum upwards ramp rate of the $j$th unit.

After receiving all power plants' bid data, a market operator determines the output power of each unit and $UMP_t$ in time slot $t$. $UMP_t$ can be calculated according to the following steps:

[Begin]

**Step1**: calculate output power of each unit $j$ for all time slots $t$ using formula (12.6);

**Step2**: compute bidding $R_{tj}$ corresponding to the generation output $P_{tj}$;

**Step3**: account $UMP_t = \max_{j=1}^{N} R_{tj}$.

[End]

## 12.3 BLML Decision Model in Competitive Electricity Markets

From the analysis above, we know that in an auction-based day-ahead electricity market, each power plant tries to maximize its own profit by strategic bidding, and each market operator tries to minimize its total electricity purchase fare. The decision of one will influence the other. This is a typical bi-level decision problem, which has multiple leaders and only one follower, with power plants as leaders and a market operator as a follower.

By combining the strategic pricing model defined in (12.5) with the generation output dispatch model defined in (12.6), we establish a BLML decision model for competitive strategic bidding-generation output dispatch in an auction-based day-ahead electricity market as follows:

$$
\max_{\alpha_{tj},\beta_{tj}\in G_i} F_i = F_i(\alpha_{t1},\beta_{t1},\ldots,\alpha_{tN},\beta_{tN},P_{t1},\ldots,P_{tN})
$$

$$
= \sum_{t=1}^{T}\left( UMP_t P_{ti} - \sum_{j\in G_i}\left(a_j P_{tj}^2 + b_{tj}P_{tj} + c_{tj}\right)\right)
$$

$$
\text{s.t.} \quad \alpha_{t\min} \leq \alpha_{tj} \leq \alpha_{t\max},
$$

$$
\beta_{t\min} \leq \beta_{tj} \leq \beta_{t\max},
$$

$$
t = 1,2,\ldots, T, \quad j = 1,2,\ldots,N, \quad j = 1,2,\ldots,L
$$

$$
\min_{P_{tj}} \quad f = f(\alpha_{t1},\beta_{t1},\ldots,\alpha_{tN},\beta_{tN},P_{t1},\ldots,P_{tN}) = \sum_{t=1}^{T}\sum_{j=1}^{N} R_{tj}P_{tj} \tag{12.7}
$$

$$
\text{s.t.} \quad \sum_{j=1}^{N} P_{tj} = P_{tD},
$$

$$
P_{j\min} \leq P_{tj} \leq P_{j\max},
$$

$$
- D_j \leq P_{tj} - P_{t-1,j} \leq U_j,
$$

$$
t = 1,2,\ldots, T.
$$

where $\alpha_{tj}$ and $\beta_{tj}$ are the bidding coefficients of unit $j$ at time $t$, $\alpha_{t\min}$, $\alpha_{t\max}$, $\beta_{t\min}$, $\beta_{t\max}$ are the lower and upper limits for $\alpha_{tj}$ and $\beta_{tj}$ respectively, $L$ is the number of power plants, $P_{ti} = \sum_{j\in G_i} P_{tj}$, $P_{j\min}$ is the minimum output power of the $j$th unit,

$P_{j\max}$ is the maximum output power of the $j$th unit, $D_j$ is the maximum downwards ramp rate of the $j$th unit, and $U_j$ is the maximum upwards ramp rate of the $j$th unit.

This model describes strategic bidding problems in competitive electricity markets from a bi-level angle. In this model, there are multiple leaders (power plants) but only one follower (a market operator). This kind of problem has been studied in Chap. 5, and we will use the developed BLML-PSO algorithm proposed in Sect. 5.6 of Chap. 5 to solve it.

## 12.4 A Case Study

In this section, we will use a real world competitive strategic bidding example to illustrate the application of bi-level decision technology on an electricity market.

### 12.4.1 Test Data

In order to test the effectiveness of the proposed BLML decision model and the BLML-PSO algorithm when solving the model defined by (12.7), a typical competitive strategic bidding case consisting of three companies with six units and twenty-four time intervals is chosen. The generation cost function can be calculated by using formula (12.1), where the cost coefficients $a_j, b_j, c_j$ of unit $j$ and other technical data are given in Table 12.1. The load demands for each time interval $t$ are given in Table 12.2.

In Table 12.1, Units 1 and 2 belong to the first power plant, Units 3 and 4 belong to the second power plant, and Units 5 and 6 belong to the third power plant.

To simplify computation, the limit of strategic bidding coefficients does not vary by different time slots and we suppose:

$$\alpha_{t\min} = 7, \alpha_{t\max} = 9, \beta_{t\min} = 0.0002, \beta_{t\max} = 0.007,$$

$$t = 1, 2, \dots T, \quad j = 1, 2, \dots, N.$$

**Table 12.1** Technical data of units

| Unit no. | $a_j$ | $b_j$ | $c_j$ | $p_{\min}$ (MW) | $p_{\max}$ (MW) | $D_j$ (MW/h) | $U_j$ (MW/h) |
|---|---|---|---|---|---|---|---|
| 1 | 0.00028 | 4.10 | 150 | 50 | 680 | 80 | 85 |
| 2 | 0.00312 | 4.50 | 80 | 30 | 150 | 45 | 60 |
| 3 | 0.00048 | 4.10 | 109 | 50 | 360 | 60 | 65 |
| 4 | 0.00324 | 3.74 | 125 | 60 | 240 | 45 | 80 |
| 5 | 0.00056 | 3.82 | 130 | 60 | 300 | 70 | 80 |
| 6 | 0.00334 | 3.78 | 100 | 40 | 160 | 55 | 40 |

**Table 12.2** Load demands in different time intervals

| $t$ | 1 | 2 | 3 | 4 | 5 | 6 |
|---|---|---|---|---|---|---|
| $P_{tD}$ | 1,033 | 1,000 | 1,013 | 1,027 | 1,066 | 1,120 |
| $t$ | 7 | 8 | 9 | 10 | 11 | 12 |
| $P_{tD}$ | 1,186 | 1,253 | 1,300 | 1,340 | 1,313 | 1,313 |
| $tt$ | 13 | 14 | 15 | 16 | 17 | 18 |
| $P_{tD}$ | 1,273 | 1,322 | 1,233 | 1,253 | 1,280 | 1,433 |
| $tt$ | 19 | 20 | 21 | 22 | 23 | 24 |
| $P_{tD}$ | 1,273 | 1,580 | 1,520 | 1,420 | 1,300 | 1,193 |

**Table 12.3** Running results for $\alpha_{tj}$ from the example

| $t$ | $j$ | | | | | |
|---|---|---|---|---|---|---|
| | 1 | 2 | 3 | 4 | 5 | 6 |
| 1 | 7.37 | 7.61 | 7.32 | 7.03 | 7.27 | 8.98 |
| 2 | 8.76 | 7.74 | 8.71 | 7.15 | 8.13 | 7.10 |
| 3 | 7.18 | 8.89 | 8.60 | 8.84 | 8.55 | 8.26 |
| 4 | 7.36 | 8.33 | 7.31 | 8.28 | 8.73 | 7.70 |
| 5 | 7.10 | 8.80 | 8.51 | 8.75 | 8.46 | 8.17 |
| 6 | 8.59 | 7.57 | 8.54 | 8.98 | 7.96 | 8.93 |
| 7 | 7.11 | 7.35 | 7.06 | 8.77 | 7.01 | 8.72 |
| 8 | 8.45 | 8.90 | 7.87 | 8.85 | 7.82 | 8.80 |
| 9 | 8.29 | 8.00 | 8.24 | 7.95 | 7.66 | 7.37 |
| 10 | 8.42 | 7.39 | 8.37 | 8.81 | 7.79 | 8.76 |
| 11 | 7.57 | 7.28 | 7.52 | 7.23 | 8.94 | 7.18 |
| 12 | 7.02 | 7.99 | 8.97 | 7.94 | 8.39 | 7.36 |
| 13 | 7.49 | 7.20 | 7.44 | 7.15 | 8.86 | 7.10 |
| 14 | 8.25 | 7.22 | 8.20 | 8.64 | 7.62 | 8.59 |
| 15 | 8.04 | 7.74 | 7.98 | 7.69 | 7.40 | 7.11 |
| 16 | 8.11 | 8.56 | 7.53 | 8.51 | 7.48 | 8.45 |
| 17 | 8.68 | 8.92 | 8.63 | 8.34 | 8.58 | 8.29 |
| 18 | 8.08 | 7.05 | 8.03 | 8.47 | 7.45 | 8.42 |
| 19 | 8.50 | 8.21 | 7.92 | 8.16 | 7.86 | 7.57 |
| 20 | 8.68 | 7.65 | 8.63 | 7.07 | 8.04 | 7.02 |
| 21 | 8.41 | 8.12 | 7.83 | 8.07 | 7.78 | 7.49 |
| 22 | 7.91 | 8.88 | 7.33 | 8.30 | 7.27 | 8.25 |
| 23 | 8.43 | 8.67 | 8.38 | 8.09 | 8.33 | 8.04 |
| 24 | 7.24 | 8.21 | 7.19 | 8.16 | 7.14 | 8.11 |

## 12.4.2 Experiment Results

This example is run by the BLML-PSO algorithm developed in Sect. 5.6. The running results are listed in Tables 12.3, 12.4, 12.5, 12.6 and 12.7, where $\alpha_{tj}$ and $\beta_{tj}$

**Table 12.4** Running results for $\beta_{tj}$ from the example

| t | j | | | | | |
|---|---|---|---|---|---|---|
| | 1 | 2 | 3 | 4 | 5 | 6 |
| 1 | 0.00089 | 0.00205 | 0.00321 | 0.00438 | 0.00554 | 0.00670 |
| 2 | 0.00208 | 0.00074 | 0.00440 | 0.00307 | 0.00673 | 0.00539 |
| 3 | 0.00641 | 0.00077 | 0.00193 | 0.00310 | 0.00426 | 0.00542 |
| 4 | 0.00278 | 0.00644 | 0.00510 | 0.00377 | 0.00063 | 0.00609 |
| 5 | 0.00048 | 0.00164 | 0.00280 | 0.00397 | 0.00513 | 0.00629 |
| 6 | 0.00382 | 0.00249 | 0.00615 | 0.00481 | 0.00348 | 0.00034 |
| 7 | 0.00350 | 0.00467 | 0.00583 | 0.00699 | 0.00135 | 0.00251 |
| 8 | 0.00022 | 0.00568 | 0.00435 | 0.00121 | 0.00667 | 0.00353 |
| 9 | 0.00687 | 0.00123 | 0.00420 | 0.00536 | 0.00653 | 0.00089 |
| 10 | 0.00556 | 0.00423 | 0.00109 | 0.00655 | 0.00522 | 0.00208 |
| 11 | 0.00060 | 0.00176 | 0.00292 | 0.00408 | 0.00525 | 0.00641 |
| 12 | 0.00626 | 0.00493 | 0.00179 | 0.00045 | 0.00411 | 0.00278 |
| 13 | 0.00147 | 0.00263 | 0.00379 | 0.00496 | 0.00612 | 0.00048 |
| 14 | 0.00051 | 0.00597 | 0.00464 | 0.00150 | 0.00696 | 0.00382 |
| 15 | 0.00449 | 0.00565 | 0.00682 | 0.00118 | 0.00234 | 0.00350 |
| 16 | 0.00551 | 0.00237 | 0.00103 | 0.00469 | 0.00336 | 0.00022 |
| 17 | 0.00106 | 0.00222 | 0.00339 | 0.00455 | 0.00571 | 0.00687 |
| 18 | 0.00406 | 0.00092 | 0.00638 | 0.00324 | 0.00191 | 0.00556 |
| 19 | 0.00658 | 0.00094 | 0.00391 | 0.00507 | 0.00624 | 0.00060 |
| 20 | 0.00295 | 0.00162 | 0.00527 | 0.00394 | 0.00080 | 0.00626 |
| 21 | 0.00065 | 0.00362 | 0.00478 | 0.00595 | 0.00031 | 0.00147 |
| 22 | 0.00580 | 0.00266 | 0.00132 | 0.00498 | 0.00365 | 0.00051 |
| 23 | 0.00368 | 0.00484 | 0.00600 | 0.00036 | 0.00333 | 0.00449 |
| 24 | 0.00220 | 0.00586 | 0.00452 | 0.00138 | 0.00684 | 0.00551 |

**Table 12.5** Running results for $UMP_t$ from the example

| t | 1 | 2 | 3 | 4 | 5 | 6 | 7 | 8 |
|---|---|---|---|---|---|---|---|---|
| UMP | 17.81 | 8.62 | 1.49 | 8.19 | 2.77 | 4.35 | 14.31 | 4.43 |
| t | 9 | 10 | 11 | 12 | 13 | 14 | 15 | 16 |
| UMP | 8.75 | 6.40 | 13.45 | 12.30 | 1.06 | 9.47 | 18.93 | 15.88 |
| t | 17 | 18 | 19 | 20 | 21 | 22 | 23 | 24 |
| UMP | 14.39 | 18.87 | 5.42 | 11.10 | 19.35 | 14.60 | 18.23 | 7.34 |

are the bidding coefficients of unit $j$ at time $t$, $P_{tj}$ is the generation output of unit $j$ at time $t$, $UMP_t$ is the uniform marginal price at time $t$.

Under these solutions, the objective values for both the leaders and the follower are listed in Table 12.7.

**Table 12.6** Running results for $P_{tj}$ from the example

| $t$ | $j$ | | | | | |
|---|---|---|---|---|---|---|
| | 1 | 2 | 3 | 4 | 5 | 6 |
| 1 | 493 | 150 | 50 | 240 | 60 | 40 |
| 2 | 445 | 145 | 63 | 232 | 70 | 45 |
| 3 | 443 | 140 | 76 | 224 | 80 | 50 |
| 4 | 442 | 135 | 89 | 216 | 90 | 55 |
| 5 | 466 | 130 | 102 | 208 | 100 | 60 |
| 6 | 505 | 125 | 115 | 200 | 110 | 65 |
| 7 | 556 | 120 | 128 | 192 | 120 | 70 |
| 8 | 608 | 115 | 141 | 184 | 130 | 75 |
| 9 | 640 | 110 | 154 | 176 | 140 | 80 |
| 10 | 665 | 105 | 167 | 168 | 150 | 85 |
| 11 | 623 | 100 | 180 | 160 | 160 | 90 |
| 12 | 593 | 110 | 193 | 152 | 170 | 95 |
| 13 | 538 | 105 | 206 | 144 | 180 | 100 |
| 14 | 478 | 108 | 231 | 165 | 210 | 130 |
| 15 | 412 | 109 | 252 | 150 | 200 | 110 |
| 16 | 333 | 130 | 285 | 180 | 210 | 115 |
| 17 | 275 | 130 | 320 | 210 | 220 | 125 |
| 18 | 268 | 150 | 360 | 240 | 260 | 155 |
| 19 | 322 | 120 | 300 | 200 | 211 | 120 |
| 20 | 390 | 150 | 360 | 240 | 290 | 150 |
| 21 | 326 | 150 | 355 | 230 | 299 | 160 |
| 22 | 266 | 143 | 356 | 240 | 270 | 145 |
| 23 | 191 | 120 | 320 | 239 | 280 | 150 |
| 24 | 207 | 100 | 300 | 200 | 254 | 132 |

**Table 12.7** Objective values for the decision makers

| The 1st power plant | The 2nd power plant | The 3rd power plant | The market operator |
|---|---|---|---|
| 73,313 | 65,799 | 46,376 | 225,272 |

## 12.4.3 Experiment Analysis

By the BLML-PSO algorithm developed in Sect. 5.6, solutions are reached for both the power plants and the market operator to help them make strategic decisions. We conclude the BLML decision model and the BLML-PSO algorithm in the experiment as follows:

1. The BLML decision model can effectively model strategic bidding problems from electricity markets. By considering the gaming and bi-level relationships between several power plants and a market operator, the BLML decision model

can better reflect the features of such real-world strategic bidding problems in electricity markets and format these problems practically.

2. The BLML-PSO algorithm is quite effective for solving strategic bidding problems defined by the BLML decision model. By making several power plants and a market operator decide sequentially, the hierarchical relation between them is fully considered. By moving the choice by power plants as close as possible to their rational reactions, the Nash equilibrium solution can be obtained.

## 12.5  Summary

Competitive strategic bidding optimization of power plants in electricity markets is in a practical sense important and it is technically implementable. This chapter applies a BLML decision model and BLML-PSO algorithm to handle the competitive strategic bidding decision-making problem in electricity markets. The proposed solution method can achieve a generalized Nash equilibrium for the BLML decision problem in an electricity market by providing power plants with competitive strategic bidding within the prevailing network security constraints.

# Chapter 13
# Bi-level Pricing and Replenishment in Supply Chains

Effective pricing and replenishment strategies in supply chain management are the keys to business success. Notably, with rapid technological innovation and strong competition in hi-tech industries such as computer and communication organizations, the upstream component price and the down-stream product cost usually decline significantly with time. As a result, effective pricing and replenishment decision models are very important in supply chain management. This chapter first establishes a bi-level pricing and replenishment strategy optimization model in hi-tech industry. Then, two bi-level pricing models for pricing problems, in which the buyer and the vendor in a supply chain are respectively designated as the leader and the follower, are presented. Experiments illustrate that bi-level decision techniques can solve problems defined by these models and can achieve a profit increase under some situations, compared with the existing methods.

This chapter is organized into four sections. After introducing the background in Sects. 13.1 and 13.2 shows a case study about hi-tech collaborative pricing and replenishment strategy making. In Sect. 13.3, we use bi-level decision techniques to develop two bi-level pricing models within another case study, one considering the buyer as the leader who has priority in deciding, and the other taking the vendor as the leader. Finally, the summary of this chapter is given in Sect. 13.4.

## 13.1 Background

Hi-tech products such as computers and communication consumer products have driven the need for globalization and massive customization, and have come to occupy a large section of the supply chain industry. Features of hi-tech products include short product life cycle time and quick response time. The lead-time from order to delivery is usually compressed from 955 (95 % order delivered within 5 days) to 1,002 (100 % order delivered within 2 days), and both component costs and product prices are declining at a rate of about 1 % per week (Sern 2003). This implies that purchasing or selling one-week earlier or later will result in an approximate loss of 1 % (Lee 2002). As a result, hi-tech products require a more

© Springer-Verlag Berlin Heidelberg 2015
G. Zhang et al., *Multi-Level Decision Making*,
Intelligent Systems Reference Library 82, DOI 10.1007/978-3-662-46059-7_13

effective optimization method to support policy-making by both the buyer and the vendor in a supply chain.

In reality, the buyer and vendor in a hi-tech product supply chain are two echelons that need to achieve a win-win business solution. Under this principle of collaboration, some collaborative pricing and replenishment optimization models are developed and both the vendor and buyer aim to reduce/optimize the purchase cost and price respectively.

To reveal well and clearly reflect the interactive and internal relationship between a vendor and a buyer, we consider both sides to be well-optimized for the supply chain: the maximum optimization to one side, such as buyer, while still considering the profit achievement of the other side, such as vendor. In fact, neither the vendor nor buyer has direct control over the strategy/policy-making of the other, but their actions affect subsequent responses of each other. Therefore, the pricing and replenishment strategy problem is naturally a bi-level optimization problem where either the vendor or buyer can be the leader based on the requirement and goal of the decision support system.

In the following sections, we will present the formulation of the hi-tech collaborative pricing and replenishment strategy problem using non-linear bi-level programming in the first case study, in which the buyer is the leader. Then, we address how bi-level pricing models are developed by using bi-level programming in the second case study. In the second case study, one bi-level pricing model considers the buyer as the leader who has the privilege of deciding first, and the vendor as the follower who makes decisions after the buyer; the other bi-level pricing model takes the vendor's profit as priority and makes the vendor the leader and the buyer the follower. These two pricing models allow the buyer and vendor to make decisions in sequence, fully considering the mutual influence of each other. To obtain solutions from these non-linear bi-level decision models, the *Fuzzy Bi-level Decision Support System* (FBLDSS) software developed in Chap. 11 is used. We also conduct experiments to illustrate the proposed models.

## 13.2  Case Study 1: Hi-tech Product Pricing and Replenishment Strategy Making

This section will handle hi-tech product pricing and replenishment strategy problem by bi-level decision techniques.

### 13.2.1  Problem Formulation

The formulation for the pricing and replenishment strategy problem is presented based on the assumptions of Yang et al. (2007):

1. Vendor and buyer's replenishment rates are instantaneous.
2. Component purchase cost and product price to the end consumer decline at a continuous rate per unit time.
3. Finite planning horizon and constant demand rate are considered.
4. Each replenishment time interval is the same.
5. No shortage is allowed.
6. Purchase lead-time is constant.

It is assumed that the purchase cost of the vendor and the market price to the end-consumer are fixed. To maximize profit through increased sales, the vendor offers a price discount rate of $r_b$ to the buyer.

The related parameters included in our model are listed in Table 13.1.

If the vendor's and buyer's costs decline at continuous rates of $r_v$ and $r_b$ respectively, their purchase costs are:

$$P_v(t) = P_{v0}(1 - r_v)^t, \quad 0 \le t \le H \tag{13.1}$$

**Table 13.1** Related parameters

| Parameter | Description |
|---|---|
| $n$ | The number of orders that a vendor places for the item from a supplier in the planning horizon |
| $m$ | The number of buyer's lot size deliveries per vendor's lot size |
| $Q$ | The buyer's lot size |
| $r_b$ | The weekly decline-rate of the buyer's purchase cost |
| $D$ | The weekly demand rate |
| $r_v$ | The weekly decline-rate of the vendor's purchase cost |
| $r_m$ | The weekly decline-rate of market price to the end-consumer |
| $H$ | The weekly length of the planning horizon |
| $F_v$ | The vendor's holding cost per dollar per week |
| $F_b$ | The buyer's holding cost per dollar per week |
| $C_v$ | The vendor's ordering cost per order |
| $C_b$ | The buyer's ordering cost per order |
| $P_{v0}$ | The vendor's unit purchase cost at the initial time |
| $P_{b0}$ | The buyer's unit purchase cost at the initial time |
| $P_{m0}$ | Market price to the end consumer at the initial time |
| $P_v(t)$ | The vendor's unit purchase cost in week $t$ |
| $P_b(t)$ | The buyer's unit purchase cost in week $t$ |
| $P_m(t)$ | The market price to the end consumer in week $t$ |
| $NP_v$ | The vendor's net profit in the planning horizon |
| $NP_b$ | The buyer's net profit in the planning horizon |
| $NP$ | The joint net profit of both the vendor and the buyer in the planning horizon |

and

$$P_b(t) = P_{b0}(1 - r_b)^t, \quad 0 \le t \le H \tag{13.2}$$

respectively.

If the market price declines at a continuous rate of $r_m$, the unit market price to the end-consumer is

$$P_m(t) = P_{m0}(1 - r_m)^t, \quad 0 \le t \le H. \tag{13.3}$$

The buyer's average inventory level is $Q/2$, that is, one half of the buyer's lot size. The unit purchase cost is

$$P_{b0}, P_{b0}(1 - r_b)^{\frac{H}{mn}}, P_{b0}(1 - r_b)^{\frac{2H}{mn}}, \ldots, P_{b0}(1 - r_b)^{\frac{(n-1+\frac{m-1}{m})H}{n}}.$$

The buyer's holding cost in the planning horizon is

$$HC_b = \frac{F_b H}{mn} \sum_{i=0}^{n-1} \sum_{j=1}^{m-1} \frac{P_{b0}(1 - r_b)^t Q}{2} = \frac{F_b H P_{b0} Q}{2mn} \frac{1 - (1 - r_b)^H}{1 - (1 - r_b)^{\frac{H}{mn}}}, \tag{13.4}$$

where $t = \left(i + \frac{j}{m}\right) H/n$, $i = 0, 1, \ldots, n - 1$, $j = 1, 2, \ldots, m - 1$.

Note that $m$ and $n$ are positive integers, $t$ is a continuous real number and is discrete valued in the analytical steps for ease of analysis.

Since the vendor-buyer-combined average inventory level is $mQ/2$, the vendor's average inventory level is $Q(m - 1)/2$ in the collaborative system. The vendor's holding cost in the planning horizon is

$$HC_v = \frac{F_v H}{n} \sum_{i=0}^{n-1} \frac{P_{v0}(1 - r_v)^t Q(m - 1)}{2} = \frac{F_v H P_{v0} Q(m - 1)}{2n} \frac{1 - (1 - r_v)^H}{1 - (1 - r_v)^{\frac{H}{n}}}, \tag{13.5}$$

where $t = \frac{iH}{n}$, $i = 0, 1, 2, \ldots, n - 1$.

The buyer's net income (sales revenue minus purchase cost) is denoted by $NI_b$ as follows:

$$NI_b = \int_0^H P_{m0}(1 - r_m)^t D dt - \sum_{i=0}^{n-1} \sum_{j=0}^{m-1} P_{b0}(1 - r_b)^t Q$$

$$= \frac{P_{m0} D}{\ln(1 - r_m)} \left( e^{H \ln(1-r_m)} - 1 \right) - \frac{P_{b0} Q \left(1 - (1 - r_b)^H\right)}{1 - (1 - r_b)^{\frac{H}{mn}}} \tag{13.6}$$

The vendor's net income (sales revenue minus purchase cost) is denoted by $NI_v$ as follows:

$$NI_v = \sum_{i=0}^{n-1}\sum_{j=0}^{m-1} P_{b0}(1 - r_b)^t Q - \sum_{i=0}^{n-1} P_{v0}(1 - r_v)^t mQ$$

$$= \frac{P_{b0}Q(1 - (1 - r_b)^H)}{1 - (1 - r_b)^{\frac{H}{mn}}} - \frac{P_{v0}mQ(1 - (1 - r_v)^H)}{1 - (1 - r_v)^{\frac{H}{n}}}, \tag{13.7}$$

where $1 > r_b > 0$, $1 > r_v > 0$.

The buyer's net profit (formula (13.6) minus formula (13.4) and the ordering cost) is denoted by $NP_b$ as follows:

$$NP_b = \frac{P_{m0}D}{\ln(1 - r_m)}\left(e^{H\ln(1-r_m)} - 1\right) - \frac{P_{b0}Q(1 - (1 - r_b)^H)}{1 - (1 - r_b)^{\frac{H}{mn}}}$$

$$- \frac{F_b H P_{b0}Q}{2mn}\frac{1 - (1 - r_b)^H}{1 - (1 - r_b)^{\frac{H}{mn}}} - mnC_b. \tag{13.8}$$

where $1 > r_b > 0$, $1 > r_m > 0$.

The vendor's net profit (formula (13.7) minus formula (13.5) and the ordering cost) is as follows:

$$NP_v = \frac{P_{b0}Q(1 - (1 - r_b)^H)}{1 - (1 - r_b)^{\frac{H}{mn}}} - \frac{P_{v0}mQ(1 - (1 - r_v)^H)}{1 - (1 - r_v)^{\frac{H}{n}}}$$

$$- \frac{F_v H P_{v0}(m - 1)Q}{2n}\frac{1 - (1 - r_v)^H}{1 - (1 - r_v)^{\frac{H}{n}}} - nC_v. \tag{13.9}$$

The joint net profit for both vendor and buyer, the sum of formula (13.8) and formula (13.9), denoted by $NP$, is

$$NP = NP_b + NP_v. \tag{13.10}$$

The relationship between the lot size and the number of deliveries is

$$Q = \frac{HD}{mn} \tag{13.11}$$

The value of $r_b$ is dependent on the net profit sharing between the two players. The relationship between the vendor's net profit and buyer's net profit is defined as

$$(NP_v) = \alpha(NP_b) \tag{13.12}$$

where $\alpha$ is a negotiation factor.

When $\alpha = 0$, it means all net profit sharing is accrued by the buyer; when $\alpha = 1$, it implies that all net profit sharing is equally distributed. A large $\alpha$ means that all net profit is accrued mainly by the vendor. The optimization problem is a constrained non-linear programming problem, stated as

$$\max \quad NP = NP_v + NP_b$$
$$\text{s.t.} \quad (NP_v) = \alpha(NP_b), \quad \alpha \ge 0, \tag{13.13}$$
$$Q = \frac{HD}{mn}$$

When there is no cost/price reduction (i.e., $r_v = 0$, $r_b = 0$ and $r_m = 0$), formulas (13.8) and (13.9) are undefined. Using L'Hospital's rule to take the derivatives of both the numerator and the denominator (Yang et al. 2007), a buyer's and a vendor's net profits in formulas (13.8) and (13.9) are derived as

$$NP_b = P_{m0}DH - P_{b0}Qmn - \frac{P_{b0}F_bHQ}{2} - mnC_b \tag{13.14}$$

and

$$NP_v = P_{b0}Qmn - P_{v0}mnQ - \frac{P_{v0}F_vH(m-1)Q}{2} - nC_v \tag{13.15}$$

respectively.

The results of formulas (13.14) and (13.15) are the same as the case for static cost and price.

In the vendor–buyer pricing system, both the vendor and buyer aim to maximize their profits but their decisions are related to each other in a hierarchical way: the buyer as the leader and the vendor as the follower, or vice versa. When making the pricing strategy, if we take the buyer's point of view as having priority over a vendor, we can set the buyer as the leader and the vendor as the follower. By combining formulas (13.14) and (13.15), we can establish a bi-level pricing and replenishment strategy optimization model in a supply chain as follows:

$$\max_{m \in M} \quad NP_b(m) = P_{m0}DH - P_{b0}Qmn - \frac{P_{b0}F_bHQ}{2} - mnC_b$$
$$\text{s.t.} \quad m > 0,$$
$$\max_{n \in N} \quad NP_v(n) = P_{b0}Qmn - P_{v0}mnQ - \frac{P_{v0}F_vH(m-1)Q}{2} - nC_v$$
$$\text{s.t.} \quad n > 0. \tag{13.16}$$

In this non-linear bi-level programming model, both the buyer and vendor adjust their own controlling variables, wishing to maximize their own profits under their specific constraints, but the buyer's objective is also subject to the vendor's optimized objective function value. That is, the buyer is the leader, who makes a decision first; and the vendor is the follower, who makes a decision based on the possible strategy of the buyer.

## 13.2.2 Experiments

The bi-level pricing and replenishment strategy optimization model can be illustrated by the following example from Yang et al. (2007).

*Example 13.1*

The demand rate per week, $D = 400$ units;
The vendor's unit purchase cost at the initial time, $P_{v0} = \$4$;
The buyer's unit purchase cost at the initial time, $P_{b0} = \$5$;
The market price to the end consumer from the buyer at the initial time, $P_{m0} = \$6$;
The buyer's ordering cost per order, $C_b = \$30$;
The vendor's ordering cost per order, $C_v = \$1,000$;
The buyer's holding cost per dollar per week, $F_b = 0.004$;
The vendor's holding cost per dollar per week, $F_v = 0.004$;
The time horizon considered, $H = 52$ weeks;
The negotiation factor, $\alpha = 1$.

After substituting the above parameters into formula (13.16), we have the following simplified bi-level pricing and replenishment strategy optimization model:

$$\max_{n \in N} \quad NP_b(n, m) = 20{,}800 - \frac{10{,}816}{mn} - 30mn$$

$$\text{s.t.} \quad n > 0$$

$$\max_{m \in M} \quad NP_v(n, m) = 20{,}800 - \frac{8{,}652.8(m-1)}{mn} - 1{,}000n \qquad (13.17)$$

$$\text{s.t.} \quad m > 0.$$

We use the developed FBLDSS in Chap. 11 to solve problem (13.17). To obtain a solution, we first input the objective functions and constraints of both the leader (buyer) and follower (vendor). We then run the software and obtain a solution $(m, n) = (6.9743, 2.7225)$, $(NP_v, NP_b) = (19{,}660.7371, 15{,}354.9525)$.

Through comparison of objective values when $m = 6$ and $m = 7$, we select $m = 6$ since it results in a bigger objective function value for the buyer in the problem. Similarly, $n = 3$ results in a bigger objective value for the vendor.

We then use the model and solution method from Yang et al. (2007) to obtain a solution for the same problem when the negotiation factor $\alpha$ is set as one. It should be noted that since both $m$ and $n$ can only be positive integers, the buyer's net profit and the vendor's net profit cannot be completely equal in most situations. In this example, the buyer's net profit and the vendor's net profit will be the closest and maximized under $m = 1$ and $n = 3$. Meanwhile, further experiments are carried out by adjusting the negotiation factor $\alpha$ in a wider range.

The results of our bi-level pricing and replenishment strategy optimization model and the model of Yang et al. (2007) are compared, as shown in Table 13.2.

**Table 13.2** Summary of results for Example 13.1

|  | A | m | n | $NP_b$ | $NP_v$ | NP |
|---|---|---|---|---|---|---|
| Yang et al. (2007) | $\alpha = 1$ | 1 | 3 | $17,104 | $17,800 | $34,904 |
|  | $\alpha > 2$ | 1 | 1 | $9,954 | $19,800 | $29,754 |
|  | $1.5 \leq \alpha \leq 2$ | 1 | 1 | $9,954 | $19,800 | $29,754 |
|  | $1 \leq \alpha \leq 1.5$ | 1 | 3 | $17,105 | $17,800 | $34,904 |
|  | $0.5 \leq \alpha \leq 1$ | 3 | 3 | $19,328 | $15,877 | $35,205 |
|  | $\alpha < 0.5$ | 2 | 10 | $19,659 | $10,367 | $30,026 |
| Our bi-level optimization model |  | 6 | 3 | $19,659 | $15,396 | $35,016 |

From the experimental results, it is noted that with our bi-level pricing and replenishment strategy optimization model, compared with Yang et al.'s (2007) original model (i.e., $\alpha = 1$), the profit for the buyer increases by about 15 % (from $17,104 to $19,659) and the profit for the vendor decreases by about 13.5 % (from $17,800 to $15,396). The total percentage increase for the buyer and the vendor is about 3.2 % (from $34,904 to $35,016) when compared with the results from Yang et al. (2007). When $\alpha$ is adjusted in a wider range, the buyer still achieves more profit in all situations of $\alpha$ with our bi-level strategy optimization model. Even if the vendor as the follower loses some profit when $\alpha \geq 0.5$, the profit sum of both the buyer and vendor are still higher in our bi-level model under most choices of $\alpha$.

The proposed bi-level pricing and replenishment optimization model can achieve more profit for a buyer in a supply chain at the price of some profit loss for the vendor. This is understandable, as bi-level optimization models always take the leader's interest as priority. The reason our results outperform others is that our bi-level pricing and replenishment strategy optimization model gives the buyer or the vendor the freedom to optimize their choices, without having to obey the heavy restrictions faced in the model by Yang et al. (2007).

## 13.3  Case Study 2: Hi-tech Product Pricing and Replenishment Strategy Making with Weekly Decline-Rates

This section takes the hi-tech product pricing and replenishment strategy problem again, where weekly demand rate and weekly decline-rates are added as extra decision variables, to carry out the second case study by bi-level decision techniques.

### 13.3.1  Problem Formulation

In this section, by switching the leader and follower roles, respectively, between a buyer and a vendor, we develop two bi-level pricing models in a supply chain.

The buyer's net profit in a buyer-vendor system can be calculated by:

$$NP_b = \frac{P_{m0}D}{\ln(1-r_m)}\left(e^{H\ln(1-r_m)}-1\right) - \frac{P_{b0}Q\left(1-(1-r_b)^H\right)}{1-(1-r_b)^{\frac{H}{mn}}}$$
$$- \frac{F_bHP_{b0}Q}{2mn}\frac{1-(1-r_b)^H}{1-(1-r_b)^{\frac{H}{mn}}} - mnC_b. \tag{13.18}$$

The vendor's net profit can be calculated by:

$$NP_v = \frac{P_{b0}Q\left(1-(1-r_b)^H\right)}{1-(1-r_b)^{\frac{H}{mn}}} - \frac{P_{v0}mQ\left(1-(1-r_v)^H\right)}{1-(1-r_v)^{\frac{H}{n}}}$$
$$- \frac{F_vHP_{v0}(m-1)Q}{2n}\frac{1-(1-r_v)^H}{1-(1-r_v)^{\frac{H}{n}}} - nC_v. \tag{13.19}$$

In (13.18), the buyer controls $m$, the number of the buyer's lot size deliveries per vendor's lot size; and $r_m$, the weekly decline-rate of market price to an end-consumer. In (13.19), the vendor controls $n$, the number of the orders that the vendor places for the item from a supplier in the planning horizon; $r_b$, the weekly decline-rate of the buyer's purchase cost; and $r_v$, the weekly decline-rate of the vendor's purchase cost. All other parameters defined in the problem are constants, which may change if other specific problems are introduced. The explanations of symbols used in the above two formulas are listed in Table 13.1.

When making the pricing strategy, if we take the buyer's point of view to make its profit a priority over the vendor, we can designate the buyer as the leader and the vendor as the follower. By combining Formulas (13.18) and (13.19), we establish a bi-level pricing model in a supply chain as follows:

$$\max_{m,r_m} \quad NP_b(m,r_m,n,r_b,r_v)$$
$$= \frac{P_{m0}D}{\ln(1-r_m)}\left(e^{H\ln(1-r_m)}-1\right) - \frac{P_{b0}Q\left(1-(1-r_b)^H\right)}{1-(1-r_b)^{\frac{H}{mn}}}$$
$$- \frac{F_bHP_{b0}Q}{2mn}\frac{1-(1-r_b)^H}{1-(1-r_b)^{\frac{H}{mn}}} - mnC_b$$

s.t. $m > 0$,
$\quad 0.0001 \le r_m \le 0.5$,

$$\max_{n,r_b,r_v} \quad NP_v(m,r_m,n,r_b,r_v) = P_{b0}Q\frac{1-(1-r_b)^H}{1-(1-r_b)^{\frac{H}{mn}}} - P_{v0}mQ\frac{1-(1-r_v)^H}{1-(1-r_v)^{\frac{H}{n}}}$$
$$- \frac{F_vHP_{v0}(m-1)Q}{2n}\frac{1-(1-r_v)^H}{1-(1-r_v)^{\frac{H}{n}}} - nC_v$$

s.t. $n > 0$,
$\quad 0.0001 \le r_b \le 0.5$,
$\quad 0001 \le r_v \le 0.5$.

$$\tag{13.20}$$

In this model, both the buyer and vendor adjust their own decision variables respectively, wishing to maximize their own profits, under specific constraints. The buyer is the leader, who makes a decision first; and the vendor is the follower, who makes a decision after the buyer.

If we take the point of view of the vendor to make its profit a priority over the buyer, we can designate the vendor as the leader and the buyer as the follower. By combining formulas (13.18) and (13.19), we establish another bi-level pricing model in a supply chain as follows:

$$\max_{n,r_b,r_v} \quad NP_v(m,r_m,n,r_b,r_v) = P_{b0}Q\frac{1-(1-r_b)^H}{1-(1-r_b)^{\frac{H}{mn}}} - P_{v0}mQ\frac{1-(1-r_v)^H}{1-(1-r_v)^{\frac{H}{n}}}$$

$$-\frac{F_vHP_{v0}(m-1)Q}{2n}\frac{1-(1-r_v)^H}{1-(1-r_v)^{\frac{H}{n}}} - nC_v$$

s.t. $n > 0$,

$0.0001 \le r_b \le 0.5$,

$0.0001 \le r_v \le 0.5$.

$$\max_{m,r_m} \quad NP_b(m,r_m,n,r_b,r_v)$$

$$= \frac{P_{m0}D}{\ln(1-r_m)}\left(e^{H\ln(1-r_m)}-1\right) - \frac{P_{b0}Q\left(1-(1-r_b)^H\right)}{1-(1-r_b)^{\frac{H}{mn}}}$$

$$-\frac{F_bHP_{b0}Q}{2mn}\frac{1-(1-r_b)^H}{1-(1-r_b)^{\frac{H}{mn}}} - mnC_b$$

s.t. $m > 0$,

$0.0001 \le r_m \le 0.5$.

$$(13.21)$$

In this model, both the buyer and vendor adjust their own decision variables respectively, wishing to maximize their own profits, under specific constraints. The vendor is the leader, who makes the first decision; and the buyer is the follower, who makes a decision after the buyer.

We will use the FBLDSS to solve problems defined by the above two bi-level pricing models.

## 13.3.2 Experiments

In this section, we illustrate the bi-level pricing models in Sect. 13.3.1 by the following numerical example where the parameters are given as follows:

*Example 13.2*

1. The demand rate per week, $D = 400$ units;
2. The vendor's unit purchase cost at the initial time, $P_{v0} = \$4$;
3. The buyer's unit purchase cost at the initial time, $P_{b0} = \$5$;
4. The market price to the end consumer from the buyer at the initial time, $P_{m0} = \$6$;
5. The buyer's ordering cost per order, $C_b = \$30$;
6. The vendor's ordering cost per order, $C_v = \$1,000$;
7. The buyer's holding cost per dollar per week, $F_b = 0.004$;
8. The vendor's holding cost per dollar per week, $F_v = 0.004$;
9. The time horizon considered, $H = 52$ weeks.

To deal with this problem, we relax the constraint of equal profit, and add $r_m$, $r_b$, and $r_v$ as decision variables. By using the FBLDSS developed in Chapter 11 to solve problems defined by Formulas (13.20) and (13.21), we obtain solutions for both the buyer and the vendor. To evaluate the results of this research, we compare these results with the results from the original model by Yang et al. (2007) under a different negotiation factor $\alpha$, which is defined as $\alpha = NP_v/Np_b$. To make the comparison fair and reasonable, besides $m$ and $n$, we add $r_m$, $r_b$, and $r_v$ as decision variables to be changeable to maximize the profit in Yang et al.'s (2007) model. Table 13.3 lists solutions from this research and solutions from the model by Yang et al. (2007).

From Table 13.3, we can see that, using the bi-level pricing model (the buyer as the leader) developed in this section, the buyer's profit will increase compared with Yang's model when $\alpha \geq 1.5$. If the vendor is taken as the leader, he or she can achieve a profit increase when $\alpha \leq 2$, which is true for most pricing problems in a supply chain. As the follower, the vendor or buyer is bound to lose, despite the range of the negotiation factor $\alpha$. This is understandable, because in a bi-level decision situation, we always take the leader's interest as priority.

**Table 13.3** Summary and comparison of running results for Example 13.2

|  | $m$ | $r_m$ | $N$ | $r_b$ | $r_v$ | $NP_b$ | $NP_v$ |
|---|---|---|---|---|---|---|---|
| Yang et al. (2007) ($\alpha \geq 2$) | 2 | 0.0001 | 9 | 0.0068 | 0.5 | 35,008 | 69,946 |
| Yang et al. (2007) ($1.5 \leq \alpha \leq 2$) | 2 | 0.0001 | 9 | 0.01 | 0.5 | 41,280 | 63,710 |
| Yang et al. (2007) ($1 \leq \alpha \leq 1.5$) | 2 | 0.0001 | 9 | 0.017 | 0.5 | 52,990 | 52,068 |
| Yang et al. (2007) ($0.5 \leq \alpha \leq 1$) | 1 | 0.0001 | 9 | 0.032 | 0.5 | 68,548 | 36,605 |
| Yang et al. (2007) ($\alpha < 0.5$) | Not applicable | | | | | | |
| This study (buyer as leader) | 5 | 0.0071 | 6 | 0.0372 | 0.0753 | 52,399 | 16,866 |
| This study (vendor as leader) | 3 | 0.0015 | 7 | 0.0026 | 0.0767 | 21,359 | 64,165 |

These results reveal that when applying bi-level decision techniques on pricing problems in supply chains, some improvements can be achieved for a player (a buyer or a vendor) if it is the leader.


## 13.4  Summary

In this chapter, the pricing and replenishment strategy making problem for hi-tech products proposed by Yang et al. (2007) is remodeled by bi-level programming. To solve problems defined by these bi-level programming models, the FBLDSS is used. Experimental results show that the bi-level pricing models can achieve profit improvements for both the buyer and vendor. In the two-stage vendor-buyer inventory system, our experimental data show that the vendor, as the leader, out-performs the buyer as the leader. This is because the vendor, as the leader, improves the actual consumption rates; the vendor making the first decision ensures that production matches demand more closely, reduces inventory and improves business performance. This is why the vendor managed inventory has become very popular in recent years.

# Chapter 14
# Bi-level Decision Making in Railway Transportation Management

Transportation management is an important application field of bi-level decision-making. For example, transportation facilities, resources planning and moving, as well as staff relocation all involve sub-optimization and optimization problems, that is, the decision entities are often at two decision levels. This chapter presents two real applications of the bi-level decision techniques in railway transportation management.

This chapter is organized by two case studies. Section 14.1 presents a case study about a bi-level decision model which is established for a railway train set organization. Section 14.2 shows a bi-level decision model for a railway wagon flow management problem. Experiments are carried out in each section to further illustrate the applications of these two bi-level decision models on them in railway transportation management.

## 14.1 Case Study 1: Train Set Organization

In this section, a decision model for *train set organization* (TSO) is developed by bi-level programming techniques. We first analyze the bi-level optimization nature of the management on TSO. A bi-level decision model for TSO is then developed, and applied in a real-world railway station to illustrate the bi-level decision model.

### 14.1.1 Background

Railway transportation, as one of the most important ways of transportation, has always been playing an irreplaceable role in social economics. For railway freight transportation, about 80 % of the whole transportation time is allotted to the operations of loading/unloading, transferring, and overhauling in railway technical stations (Li and Du 2002). The working state of technical stations, therefore, will

© Springer-Verlag Berlin Heidelberg 2015
G. Zhang et al., *Multi-Level Decision Making*,
Intelligent Systems Reference Library 82, DOI 10.1007/978-3-662-46059-7_14

influence the whole overpass ability of the railway network. Thus the research on the railway transportation optimization will be bound to focus on the operation of technical stations.

Train set organization, aiming at arranging the train set in railway freight transportation and with extraordinary professional and technical specialties, is one of the main subjects in railway transportation management. The objectives of TSO include: to make the transportation efficient and even; to use the transporting device reasonably and to promote the cooperation among different departments involved in the freighting procedure. The term of *organizing* here means arranging, deciding and managing, while *train set organizing* acts to arrange the train set, make decisions on related issues, and manage the procedure in railway transportation.

There exist multiple levels among the running of TSO: (1) the *national railway network level* the top, (2) the *local bureau railway network level* the second, (3) the *stations* the third, and (4) the *operating group* the bottom. However, as the operating objects of both the national railway network and the local bureau railway network are train sets, while those of the two lower levels are trains, the organization of TSO can be generalized into two levels: the railway network as the leader and the stations as the followers. Thus bi-level programming techniques can be used to analyze the problem.

The main concerns of the railway network are to decide the train type (pick-up-and-drop-train, district-train, transit-train, or through-train), the train constitution, the train number, and the detailed route of the departing train set. The objectives of the railway network include: improving the transportation capacity and service speed, reducing the cost, balancing the working rhythm among divisions, and assigning the break-up and make-up jobs among different stations rationally.

The tasks assigned to a station are to constitute a normative train set required by the railway network from all kinds of freight wagons that stop by this station. Involved with these tasks, there also include a series of relevant operations, such as: collecting or delivering, shunting, loading/unloading, and wagon checking.

The main concerns of stations include: making the operating efficient, economical and safe; rationally using the transportation devices such as track, shunting locomotive, and hump; deciding the operation steps together with its schedules; and cooperating among steps within the schedule-frame of the railway network.

The TSO can be divided into two levels, even though the separate levels still share intrinsic consistency. For the upper level, when making a TSO plan, the railway network must consider the influence from the specific operating ability and device conditions of stations, while calculating the influence factors from itself such as the amount and destinations of trains and the track conditions. For stations located at the lower level, when implementing the working goals, they should try their best to harmonize between their own operation abilities and the working arrangement from their top counterpart.

Railway stations can be grouped into two classes: *through stations* and *technical stations*. Compared with *technical stations*, *through stations* are small sized and their daily works, mainly on helping trains go through or two train set from opposite directions meet, are simple and the workload is small. Except for all the functions of *through stations*, *technical stations* are to make a new train set by breaking up the old ones and adding transship trains and trains originated there. Related tasks also include: arrival/departure operating, collection-and-delivering operating, shunting, loading/unloading, and wagon checking. We generalize these operations at *technical stations* as *shunting and transship operations*.

For the reason of facilitating the modeling, we simplify the tasks of TSO by the following assumptions:

1. The railway transportation supply is less than the demand; the aim of the TSO is to fully use the transportation ability to provide as much transportation as possible.
2. The topological structure of a railway network is a circle formed by train lines. This is to embody the continuous nature of the net and transportation circulation.
3. The main line is double-track with every track direction fixed, which means there allows two train sets running in opposite directions between two stations simultaneously. This is to avoid the meeting problem of two train set with opposite running directions.
4. Within a railway network, there are located only *technical stations*, and only one type of trains run, the *district trains*, which are from one *technical station* A and to another *technical station* B. Between *technical stations* A and B there are no other *technical stations*.
5. The unit workload of *shunting and transship operation* for all *technical stations* are the same. In other words, every *technical station* shares the identical amount of operating time for the same train set.

Based on these assumptions above, the decision maker on a railway network wishes that the density of train sets (calculated by the time intervals between any two side-by-side running train sets) and the length of a train set (the number of trains of any train set) as large as possible to obtain the maximal transport capacity. However, for the sake of safety, the density has its upper limit set by the railway network. Restricted by the motive power of the locomotive and the useful length limit of arrival-departure track, the train set length has its upper limit as well.

Ignoring the constraints by a railway network, the stations, on one hand, wish the length of train set to be large because the larger the length, the more efficient the operating and the lower the unit operating cost. The operating efficiency is the amount of trains shunted and transshipped per unit time, while the unit operating cost is the cost for every single train. On the other hand, the operating time for shunting and transshipping, which influences the cost, will increase if the length of the train set increases. However, the overall effect of the train set length is that the general unit operating cost will decrease with the increase of the train set length.

From the analysis above, we can conclude that:

1. For the variable of the length of a train set, the two levels share the same objective: the larger the length of the train set the better.
2. For the variable of the density of a train set, the decision makers at the upper level pursue its minimum while those at the lower level wish it to change with the train set length in the same direction of travel.

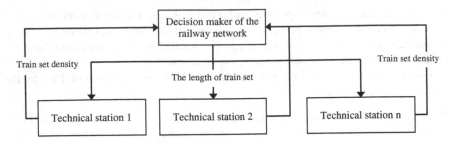

**Fig. 14.1** The relationship between the railway network and technical stations

Generally speaking, the shunting and transshipping time in stations is larger than the safe time intervals of any two side-by-side running train sets, so the variable of the density of train set is determined by the lower level, the stations, while the variable of the length of train set is controlled by the top level, the railway network. The relationship between decision makers of the railway network and technical stations is illustrated in Fig. 14.1.

## 14.1.2 Problem Formulation

Based on the analysis above, a bi-level decision model of TSO is built as:

For $x = (x_1, x_2, \ldots, x_n) \in X \subset R^n$, $y \in Y \subset R^m$, $F, f : X \times Y \to F(R)$, Leader: decision-maker of the railway network

$$\max_{x \in X} \quad F(x, y) = \max_{x \in X} \quad F(x, y) = \frac{a_1 \cdot \sum_{i=1}^{n} w_i \cdot x_i}{\sum_{i=1}^{n} w_i \cdot y_i} \tag{14.1a}$$

$$\text{s.t.} \quad \sum_{i=1}^{n} w_i \cdot x_i < m, \tag{14.1b}$$

$$\sum_{i=1}^{n} w_i \cdot y_i < c_1, \tag{14.1c}$$

The $i$th follower: the $i$th technical station

$$\min_{y_i} \quad f_i(x_i, y_i) = -b_1 \cdot x_i - b_2 \cdot y_i \tag{14.1d}$$

$$\text{s.t.} \quad c_2 \le \frac{x_i}{y_i} \le c_3, \tag{14.1e}$$

$$y_i > c_4. \tag{14.1f}$$

Explanation:

1. Variables:

$x_i$: the length of a train set for the $i$th station, which is the number of trains of any train set controlled by the leader, the decision maker of the railway network.

$y_i$: the density of train sets for the $i$th station, which is the time interval between any two side-by-side running train sets, controlled by the $i$th follower, the $i$th technical station.

2. Coefficients and constants:

$n$: the number of *technical stations* in the railway network.

$w_i$: the relative weight for the $i$th station in the railway network.

$a_1$: the time interval. If $a_1 = 24$, then $a_1 / \sum_{i=1}^{n} w_i \cdot y_i$ means the number of train sets going through the network within 24 h. $a_1 \cdot \sum_{i=1}^{n} w_i \cdot x_i / \sum_{i=1}^{n} w_i \cdot y_i$ is the number of trains going through the network per day, and $a_1 > 0$.

$m$: the maximum number of trains of any train set regulated by the *Safety Terms*. When the trains are empty, the main concern is not to exceed the length limit. When the trains are loaded, the weight limit becomes the decisive factor. However, for the sake of safety, when computing, both the length and weight must meet the requirements. No matter whether it is the weight or length, the ultimate limit is put on the number of trains.

$c_1$: the minimum time interval between any trains list regulated by the *Safety Terms*.

$b_1$ and $b_2$: the weights set for the influencing power by the length and density to the unit cost.

$c_2$ and $c_3$: the lower and upper number limits of the trains for *technical stations* to shunt and transship per time unit.

$c_4$: the least time for the technical stations to complete the shunting and transshipping.

3. Formula:

(14.1a) means that the leader aims at obtaining the maximum throughput capacity within a certain period of time. $\sum_{i=1}^{n} w_i \cdot x_i / \sum_{i=1}^{n} w_i \cdot y_i$ means the number of trains shunted and transshipped per time unit.

(14.1b) means that the length of a train set has its upper limit imposed by the locomotive's motive power and the arrival-departure track's useful length. When the trains are loaded, except for the length limit, there is still weight restriction set upon the train set, which means, the weights of goods loaded together with the weights of trains cannot exceed its upper limit.

(14.1c) means that any two adjacent running train sets cannot be too close for the sake of safety.

(14.1d) means that the followers wish that the cost is as low as possible. The first part of (14.1d) means that the more the length of trains sets results in the more efficient of the shunting and the lower the unit cost. The second part means that the longer of the time the train set remains in the station the higher the cost.

(14.1e) means that *technical stations* have their own lower and upper time limits to shunt and transship trains.

(14.1f) means that there exists a least period of time for the *technical station* to complete the operation.

### 14.1.3 Experiments

In this section, we take the railway freight operation in a railway station: Station A into consideration. Station A is a railway *technical station* with the duty of managing both passenger transportation and freight transportation within the precinct of its Railway Bureau. The data collected from Station A cover the duration between November 1, 2006 and December 31, 2006.

Suppose that the trains shunted and transshipped are to the direction of Station B, which is another station located next to Station A along its downlink. The weight distribution of trains is listed in Table 14.1, with the locomotive being SS1 (137 ton, 1.9 unit length).

The terms in Table 14.1 are explained as below:

1. Wagon Type: the type of wagon used.
2. Wagon Suttle: the weight of the empty wagon.
3. Load: the weight of the goods loaded.
4. Equivalent Length: the equivalent length of a wagon is calculated from the front clasp to the rear clasp, with the unit length as 11 m. If the equivalent length is 1.1, then its actual length is $11 \times 1.1 = 12.1$ m.

According to the model defined by (14.1a–14.1f), the coefficients are calculated and discussed below:

1. $a_1$: as the computation is within the *Basal Daily Working Plan*, which is to arrange wagon assignment and schedule necessary operations based on the *Trains Running Chart*, *Trains Shunting Plan*, *Detailed Rules on Technical Station Management*, and constraints set by operating spots; the computing of the freighting wagon organization is limited within a working day of 24 h. So $a_1$ is set to 24.

**Table 14.1** Train set distribution

| Wagon type | Wagon Suttle (ton) | Load (ton) | % | Equivalent length |
|---|---|---|---|---|
| B23 | 38 | 40 | 3 | 2.1 |
| P64A | 26 | 58 | 3 | 1.5 |
| G70 | 23 | 58 | 9 | 1.1 |
| G60 | 23 | 50 | 59 | 1.1 |
| G70 | 23 | 55 | 35 | 1.1 |

2. $m$: limited by the pulling ability of the locomotive and the territorial landform, such as grading, within Station A's precinct, the weight of the train set must not be larger than 3,500 tons. The departure track used for train sets to the direction to Station B is Track IV, Filed II, whose effective length is 890 m. By 30 m of braking distance, which is left for trains to stop safely, the maximum length for the trains sets is 860 m.

Taking the constitution of the trains listed in Table 14.1, we set 1 *unit train* as a virtual train whose equivalent length, denoted by $l_1$ (meters), and weight, denoted by $w_1$ (ton), are calculated below:

$$l_1 = 2.1 \times 0.03 + 1.5 \times 0.03 + 1.1 \times 0.09$$
$$+ 1.1 \times 0.5 + 1.1 \times 0.35 = 1.142$$
$$w_1 = (38 + 40) \times 0.03 + (26 + 58) \times 0.03 + (23 + 58) \times 0.09$$
$$+ (23 + 50) \times 0.5 + (23 + 55) \times 0.35 = 66.95$$

The maximum number of such empty *unit train*, denoted by $m_e$, is $(860 - 1.9 \times 11)/(1.142 \times 11) = 66$, and the maximum number of such loaded *unit train*, denoted by $m_l$, is $(3,500 - 137)/66.95 = 50$.

From above analyzing and computing, we obtain:

$$m = \min(m_e, m_l) = \min(66, 50) = 50.$$

3. $c_1$: for the sake of safety, the pursuing distance, the minimum distance interval between any side-by-side running trains list, is 10 km, which costs about 0.2 h in the journey from Station A to Station B. So $c_1$ is set to 0.2.
4. $b_1$ and $b_2$: we set the weights of length and density of trains set on the cost of the station as 0.4 and 0.6 respectively.
5. $c_2$ and $c_3$: the least number of trains Station A can shunt and transship is 30 per hour, while the max number is 150.
6. $c_4$: the least time for Station A to complete the shunting and transshipping for a train set is 0.68 h.

Thus, the bi-level decision problem defined by (14.1a–14.1f) is specialized as (14.2a–14.2f) in Station A.

Leader: decision-maker of the railway network

$$\max_x \quad F(x, y) = \frac{24x}{y} \tag{14.2a}$$

$$\text{s.t.} \quad x < 50, \tag{14.2b}$$

$$y > 0.2. \tag{14.2c}$$

Follower: Station A

$$\min_{y} \quad f(x,y) = -0.4x - 0.6y \tag{14.2d}$$

$$\text{s.t.} \quad 30 \le \frac{x}{y} \le 150, \tag{14.2e}$$

$$y > 0.68. \tag{14.2f}$$

To solve the problem in (14.2a–14.2f), we use the *fuzzy bi-Level decision support system* (FBLDSS) software presented in Chap. 11, and come to the solutions of $(x^*, y^*) = (50, 1.67)$ with $F^* = 718.6$ and $f^* = -21.002$, which means, the railway network will obtain its maximum throughput capacity of 718.6 trains per day, if the decision makers of the railway network set the average number of trains to 50, followed by Station A setting the time interval between every two side-by-side train sets to 1.67 h.

## 14.2 Case Study 2: Railway Wagon Flow Management

This section presents a bi-level decision model for *railway wagon flow management* (RWFM). We first analyze the multi-level nature of RWFM and then develop a bi-level decision model for it. Experiments are then carried out to illustrate its applications.

### 14.2.1 Background

*Railway wagon flow management* (RWFM) is to arrange wagon flows in railway freight transportation. One of the key issues faced by RWFM is how to arrange wagons generated or transferred in *technical stations* to form new wagon flows, while aiming at making transportation cost minimum and under constraints from both technical stations and rail tracks. An optimal solution to this problem can not only ensure freight to be sent to the destinations economically, but also make full use of all transportation facilities, thus reduce jamming probabilities and improve the transportation ability as a whole.

Due to the difficulties arising from both wagon routing and marshalling plan optimization, it is even more difficult to integrate these two issues. The most popular way is to choose wagon rout first, and then optimize the marshalling plans in every *technical station*. Although this strategy can decrease the problem solving difficulties, it still cannot reach global solutions as the benefits from the best routing can be offset by some extra workload brought in stations (Lin and Zhu 1996).

Most current research by bi-level decision techniques on traffic controlling focuses on the transformation network design and layout (Feng and Wen 2005). Little research has been conducted towards wagon flow management problems from the multi-level angle. In this section, we use a bi-level method to study the problem of RWFM.

## 14.2.2 Problem Formulation

Before establishing the bi-level decision model for RWFM, we list some terms used in following content.

1. *Local wagon flow*: wagons that are loaded/uploaded or repaired in one *technical station* are called local wagon flow for this station.
2. *Local district wagon flow*: some wagons are loaded/uploaded or repaired in intermediate stations between two *technical stations*. This kind of wagon flow is called *local district wagon flow* for the two *technical stations*.
3. *Long-distance wagon flow*: for a *technical station*, if a wagon flow is not its *local wagon flow* or *local district wagon flow* but belongs to another *technical station* (*local wagon flow* or *local district wagon flow*), this wagon flow is called *long-distance wagon flow* for this *technical station*.
4. *Service operation*: to assist on the marshalling operation within one station, some auxiliary operations must be made, including: taking-out and placing-in of cars, picking-up and dropping trains, loading/uploading goods, and repairing. We call this kind of auxiliary operation as *service operation*.

Railway wagon flow management characterized by monopolization, is usually run by three levels, i.e. railway ministry level, railway bureau level, and station level. However, when carrying out tasks assigned by its corresponding superior, a lower level can arrange its own resources to achieve as much profit as possible. The communication among levels is through marshalling plans which are designed by the upper level but implemented by the lower counterparts. Marshalling plans are regulations on organizing vans which may be destined to different destinations to form van lists. Optimization on marshalling plans aims at minimizing the time spent for centralizing and detention in *technical stations*.

In this section, we take *railway bureaus as leaders and stations as followers*. A *railway bureau* controls the workload and working rhythm of the stations in its administration area. A station, while controlling its own producing resources, decides which specific method it will use to achieve tasks to be carried in this station. Thus, the cost in a station is determined by both the station and its upper administrator, the *railway bureau*. However, the optimal cost level for a station does not necessarily produce the most ideal cost status for the *railway bureau* who seeks equilibrium with traffic and cost. Although the *railway ministry* is located above *railway bureaus* and *technical stations*, this study, while not focusing on the reciprocal decision relation between a *railway ministry* and its *bureaus*, only takes the decision from the *railway ministry* as input constraints.

Once a *railway bureau* selects a marshalling plan, it means two kinds of data are determined. One is the *technical station sequences* where some marshalling operation will be carried for every *long-distance wagon flow*. The other is the number of vans to be marshaled in every station. For the leader, a decision involves whether accepting a carriage and the way to deliver it. The marshalling plans made by a *railway bureau* involve only *long-distance wagons*. In some *technical stations*, some *long-distance wagons* should be merged or separated to decrease cost in stations and increase traffic efficiencies.

*Technical stations* perform marshalling operations as well as relevant following services, such as collecting, delivering, shunting, loading/unloading, and wagon checking. The facilities of these services depend on the quality of the marshalling operation which is performed beforehand. Having *local wagons*, *local district wagons* and *long-distance wagons* as three kinds of marshalling objectives, marshalling operations with *local wagons* and *local district wagons* are flexible in *technical stations*. Stations can determine the extent and depth of the marshalling operation for *local wagons* and *local district wagons*. The better performance of marshalling operation results in the easier the following services and the lower the cost. With the objective of making the costs as low as possible, *technical stations* reasonably marshal *local wagons* and *local district wagons* as thoroughly as possible. However, profound marshalling operation will inevitably raise the cost and the time allocated for marshalling in a *technical station* within some limitations. Thus a *technical station* will seek a best point where its marshalling operations can bring itself the lowest cost. The decision on how to marshal local wagons and local district wagons becomes key content for *technical stations*.

Among 1,440 min a day, some time is allocated for operations other than marshalling. Also, some marshalling operations are fixed so that a *technical station* cannot adjust it. Thus, a station can only decide on flexible wagon flows within available working time. A station needs first to distribute working time between *local wagons* and *local district wagons*, then divide it among different sections of a *local district wagon flow*. Based on this distribution, a station will decide the amount of marshalling a day, the amount of wagons and time for every marshalling. Generally speaking, *technical stations* make decisions from the following aspects.

1. Marshalling percentage: Influenced by time limitation, some marshalling operations can be executed to only some wagons while others must be treated as if they had the same sequence number (the same destination station) to reduce marshalling load. Thus, the percentage of wagons which will be marshaled is a decision made by a *technical station*.

2. Shunting choice: within limited working time, a *technical station* can decrease the shunting precision to finish a marshalling operation on time. Different shunting precisions occur in both *sort-shunting* and *group-shunting*. For *sort-shunting*, every wagon should be placed sequentially by their destinations. For *group-shunting*, marshalling is supposed to be finished as long as wagons with

the same destination are placed together. *Group-shunting* takes less working time than *sort-shunting*.

3. Marshalling precision: marshalling can be divided into different precise degrees. Actually, the destination of a wagon can be defined from generality to specificity by stations, operation areas, operation lines, or operation spots. The more precise, the more working time will be needed.

To facilitate modeling the RWFM problem, we have the following assumptions:

1. Marshalling difficulty is decided by the disorder degree of the wagon flow to be marshaled. Disorder degree depends on the destination stations of every wagon and the relationship among them, which occurs randomly. In this research, we hold that the disorder degrees for wagon flows have no difference.
2. Marshalling costs from two train flows, one of which is from Station A to Station B and the other is from Station B to Station A, may have trivial difference on marshalling cost. When making plans and calculating the cost, decision makers sometimes need to consider the influence from these differences. However, compared with other influencing factors, the influence from different directions is trivial and can be ignored. In this research, we hold that the marshalling costs with two train flows with different directions are exactly the same.

From the analysis above, a bi-level decision model for RWFM is built as follows:

For $x = (x_1, x_{21}, x_{22}, \ldots, x_{2m}) \in X \subset R^{m+1}, y_i = (\eta_{li}, \eta_{di}, y_{1Gil}, y_{1Sil}, y_{1Sid},$

$y_{wGil}, y_{2Sil}, y_{2Gid}, y_{2Sid}, y_{2ik}) \in Y_i \subset R^{10}, \quad i = 1, 2, \ldots,$

$$\max_{x \in X} \quad \sum_{p_j, l_j \in D} (p_j \times l_j) \left[ J_w (1 - r_j) - \bar{C}_{wj}(x) \right]$$
$$+ \sum_{s_i \in S} [q_{di}(x_1) \times \xi_i \times J_s - C_{si}(x, y_i)] \tag{14.3a}$$

$$\text{s.t.} \quad p_{ju\_min} \le p_{ju} \le p_{ju\_max}, \quad j = 1, 2, \ldots, m \tag{14.3b}$$

$$p_{jd\_min} \le p_{jd} \le p_{jd\_max}, \quad j = 1, 2, \ldots, m \tag{14.3c}$$

$$m_{j\_min} \le x_{2j} \le m_{j\_max}, \quad j = 1, 2, \ldots, m \tag{14.3d}$$

$$m_{j\_min} = \max \{ I_{ju} \times t_{T_jd}, I_{jd} \times t_{T_jd} \}, \quad j = 1, 2 \tag{14.3e}$$

$$0 \le q_{di} \le v_i \tag{14.3f}$$

$$0 \le q_{2i} + q_{di} \le u_i \tag{14.3g}$$

$$\min_{y_i \in Y_i} \sum_{s_i \in S} \left[ \left( C'_{z1i} + C''_{z1i} + C'''_{z1i} + C'_{z2i} + C''_{z2i} + C_{z3i} \right) \left( 1 + A_i \sigma_i^{B_i} \right) \right. \tag{14.4a}$$

$$\left. + \Delta C_{2i} \right]$$

$$\text{s.t. } q_i = q'_{d_i} + \sum_{d_k \in D_i} q''_{dik} + q_{zi} \tag{14.4b}$$

$$T_{il\_min} \le \eta_{li} \times q_{di} \times (y_{1Gil} \times y_{2Gil} \times T_{Gi} + y_{1Sil} \tag{14.4c}$$
$$\times y_{2Sil} \times T_{Si}) + q_{di} \times S_{il} \times T_{si} \le T_{il\_max}$$

$$T_{id\_min} \le \eta_{di} \times q_{di} \times (y_{1Gil} \times y_{2Gil} \times T_{Gi} + y_{1Sil} \tag{14.4d}$$
$$\times y_{2Sil} \times T_{Si}) + q_{di} \times S_{id} \times T_{si} \le T_{id\_max}$$

$$i = 1, 2, \ldots, m \tag{14.4e}$$

where

$$\bar{C}_{wj}(x) = \bar{C}_{w0} + \frac{\bar{C}_{wj}}{p_j} + \frac{\bar{C}_{2wj}}{x_{2j}} + \Delta \bar{C}_{wj},$$

$$\Delta \bar{C}_{wj} = \left| p_{ju} - p_{jd} \right| \times \frac{\bar{C}_{w2j}}{x_{2j} \times p_j},$$

$$I_{ju} = \frac{p_{ju}}{1,440 \times \psi_{ju}},$$

$$I_{jd} = \frac{p_{jd}}{1,440 \times \psi_{jd}},$$

$$C'_{z1i} = \bar{C}'_{z1i} \times \bar{q}'_{d_i},$$

$$\bar{C}'_{z1i} = z'_{10i} + \frac{Z_{11i}}{a'_{1i} \times \bar{q}_{d_i}^{-b'_{1i}}} + Z_{12i} \times \left( (y_{1Gil} + 2 \times y_{1Sil} + y_{1i})^{b'_{2i}} \right.$$

$$\left. + (y_{2Gil} + y_{2Sil})^{b'_{3i}} + \eta_{li}^{b'_{4i}} \right),$$

$$C''_{z1i} = \bar{C}''_{z1i} \times \bar{q}''_{d_i},$$

$$\bar{C}''_{z1i} = z''_{10i} + \frac{Z_{11i}}{\sum_{d_k \in D_i} a''_{1ik} \times x_{2ik}^{b''_{1ik}}}$$

$$+ Z_{12i} \times \sum_{d_k \in D_i} a''_{1ik}$$

$$\times \left[ a_{2ik} \times \left( (y_{1Gil} + 2 \times y_{1Sil} + y_{1i})^{b''_{2ik}} + y_{2ik}^{b''_{3ik}} + \eta_{dik}^{b''_{4ik}} \right) \right]$$

$$C'''_{z1i} = \bar{C}'''_{z1i} \times q_{zi},$$

$$\bar{C}'''_{z1i} = z'''_{10i} + \frac{Z_{11i}}{\sum_{d_k \in D_i} a'''_{1ik} \times x_{2ik}^{b'''_{1ik}}},$$

$$C'_{z2i} = \bar{C}'_{z2i} \times \bar{q}'_{d_i},$$

$$\bar{C}'_{z2i} = z'_{20i} + \frac{Z'_{21i}}{\alpha'_{21i} \times ((y_{1Gil} + y_{1Sil} + y_{1i})^{\beta'_{21i}} + (y_{2Gil} + y_{2Sil})^{\beta'_{22i}} + \eta_{li}^{\beta'_{23i}})},$$

$$C''_{z2i} = \bar{C}''_{z2i} \times \bar{q}''_{d_i},$$

$$\bar{C}''_{z2i} = z''_{20i} + \frac{Z''_{21i}}{\sum_{d_k \in D_i} \times \alpha''_{21i} \times ((y_{1Gil} + 2 \times y_{1Sil} + y_{1i})^{\beta''_{21i}} + y_{2ik}^{\beta''_{22i}} + \eta_{dik}^{\beta''_{23i}})},$$

$$C_{z3i} = N_i \times \bar{C}_{cx},$$

$$N_i = \sum_{d_k \in D_i} [(C_{di} + C_{cHi} \times \varsigma_{ik}) \times x_{2ik} + C''_{fik} \times q''_{dik}] + C'_{fi} \times q'_{di},$$

$$C'_{fi} = \frac{Z'_{31i}}{\alpha'_{31i} \times ((y_{1Gil} + y_{1Sil} + y_{1i})^{\beta'_{31i}} + (y_{2Gil} + y_{2Sil})^{\beta'_{32i}} + \eta_{li}^{\beta'_{33i}})},$$

$$C''_{fi} = \frac{Z''_{31i}}{\alpha''_{3iki} \times ((y_{1Gil} + y_{1Sil} + y_{1i})^{\beta''_{31ik}} + y_{2ik}^{\beta''_{32ik}} + \eta_{dik}^{\beta''_{33ik}})},$$

$$\Delta C_{2i} = \left( \sum_{d_k, d_l \in D_{i,k \neq i}} |x_{2ik} - x_{2il}| \right) \times Z_{4i},$$

$$Z_{4i} = Z_{40i} + \frac{Z_{41i}}{\sum_{d_k, d_l \in D_{i,k \neq i}} \alpha_{4ik} |x_{2ik} - x_{2il}|}.$$

The explanations for the above formulas are listed below:

1. Controlling variables:

$x_1$: Assignment of wagons which will go through the area administrated by a *railway bureau* and have more than one shunting operation in some *technical station* in this area.

$x_{2j}$: The number of vans within a shunted wagon list from the $j$th section, which is from one *technical station* to another in a *railway bureau*.

$x_{2ik}$: The number of wagons in a wagon list which is to the $k$th section in the $i$th station.

$\eta_{li}$: The percentage of wagons to be marshaled for local wagon list in the $i$th station.

$\eta_{di}$: The percentage of wagons to be marshaled for *local district wagon list* in the $i$th station.

$\eta_{dik}$: The percentage of wagons to be marshaled for *local district wagon list* to the $k$th direction in the $i$th station.

$y_{1Gil}$: The percentage of wagons to be marshaled by *group-shunting* of *local wagons* in the $i$th station.

$y_{1Sil}$: The percentage of wagons to be marshaled by *sort-shunting* of *local wagons* in the $i$th station.

$y_{1Gid}$: The percentage of wagons to be marshaled by *group-shunting* of *local district wagons* in the $i$th station.

$y_{1Sid}$: The percentage of wagons to be marshaled by *sort-shunting* of *local district wagons* in the $i$th station.

$y_{2Gil}$: The shunting precision for *local wagons* to be marshaled by *group-shunting* in the $i$th station.

$y_{2Sil}$: The shunting precision for *local wagons* to be marshaled by *sort-shunting* in the $i$th station.

$y_{2Gid}$: The shunting precision for *local district wagons* to be marshaled by *group-shunting* in the $i$th station.

$y_{2Sid}$: The shunting precision for *local district wagons* to be marshaled by *sort-shunting* in the $i$th station.

$y_{2ik}$: The shunting precision for *local district wagon flow* marshaled in the $i$th station to the $k$th direction.

2. Other variables: while decision makers from both the upper and lower levels directly control variables of $x$ and $y$, there are some other variables whose values are influenced by $x$ and $y$ directly or indirectly. These variables are summarized below:

**Variables influenced by $x_1$:**

$p_j$: The average wagon flow in the $j$th section, $p_j = p_{ju} + p_{jd}$.

$p_{ju}$: The average wagon flow in the $j$th section in the up-direction, which fluctuates with the change of $x_1$.

$p_{jd}$: The average wagon flow in the $j$th section in the down-direction, which fluctuates with the change of $x_1$.

$q_{di}$: The number of *local district wagons* and *local wagons* operated per day in the $i$th station.

$q'_{d_i}$: The number of local wagons operated in the $i$th station.

$q''_{d_ik}$: The number of local district wagons operated to the $k$th direction in the $i$th station.

$q''_{d_i}$: The number of local district wagons operated in the $i$th station.

$q_{2i}$: The number of long-distance wagons operated in the $i$th station.

$\xi_i$: The loading percentage in the $i$th station.

$r_j$: The percentage of empty to loaded wagon kilometers in the $j$th section. It fluctuates with the change of $x_1$.

**Variables influenced by $x$:**

$q_{zi}$: The number of wagons marshaled in the $i$th station a day.

$\varsigma_{ik}$: The number of *long distance wagons* to the $k$th direction in the $i$th station.

$n_{kl}$: The number of wagon lists some of whose wagons have been added/removed from the $k$th section to the $l$th section.

**Variables influenced by $y$:**

$y_{li}$: Marshalling degree, which is determined by different operating depth, such as *group shunting*, *sort-shunting* and the fit degree of the regulation of *Safety terms*, for the *local district wagon* in the $i$th station.

$\sigma_i$: The average time difference among the operations for *local wagon flow*, *local district wagon flow*, and *long-distance wagon flow* in the $i$th station.

**Variables influenced by $x$ and $y$:**

$q_i$: The number of wagons operated in the $i$th station.

3. Coefficients and constants:

$S = \{s_i, \ i = 1, 2, \ldots, n\}$: The set of the *technical stations* administrated by a *railway bureau*.

$D = \{d_j, \ j = 1, 2, \ldots, m\}$: The set of the train running sections administrated by a *railway bureau*.

$D_i = \{d_k, \ k = 1, 2, \ldots, l\}$: The set of train running sections which are adjacent to the $i$th station.

$l_j$: The hauling distance in the $j$th section, which is a constant.

$J_w$: Railway average tariff, which is a constant.

$\bar{C}_{w0}$: Freight traffic fixed unit cost.

$\bar{C}_{w1j}$: The freight traffic unit cost in the $j$th section per day per kilometer.

$\bar{C}_{w2j}$: Hauling cost in the $j$th section per wagon per kilometer.

$\bar{C}'_{z2j}$: The locomotive cost in the $j$th section per kilometer when there is no wagon hauled by the locomotive.

$J_s$: Fees charged per wagon.

$p_{ju\_min}$: The minimum wagon flow which can meet the requisite traffic demand required for the $j$th section in the up-direction.

$p_{ju\_max}$: The maximum wagon flow which can be run for the $j$th section in the up-direction.

$p_{jd\_min}$: The minimum wagon flow which can meet the requisite traffic demand required for the $j$th section in the down-direction.

$p_{jd\_max}$: The maximum wagon flow which can be run for the $j$th section in the down-direction.

$m_{j\_max}$: The maximum number of wagons to form a wagon list in the $j$th section. It is determined by the locomotive hauling limit and the useful length of the receiving and departure tracks in the $j$th section.

$v_i$: The maximum possible number of wagons that can be operated by *service operation* in the $i$th station.

$u_i$: The maximum possible number of wagons that can be marshaled in the $i$th station.

$\varphi_{ju}$: The percentage of time that can be used a day (1,440 min) for freight transportation in the $j$th section in the up direction.

$\varphi_{jd}$: The percentage of time that can be used a day (1,440 min) for freight transportation in the $j$th section in the down direction.

$t_{Tju}$: The minimum time interval between two wagon lists of the $j$th section in the up direction regulated by the train working diagram.

$t_{Tjd}$: The minimum time interval between two wagon lists of the $j$th section in the down direction regulated by the train working diagram.

$Z'_{10i}$: The minimum cost for marshalling one *local wagon* in the $i$th station. This cost happens in an ideal situation when the number of wagons to be marshaled is large enough and the marshalling degree is deep enough for one marshalling operation.

$Z_{11i}$: Coefficient for the effect from centralized marshalling operation.

$\bar{q}_{di}$: The number of local wagons to be marshaled for one marshalling operation. It is determined by the loading/uploading capacity in the $i$th station.

$Z_{12i}$: Coefficient for the effect from deepened marshalling operation. It is an average additional cost spent for one wagon for marshalling operation.

$A_i, B_i, a'_{1i}, b'_{1i}, b'_{2i}, b'_{3i}, b'_{4i}, a''_{1ik}, b''_{1ik}, \alpha'_{21i}, \beta'_{21i}, \beta'_{22i}, \beta'_{23i}, \alpha''_{21i}, \beta''_{21i}, \beta''_{22i},$
$\beta''_{23i}, \alpha'_{31k}, \beta'_{31k}, \beta'_{32k}, \beta'_{33k}, \alpha''_{31k}, \beta''_{31k}, \beta''_{32k}, \beta''_{33k}$: Coefficients which are to be obtained through statistic data.

$Z''_{10i}$: The minimum cost for marshalling one *local district wagon* in the $i$th station. This cost happens in an ideal situation when the number of wagons to be marshaled is large enough and the marshalling degree is deep enough for one marshalling operation.

$Z'''_{10i}$: The minimum cost for marshalling one *long distance wagon* in the $i$th station. This cost happens in an ideal situation when the number of wagons to be marshaled is large enough and the marshalling degree is deep enough for one marshalling operation.

$Z'_{20i}$: The minimum cost for *service operation* for one local wagon in the $i$th station. This cost happens in an ideal situation when marshalling is deep enough such that service operation can be operated easily and conveniently.

$Z'_{21i}$: The additional cost for service operation for one *local wagon* in the $i$th station. This cost happens when marshalling is superficial thus service operation become unhandy.

$Z''_{20i}$: The minimum cost for service operation for one *local district wagon* in the $i$th station. This cost happens in an ideal situation when marshalling is deep enough such that service operation can be operated easily and conveniently.

$Z''_{21i}$: The additional cost for service operation for one local district wagon in the $i$th station. This cost happens when marshalling is superficial thus service operation become unhandy.

$C_{di}$: The coefficient on centralisation and detention for *local wagon flow*, which is a number between eight and twelve. The number is decided by specialties from different wagon flows.

$C_{cHi}$: The coefficient on centralisation and detention for *long distance wagon flow*, which is a number between eight and twelve.

$Z'_{3i}$: Coefficient on service facilitation for *local wagons*, which equals the additional halting time when service operation is totally inconvenient.

$Z''_{3i}$: Coefficient on service facilitation for *local district wagons*, which equals the additional halting time when service operation is totally inconvenient.

$Z_{40i}$: The basic cost for adding/removing one wagon to/from a wagon list in the *i*th station. This cost happens when the number of wagons to be added/removed is large enough.

$Z_{41i}$: The additional cost for adding/removing one wagon to/from a wagon list in the *i*th station. This cost happens when the number of wagons to be added/removed is small enough (equaling one).

$\bar{C}_{cx}$: The cost for one wagon to halt one hour, which is caused by wagon depreciation.

$S_{il}$: The percentage of wagons which have special safety requirement of *local wagons* in the *i*th station.

$S_{id}$: The percentage of wagons which have special safety requirement of *local district wagons* in the *i*th station.

$T_{si}$: Average time to marshal a wagon which has special safety requirement in the *i*th station.

$T_{il\_min}$: The least time for marshalling one *local wagon list* in the *i*th station.

$T_{il\_max}$: The time spent for marshalling one *local wagon list* with the highest specification and completeness in the *i*th station.

$T_{id\_min}$: The least time for marshalling one *local district wagon list* in the *i*th station.

$T_{il\_max}$: The time spent for marshalling one *local district wagon list* with the highest specification and completeness in the *i*th station.

4. Formula:

(14.3a) describes a *railway bureau*'s objective which aims at achieving the maximum profit for freight operation administrated by this *railway bureau*. It has two parts: the profit from the railway network and technical stations in this bureau.

(14.3b) and (14.3c) mean that wagon flows in both the up-direction and down-direction have their minimum and maximum limits. Thus the total number of vans from one station to another has its limits too.

(14.3d) and (14.3e) tell how the limits set for wagon flows in the up-direction and down-direction are determined and calculated.

(14.3f) and (14.3g) mean the number of long-distance wagon flows to be marshaled in *technical stations* cannot exceed their operating abilities.

(14.4a) is the objective for a *technical station*, which aims at lowering its operation cost. The operation cost is from three parts: *local wagon flow, local district wagon flow*, and *long-distance wagon flow*.

(14.4b) denotes how the operation cost for *local wagon flow, local district wagon flow*, and *long-distance wagon flow* are calculated.

(14.4c) and (14.4d) mean that there exist minimum and maximum limits for marshalling both *local wagon flow* and *local district wagon flow*.

5. Symbols:

$F_1$: The economical benefit of the railway network within the area administrated by a *railway bureau*.

$F_2$: The economical benefit obtained by all of the *technical stations* administrated by the *railway bureau*.

$\bar{C}_{w_j}$: The freight traffic unit cost in the $j$th section.

$\Delta \bar{C}_{w_j}$: Additional unit cost in the $j$th section.

$C_{s_i}$: The operating cost in the $i$th station. It fluctuates with the change of controlling variables from both the leader and the followers.

$I_{ju}$: The number of wagons that go through the $j$th section in the up direction per minute.

$m_{j\_min}$: The minimum number of wagons to form a wagon list in the $j$th section.

$I_{jd}$: The number of wagons that go through the $j$th section in the down direction per minute.

$C'_{z1i}$: Daily cost spent for marshalling *local wagon flow* for the $i$th station.

$C''_{z1i}$: Daily cost spent for marshalling *local district wagon flow* for the $i$th station.

$C'''_{z1i}$: Daily cost spent for marshalling *long-distance wagon flow* for the $i$th station.

$C'_{z2}$: Daily cost spent for *service operation* made for *local wagon flow* for the $i$th station.

$C''_{z2i}$: Daily cost spent for *service operation* made for *local district wagon flow* for the $i$th station.

$C_{z3i}$: Daily cost spent for centralizing and detention of wagons in the $i$th station.

$\Delta C_{2i}$: Different sections may have different requests on the number of wagon lists to be run in that section. Thus adding/reducing wagons may be needed in *technical stations* to meet the requirements of its adjacent sections. $\Delta C_{2i}$ is the daily cost spent for adding/reducing wagons in the $i$th station.

$\bar{C}'_{z1i}$: Average daily cost spent for marshalling one *local wagon* for the $i$th station.

$\bar{C}''_{z1i}$: Average daily cost spent for marshalling one *local district wagon* for the $i$th station.

$\bar{C}'_{z2i}$: Average daily cost spent for *service operation* made for *local wagon flow* for the $i$th station.

$\bar{C}''_{z2i}$: Average daily cost spent for *service operation* made for *local district wagon flow* for the $i$th station.

$N_i$: The total hours spent by all wagons which are halted in the $i$th station.

$C'_{fi}$: The number of additional hours spent for *service operation* for a *local wagon* in the $i$th station.

$C''_{fik}$: The number of additional hours spent for *service operation* for a *local district wagon* to the $k$th direction in the $i$th station.

$Z_{4i}$: The cost spent for adding/removing one wagon.

### 14.2.3 Experiments

In this section, we consider the RWFM problem in a railway bureau: Bureau U. Within the area administrated by Bureau U, there are three *technical stations*: Station A, Station B, and Station C. Connecting these stations, we have three sections: Section 1 that connects Station A and Station B, Section 2 that connects Station B and Station C, and Section 3 that connects Station C and Station A. We list the values of some of the main coefficients, which are used to build the bi-level decision model for this RWFM problem in Tables 14.2 and 14.3.

To help the decision maker in Bureau U make an optimal RWFM plan, we use the FBLDSS software presented in Chap. 11 to reach a solution. By 342 s running, the solutions for Bureau U are reached and summarized in Table 14.4.

To test the stability of the FBLDSS, this example has been run six times by the FBLDSS software. The solution variances are summarized in Table 14.5.

**Table 14.2**  Summary of the coefficient values in the experiments 1

| Station | $P_{ju\_min}$ | $P_{ju\_max}$ | $P_{jd\_min}$ | $P_{jd\_max}$ | $u_i$ | $v_i$ |
|---|---|---|---|---|---|---|
| A | 10 | 29 | 10 | 29 | 19 | 19 |
| B | 10 | 29 | 10 | 29 | 19 | 19 |
| C | 10 | 29 | 10 | 29 | 19 | 19 |

**Table 14.3**  Summary of the coefficient values in the experiments 2

| Section | $T_{il\_min}$ (min) | $T_{il\_max}$ (min) | $T_{id\_min}$ (min) | $T_{id\_max}$ (min) |
|---|---|---|---|---|
| 1 | 15 | 20 | 15 | 22 |
| 2 | 15 | 23 | 15 | 25 |
| 3 | 15 | 22 | 15 | 21 |

**Table 14.4**  Summary of the solutions for Bureau U and the stations

| Station | $P_{ju}$ | $P_{jd}$ | $q_{di} + q'_{di}$ | $q_{2i}$ | $x_{2k}$ | $y_{1Gil}$ | $y_{1Sil}$ | $y_{1i}$ |
|---|---|---|---|---|---|---|---|---|
| A | 24 | 28 | 42 | 10 | 18 | 0.2 | 0.78 | 0.41 |
| B | 27 | 23 | 38 | 12 | 11 | 0.55 | 0.67 | 0.35 |
| C | 24 | 17 | 25 | 16 | 7 | 0.53 | 0.98 | 0.48 |

**Table 14.5**  Summary of the solutions for Bureau U and the stations

| Station | $P_{ju}$ | $P_{jd}$ | $q_{di} + q'_{di}$ | $q_{2i}$ | $x_{2k}$ | $y_{1Gil}$ | $y_{1Sil}$ | $y_{1i}$ |
|---|---|---|---|---|---|---|---|---|
| A | 0.002 | 0.12 | 0.1 | 0.05 | 0.37 | 0.16 | 0.00043 | 0.1 |
| B | 0.04 | 0.29 | 0.02 | 0.27 | 0.42 | 0.33 | 0.01 | 1.23 |
| C | 0.17 | 0.37 | 0.42 | 0.07 | 0.57 | 0.02 | 0.27 | 0.09 |

In Table 14.5, we can see that, there is no very large diversion among the solutions obtained. For every running, the solution has been obtained within 400 s. Thus, we can come to the conclusion that the FBLDSS could explore veracious solutions for RWFM problems with quite effective and stable performance.

## 14.3 Summary

In this chapter, the bi-level optimization natures in *train set organization* (TSO) and *railway wagon flow management* (RWFM) have been put forward by abstracting and simplifying railway trains management. First, two bi-level decision models are established for the two problems respectively. Then, these two decision models are applied to technical stations for real case studies. The experiment results obtained from these two case studies could be helpful for the tasks of train set organization and railway wagon flow management.

# References

Aamodt, A., Plaza, E.: Case-based reasoning: fundamental issues, methodological variations, and system approaches. AI Commun. **7**, 39–59 (1994)

Abo-Sinna, M.A., Baky, I.A.: Interactive balance space approach for solving multi-level multi-objective programming problems. Inf. Sci. **177**(16), 3397–3410 (2007)

Aboussoror, A., Mansouri, A.: Existence of solutions to weak nonlinear bilevel problems via MinSup and d.c. problems. Rairo-Oper. Res. **42**(2), 87–102 (2008)

Adewumi, A.O., Ali, M.M.: A multi-level genetic algorithm for a multi-stage space allocation problem. Math. Comput. Model. **51**(1–2), 109–126 (2010)

Agrawal, R., Imielinski, T., Swami, A.: Mining association rules between sets of items in large databases. The ACM SIGMOD Conference on Management of Data Washington, D.C., pp. 207–216. (1993)

Agrawal, R., Srikant, R.: Fast algorithms for mining association rules. The 20th International Conference of Very Large Data Bases, San Francisco, pp. 487–499. (1994)

Aiyoshi, E., Shimizu, K.: Hierarchical decentralized systems and its new solution by abarrier method. IEEE Trans. Syst. Man Cybern. B Cybern. **11**, 444–449 (1982)

Akartunali, K., Miller, A.J.: A heuristic approach for big bucket multi-level production planning problems. Eur. J. Oper. Res. **193**(2), 396–411 (2009)

Almeder, C.: A hybrid optimization approach for multi-level capacitated lot-sizing problems. Eur. J. Oper. Res. **200**(2), 599–606 (2010)

Amador, F., Romero, C.: Redundancy in lexicographic goal programming: an empirical approach. Eur. J. Oper. Res. **41**(3), 347–354 (1989)

Amailef, K., Lu, J.: Ontology-supported case-based reasoning approach for intelligent m-Government emergency response services. Decis. Support Syst. **55**(1), 79–97 (2013)

Amat, J.F., McCarl, B.: A representation and economic interpretation of a twolevel programming problem. J. Oper. Res. Soc. **32**, 783–792 (1981)

Amati, G.: A multi-level filtering approach for fairing planar cubic B-spline curves. Comput. Aided Geom. Des. **24**(1), 53–66 (2007)

Amouzegar, M.A.: Nonlinear Bilevel Programming: Analysis and Application to Regional Hazardous Waste Management. University of California, Los Angeles (1995)

Amouzegar, M.A.: A global optimization method for nonlinear bilevel programming problems. IEEE Trans. Syst. Man Cybern. Part B Cybern. **29**(6), 771–777 (1999)

Amouzegar, M.A., Moshirvaziri, K.: Strategic management decision support system: an analysis of the environmental policy issues. Environ. Model. Assess. **6**, 297–306 (2001)

An, L.T.H., Tao, P.D., Canh, N.N., Van Thoai, N.: DC programming techniques for solving a class of nonlinear bilevel programs. J. Glob. Optim. **44**(3), 313–337 (2009)

Anandalingam, G., Friesz, T.: Hierarchical optimization: an introduction. Ann. Oper. Res. **34**, 1–11 (1992)

© Springer-Verlag Berlin Heidelberg 2015
G. Zhang et al., *Multi-Level Decision Making*,
Intelligent Systems Reference Library 82, DOI 10.1007/978-3-662-46059-7

Andreani, R., Castro, S.L.C., Chela, J.L., Friedlander, A., Santos, S.A.: An inexact-restoration method for nonlinear bilevel programming problems. Comput. Optim. Appl. **43**(3), 307–328 (2009)

Angern, A.A., Luthi, H.J.: Intelligent decision support systems: a visual interactive approach. Interfaces **20**, 17–28 (1990)

Ankhili, Z., Mansouri, A.: An exact penalty on bilevel programs with linear vector optimization lower level. Eur. J. Oper. Res. **197**(1), 36–41 (2009)

Areibi, S., Yang, Z.: Effective memetic algorithms for VLSI design = genetic algorithms + local search +multi-level clustering. Evol. Comput. **12**(3), 327–353 (2004)

Arora, S.R., Gupta, R.: Interactive fuzzy goal programming approach for bilevel programming problem. Eur. J. Oper. Res. **194**(2), 368–376 (2009)

Arroyo, J.M.: Bilevel programming applied to power system vulnerability analysis under multiple contingencies. IET Gener. Transm. Distrib. **4**(2), 178–190 (2010)

Arroyo, J.M., Galiana, F.D.: On the solution of the bilevel programming formulation of the terrorist threat problem. IEEE Trans. Power Syst. **20**(2), 789–797 (2005)

Audet, C., Haddad, J., Savard, G.: A note on the definition of a linear bilevel programming solution. Appl. Math. Comput. **181**(1), 351–355 (2006)

Audet, C., Haddad, J., Savard, G.: Disjunctive cuts for continuous linear bilevel programming. Optim. Lett. **1**(3), 259–267 (2007a)

Audet, C., Savard, G., Zghal, W.: New branch-and-cut algorithm for bilevel linear programming. J. Optim. Theory Appl. **134**(2), 353–370 (2007b)

Aviso, K.B., Tan, R.R., Culaba, A.B., Cruz, J.B.: Bi-level fuzzy optimization approach for water exchange in eco-industrial parks. Process Saf. Environ. Prot. **88**(1), 31–40 (2010)

Babahadda, H., Gadhi, N.: Necessary optimality conditions for bilevel optimization problems using convexificators. J. Glob. Optim. **34**(4), 535–549 (2006)

Baky, I.A.: Fuzzy goal programming algorithm for solving decentralized bi-level multi-objective programming problems. Fuzzy Sets Syst. **160**(18), 2701–2713 (2009)

Ballou, R.H.: Business Logistics/Supply Chain Management, 5th edn. Pearson Prentice Hall, Upper Saddle River (2004)

Bard, J., Falk, J.: Necessary conditions for the linear three level programming problem. The 21st IEEE Conference on Decision and Control, DEC, Orlando, pp. 642–646. (1982a)

Bard, J.F.: Optimality conditions for the bilevel programming problem. Nav. Res. Logistics Q. **31**, 13–26 (1984)

Bard, J.F.: Geometric and algorithmic developments for a hierarchical planning problem. Eur. J. Oper. Res. **19**(3), 372–383 (1985)

Bard, J.F.: Practical Bilevel Optimization: Algorithms and Applications. Kluwer Academic Publishers, Dordrecht (1998)

Bard, J.F., Falk, J.E.: An explicit solution to the multi-level programming problem. Comput. Oper. Res. **9**, 77–100 (1982b)

Bard, J.F., Moore, J.T.: A branch and bound algorithm for the bilevel programming problem. SIAM J. Sci. Stat. Comput. **11**, 281–292 (1990)

Bard, J.F., Moore, J.T.: An algorithm for the discrete bi-level programming problem. Nav. Res. Logistics **39**, 419–435 (1992)

Basener, F.: Topology and Its Applications. Wiley, Hoboken (2006)

Bazaraa, M.S., Sherali, H.D., Shetty, C.M.: Nonlinear programming: theory and algorithms. Wiley, Hoboken (2013)

Bedeschi, F., Fackenthal, R., Resta, C., Donze, E.M., Jagasivamani, M., Buda, E.C., Pellizzer, F., Chow, D.W., Cabrini, A., Calvi, G.M.A., Faravelli, R., Fantini, A., Torelli, G., Mills, D., Gastaldi, R., Casagrande, G.: A bipolar-selected phase change memory featuring multi-level cell storage. IEEE J. Solid-State Circuits **44**(1), 217–227 (2009)

Begnaud, J., Benjaafar, S., Miller, L.A.: The multi-level lot sizing problem with flexible production sequences. IIE Trans. **41**(8), 702–715 (2009)

Bellman, R.E., Zadeh, L.A.: Decision-making in a fuzzy environment. Manag. Sci. **17**, 141–164 (1970)

Ben-Ayed, O.: Bilevel Linear Programming: Analysis and Application to the Network Design Problem. University of Illinois at Urbana-Champaign, Champaign (1988)

Ben-Ayed, O.: Bilevel linear programming. Comput. Oper. Res. **20**, 485–501 (1993)

Benayoun, R., Montogolfier, J., Tergny, J., Larichev, O.: Linear programming with multiple objective functions: step method (STEM). Math. Program. **1**, 366–375 (1971)

Benson, H.P.: On the structure and properties of a linear multilevel programming problem. J. Optim. Theory Appl. **60**(3), 353–373 (1989)

Berezinski, M., Holubiec, J., Petriczek, G.: Multilevel models for railway traffic control. Int. Conf. Control **1**, 54–56 (1994)

Bergounioux, M., Haddou, M.: A regularization method for ill-posed bilevel optimization problems. Rairo-Oper. Res. **40**(1), 19–35 (2006)

Bhargava, H.K., Krishnan, R., Müller, R.: Decision support on demand: emerging electronic markets for decision technologies. Decis. Support Syst. **19**, 193–214 (1997)

Bialas, W., Karwan, M., Shaw, J.: A parametric complementary pivot approach for twolevel linear programming. In: Operations Research Program. Technical report 802, State University of New York, Buffalo, (1980a)

Bialas, W., Karwan, M.H.: Two-level linear programming. Manag. Sci. **30**, 1004–1020 (1984)

Bialas, W.F., Karwan, M.: Multilevel linear programming. Operations Research Program. Technical report, vol. 78-1, State University of New York at Buffalo (1978)

Bialas, W.F., Karwan, M., Shaw, J.: A parametric complementary pivot approach for twolevel linear programming. Operations Research Program. State University of New York at Buffalo, (1980b)

Bialas, W.F., Karwan, M.H.: On two-level optimization. IEEE Trans. Autom. Control **AC-26**, 211–214 (1982)

Bianco, L., Caramia, M., Giordani, S.: A bilevel flow model for hazmat transportation network design. Transp. Res. Part C Emerg. Technol. **17**(2), 175–196 (2009)

Bilgili, E., Goknar, I.C., Albora, A.M., Ucan, O.N.: Potential anomaly separation and archeological site localization using genetically trained multi-level cellular neural networks. ETRI J. **27**(3), 294–303 (2005)

Biswal, B., Dash, P.K., Panigrahi, B.K.: Power quality disturbance classification using fuzzy c-means algorithm and adaptive particle swarm optimization. Trans. Ind. Electron. **56**, 212–220 (2008)

Bjondal, M., Jornsten, K.: The deregulated electricity market viewed as a bilevel programming problem. J. Glob. Optim. **33**(3), 465–475 (2005)

Blair, C.: The computational complexity of multi-level linear programs. Ann. Oper. Res. **34**(1), 13–19 (1992). doi:10.1007/BF02098170

Bollas, G.M., Barton, P.I., Mitsos, A.: Bilevel optimization formulation for parameter estimation in vapor-liquid(-liquid) phase equilibrium problems. Chem. Eng. Sci. **64**(8), 1768–1783 (2009)

Bonczek, R.H., Holsapple, C.W., Whinston, A.B.: Foundations of Decision Support Systems. Academic Press, New York (1981)

Bonnel, H., Morgan, J.: Semivectorial bilevel optimization problem: penalty approach. J. Optim. Theory Appl. **131**(3), 365–382 (2006)

Boyce, D.: Practical bilevel optimization: Algorithms and applications. Transp. Sci. **36**(1), 144–146 (2002)

Bracken, J., McGill, J.: Mathematical programs with optimization problems in the constraints. Oper. Res. **21**, 37–44 (1973)

Breiner, A., Avriel, M.: Two-stage approach for quantitative policy analysis using bilevel programming. J. Optim. Theory Appl. **100**(1), 15–27 (1999)

Broderick, A.J., Greenley, G.E., Mueller, R.D.: The behavioural homogeneity evaluation framework: multi-level evaluations of consumer involvement in international segmentation. J. Int. Bus. Stud. **38**(5), 746–763 (2007)

Brotcorne, L.., Hanafi, S., Mansi, R.: A dynamic programming algorithm for the bilevel knapsack problem. Oper. Res. Lett. **37**(3), 215–218 (2009)

Brotcorne, L., Labbe, M., Marcotte, P., Savard, G.: A bilevel model and solution algorithm for a freight tariff-setting problem. Transp. Sci. **34**(3), 289–302 (2000)

Brotcorne, L., Labbe, M., Marcotte, P., Savard, G.: A bilevel model for toll optimization on a multicommodity transportation network. Transp. Sci. **35**(4), 345–358 (2001)

Brotcorne, L., Marcotte, P., Savard, G.: Bilevel programming: the montreal school. Inf. Syst. Oper. Res. **46**(4), 231–246 (2008)

Bui, T., Lee, J.: An agent-based framework for building decision support systems. Support Syst. **25**, 225–237 (1999)

Bulinskaya, E.: Multi-level control and asymptotic behaviour of some inventory systems. Int. J. Prod. Econ. **59**(1–3), 289–295 (1999)

Calvete, H.I., Gale, C.: On the quasiconcave bilevel programming problem. J. Optim. Theory Appl. **98**(3), 613–622 (1998)

Calvete, H.I., Gale, C.: A note on 'bilevel linear fractional programming problem'. Eur. J. Oper. Res. **152**(1), 296–299 (2004a)

Calvete, H.I., Gale, C.: Solving linear fractional bilevel programs. Oper. Res. Lett. **32**(2), 143–151 (2004b)

Calvete, H.I., Gale, C.: Linear bilevel multi-follower programming with independent followers. J. Glob. Optim. **39**(3), 409–417 (2007)

Calvete, H.I., Gale, C., Mateo, P.M.: A genetic algorithm for solving linear fractional bilevel problems. Ann. Oper. Res. **166**(1), 39–56 (2009)

Calvete, H.I., Galee, C., Mateo, P.M.: A new approach for solving linear bilevel problems using genetic algorithms. Eur. J. Oper. Res. **188**(1), 14–28 (2008)

Cambridge, U.O. http://thesaurus.maths.org/dictionary/map/word/10037 (2001).

Campelo, M., Dantas, S., Scheimberg, S.: A note on a penalty function approach for solving bilevel linear programs. J. Glob. Optim. **16**(3), 245–255 (2000)

Campelo, M., Scheimberg, S.: A simplex approach for finding local solutions of a linear bilevel program by equilibrium points. Ann. Oper. Res. **138**(1), 143–157 (2005a)

Campelo, M., Scheimberg, S.: A study of local solutions in linear bilevel programming. J. Optim. Theory Appl. **125**(1), 63–84 (2005b)

Candler, W., Norton., R.: Multilevel programming and development policy. In: World Bank Staff. vol. 258. Washington D.C. (1977)

Candler, W., Townsley, R.: A linear two-level programming problem. Comput. Oper. Res. **9**, 59–76 (1982)

Cao, D., Chen, M.Y.: Capacitated plant selection in a decentralized manufacturing environment: a bilevel optimization approach. Eur. J. Oper. Res. **169**(1), 97–110 (2006)

Cao, D., Leung, L.C.: A partial cooperation model for non-unique linear two-level decision problems. Eur. J. Oper. Res. **140**(1), 134–141 (2002)

Carlin, U.S., Komorowski, J., Ohrn, A.: Rough set analysis of patients with suspected of acute appendicitis. The 7th Conference on Information Processing and Management of Uncertainty in Knowledge-Based Systems, Paris, France, pp. 1528–1533. (1998)

Carrion, M., Arroyo, J.M., Conejo, A.J.: A bilevel stochastic programming approach for retailer futures market trading. IEEE Trans. Power Syst. **24**(3), 1446–1456 (2009)

Carvalho, M.C.A., Pinto, J.M.: A bilevel decomposition technique for the optimal planning of offshore platforms. Braz. J. Chem. Eng. **23**(1), 67–82 (2006)

Castellazzi, A., Ciappa, M.: Multi-domain multi-level abstraction modelling of integrated power devices. Solid-State Electron. **53**(11), 1202–1208 (2009)

Chan, C.M., Makino, S., Isobe, T.: Interdependent behavior in foreign direct investment: the multi-level effects of prior entry and prior exit on foreign market entry. J. Int. Bus. Stud. **37**(5), 642–665 (2006)

Chang, J.S., Mackett, R.L.: A bi-level model of the relationship between transport and residential location. Transp. Res. Part B Methodol. **40**(2), 123–146 (2006)

Charnes, A., Cooper, W.W.: Management Models and Industrial Applications of Linear Programming. Manag. Sci. **4**(1), 38–91 (1957)

Charnes, A., Cooper, W.W.: Goal programming and multiple objective optimizations. Eur. J. Oper. Res. **1**, 39–54 (1977)

Chen, C.I., Gruz, J.B.: Stackelberg solution for two person games with biased information patterns. IEEE Trans. Autom. Control **AC-17**, 791–798 (1972)

Chen, C.N.: Non-linear multi-level finite element analysis based on multiplicative and additive corrections. Finite Elem. Anal. Des. **37**(6–7), 559–573 (2001)

Chen, J.Q., Lee, S.M.: An exploratory cognitive DSS for strategic decision making. Decis. Support Syst. **36**, 147–160 (2003)

Chen, K.J., Ji, P.: Development of a genetic algorithm for scheduling products with a multi-level structure. Int. J. Adv. Manuf. Technol. **33**(11–12), 1229–1236 (2007)

Chen, Y.: Bilevel programming problems: analysis, algorithms and applications. Ph.D. thesis, University of Saskatchewan, Sask,Canada (1994)

Chen, Y., Florian, M., Wu, S.: A descent dual approach for linear bilevel programs. Centre de Recherche sur les Transports (1992)

Chidambaram, L.: Relational development in computer-supported groups. MIS Quart. **20**, 143–163 (1996)

Chiou, S.-W.: A bi-level programming for logistics network design with system-optimized flows. Inf. Sci. **179**(14), 2434–2441 (2009a)

Chiou, S.W.: Bilevel programming for the continuous transport network design problem. Transp. Res. Part B Methodol. **39**(4), 361–383 (2005)

Chiou, S.W.: A bi-level programming for logistics network design with system-optimized flows. Inf. Sci. **179**(14), 2434–2441 (2009b)

Christiansen, S., Patriksson, M., Wynter, L.: Stochastic bilevel programming in structural optimization. Struct. Multi. Optim. **21**(5), 361–371 (2001)

Chung, K.S., Kim, T.W., Liu, C.L.: Power optimization of multi-level logic circuits utilizing circuit symmetries. Int. J. Electron. **87**(7), 853–864 (2000)

Clegg, J., Smith, M., Xiang, Y.L., Yarrow, R.: Bilevel programming applied to optimising urban transportation. Transp. Res. Part B Methodol. **35**(1), 41–70 (2001)

Coelho, R.F., Breitkopf, P., Knopf-Lenoir, C., Villon, P.: Bi-level model reduction for coupled problems. Struct. Multi. Optim. **39**(4), 401–418 (2009)

Cohen, G., Quadrat, J.P., Wynter, L.: On the convergence of the algorithm for bilevel programming problems by Clegg and Smith. Transp. Res. Part B Methodol. **36**(10), 939–944 (2002)

Cohen, M., Kelly, C.B., Medaglia, A.L.: Decision support with web-enabled software. Interfaces **31**, 109–128 (2001)

Colson, B., Marcotte, P., Savard, G.: A trust-region method for nonlinear bilevel programming: algorithm and computational experience. Comput. Optim. Appl. **30**(3), 211–227 (2005)

Colson, B., Marcotte, P., Savard, G.: An overview of bilevel optimization. Ann. Oper. Res. **153**(1), 235–256 (2007)

Courtney, J.F.: Decision making and knowledge management in inquiring organizations: toward a new decision-making paradigm for DSS. Decis. Support Syst. **31**, 17–38 (2001)

Cutello, V., Montero, J.: Hierarchies of aggregation operators. Int. J. Intell. Syst. **9**(11), 1025–1045 (1994)

Cutello, V., Montero, J.: Recursive connective rules. Int. J. Intell. Syst. **14**(1), 3–20 (1999)

Dantzig, G.B.: Linear programming. Oper. Res. **50**(1), 42–47 (2002)

del Amo, A., Montero, J., Molina, E.: Representation of consistent recursive rules. Eur. J. Oper. Res. **130**(1), 29–53 (2001)

Dellaert, N., Jeunet, J., Jonard, N.: A genetic algorithm to solve the general multi-level lot-sizing problem with time-varying costs. Int. J. Prod. Econ. **68**(3), 241–257 (2000)

DeMiguel, V., Murray, W.: A local convergence analysis of bilevel decomposition algorithms. Optim. Eng. **7**(2), 99–133 (2006)

DeMiguel, V., Xu, H.: A stochastic multiple-leader stackelberg model: analysis, computation, and application. Oper. Res. **57**(5), 1220–1235 (2009)

Dempe, S.: A simple algorithm for the linear bilevel programming problem. Optimization **18**, 373–385 (1987)

Dempe, S.: First-order necessary optimality conditions for general bilevel programming problems. J. Optim. Theory Appl. **95**(3), 735–739 (1997)

Dempe, S.: A bundle algorithm applied to bilevel programming problems with non-unique lower level solutions. Comput. Optim. Appl. **15**(2), 145–166 (2000)

Dempe, S.: Foundations of Bilevel Programming. Kluwer Academic Publishers, Dordrecht (2002)

Dempe, S., Bard, J.F.: Bundle trust-region algorithm for bilinear bilevel programming. J. Optim. Theory Appl. **110**(2), 265–288 (2001)

Dempe, S., Gadhi, N.: Necessary optimality conditions for bilevel set optimization problems. J. Glob. Optim. **39**(4), 529–542 (2007)

Dempe, S., Kalashnikov, V., Rios-Mercado, R.Z.: Discrete bilevel programming: application to a natural gas cash-out problem. Eur. J. Oper. Res. **166**(2), 469–488 (2005)

DeSanctis, G., Gallupe, B.: A foundation for the study of group decision support systems. Manag. Sci. **33**, 1589–1609 (1987)

Dewez, S., Labbe, M., Marcotte, P., Savard, G.: New formulations and valid inequalities for a bilevel pricing problem. Oper. Res. Lett. **36**(2), 141–149 (2008)

Dolgui, A., Guschinsky, N., Levin, G.: Optimization of power transmission systems using a multi-level decomposition approach. Rairo Oper. Res. **41**(2), 213–229 (2007)

Dong, Y., Xu, K., Dresner, M.: Environmental determinants of VMI adoption: an exploratory analysis. Transp. Res. Part E Logistics Transp. Rev. **43**(4), 355–369 (2007)

Dotoli, M., Fanti, M.P., Meloni, C., Zhou, M.C.: A multi-level approach for network design of integrated supply chains. Int. J. Prod. Res. **43**(20), 4267–4287 (2005)

Dussault, J.P., Marcotte, P., Roch, S., Savard, G.: A smoothing heuristic for a bilevel pricing problem. Eur. J. Oper. Res. **174**(3), 1396–1413 (2006)

Dyer, J.S.: Interactive goal programming. Manag. Sci. **19**(1), 62–70 (1972)

Eberhart, R.C., Simpson, P.K., Dobbins, R.W.: Computational Intelligence PC Tools. Academic Press, New York (1996)

Ehtamo, H., Raivio, T.: On applied nonlinear and bilevel programming for pursuit-evasion games. J. Optim. Theory Appl. **108**(1), 65–96 (2001)

Eichfelder, G.: Multiobjective bilevel optimization. Math. Program. **123**(2), 419–449 (2007)

Eiselt, H.A., Sandblom, C.-L.: Linear Programming and its Applications. Springer, New York (2007)

Emam, O.E.: A fuzzy approach for bi-level integer non-linear programming problem. Appl. Math. Comput. **172**(1), 62–71 (2006)

Faísca, N., Saraiva, P., Rustem, B., Pistikopoulos, E.: A multi-parametric programming approach for multilevel hierarchical and decentralised optimisation problems. Comput. Manag. Sci. **6**(4), 377–397 (2009)

Faisca, N.P., Dua, V., Rustem, B., Saraiva, P.M., Pistikopoulos, E.N.: Parametric global optimisation for bilevel programming. J. Glob. Optim. **38**(4), 609–623 (2007)

Fampa, M., Barroso, L.A., Candal, D., Simonetti, L.: Bilevel optimization applied to strategic pricing in competitive electricity markets. Comput. Optim. Appl. **39**(2), 121–142 (2008)

Fan, H., Cheng, H.Z.: Transmission network expansion planning with security constraints based on bi-level linear programming. Eur. Trans. Electr. Power **19**(3), 388–399 (2009)

Fan, S.K.S., Lin, Y.: A multi-level thresholding approach using a hybrid optimal estimation algorithm. Pattern Recogn. Lett. **28**(5), 662–669 (2007)

Feng, C., Wen, C.: A fuzzy bi-level and multi-objective model to control traffic flow into the disaster area post earthquake. J. East. Asia Soc. Transp. Stud. **6**, 4253–4268 (2005)

Fliedner, M., Boysen, N., Scholl, A.: Solving symmetric mixed-model multi-level just-in-time scheduling problems. Discrete Appl. Math. **158**(3), 222–231 (2010)

Fliege, J., Vicente, L.N.: Multicriteria approach to bilevel optimization. J. Optim. Theory Appl. **131**(2), 209–225 (2006)

Fong, A.C.M., Hui, S.C., Jha, G.: Data mining for decision support. IT Prof. **4**(2), 9–17 (2002)

Frambach, R.T., Schillewaert, N.: Organizational innovation adoption—a multi-level framework of determinants and opportunities for future research. J. Bus. Res. **55**(2), 163–176 (2002)

Frishman, Y., Tal, A.: Multi-level graph layout on the GPU. IEEE Trans. Visual Comput. Graphics **13**(6), 1310–1317 (2007)

Fritz, M., Schiefer, G.: Tracking, tracing, and business process interests in food commodities: a multi-level decision complexity. Int. J. Prod. Econ. **117**(2), 317–329 (2009)

Fujita, M., Matsunaga, Y., Ciesielski, M.: Multi-level logic optimization. Hassoun, S., Sasao, T. (eds.), Logic Synthesis and Verification, vol. 654. Kluwer International Series in Engineering and Computer Science, Kluwer Academic Publishers, Dordrecht, pp. 29–63. (2002)

Gao, J., Liu, B.: Fuzzy multilevel programming with a hybrid intelligent algorithm. Comput. Math. Appl. **49**(9–10), 1539–1548 (2005)

Gao, Y., Zhang, G., Lu, J.: A particle swarm optimization based algorithm for fuzzy bilevel decision making. The 2008 International Conference on Fuzzy Systems, Hong Kong, China, pp. 1452–1457. (2008a)

Gao, Y., Zhang, G., Lu, J.: A decision support system for fuzzy multi-objective bilevel decision making. Int. J.Inf. Technol. Decis. Mak. **8**(1), 93–108 (2009)

Gao, Y., Zhang, G., Lu, J.: Particle swarm optimization for bi-level pricing problems in supply chains. J. Glob. Optim. **51**(2), 245–254 (2010a)

Gao, Y., Zhang, G., Lu, J., Dillon, T., Zeng, X.: A λ-cut-approximate algorithm for goal-based bilevel risk management systems. Int. J. Inf. Technol. Decis. Mak. **7**(4), 589–610 (2008b)

Gao, Y., Zhang, G., Ma, J., Lu, J.: A λ-cut and goal programming based algorithm for fuzzy linear multiple objective bi-level optimization. IEEE Trans. Fuzzy Syst. **18**(1), 1–13 (2010b)

Gao, Z.Y., Wu, J.J., Sun, H.J.: Solution algorithm for the bi-level discrete network design problem. Transp. Res. Part B Methodol. **39**(6), 479–495 (2005)

Garces, L.P., Conejo, A.J., Garcia-Bertrand, R., Romero, R.: A bilevel approach to transmission expansion planning within a market environment. IEEE Trans. Power Syst. **24**(3), 1513–1522 (2009)

Geels, F.W.: Analysing the breakthrough of rock 'n' roll (1930–1970) Multi-regime interaction and reconfiguration in the multi-level perspective. Technol. Forecast. Soc. Change **74**(8), 1411–1431 (2007)

Geoffrion, A.M., Dyer, J.S., Feinberg, A.: An interactive approach for multi-criterion optomization with an application to the operation of an academic department. Manag. Sci. **19**(4), 357–368 (1972)

Ginzberg, M.J., Stohr, E.A.: Decision support systems: issues and perspectives. Decision Support Systems. North-Holland, Amsterdam (1982)

Glackin, J., Ecker, J.G., Kupferschmid, M.: Solving bilevel linear programs using multiple objective linear programming. J. Optim. Theory Appl. **140**(2), 197–212 (2009)

Gorry, G.A., Scott-Morton, M.S.: A framework for management information system. Sloan Manag. Rev. **13**, 55–70 (1971)

Govindan, K.: Vendor-managed inventory: a review based on dimensions. Int. J. Prod. Res. **51**(13), 3808–3835 (2013)

Gray, P.: Group decision support systems. Decis. Support Syst. **3**(3), 233–242 (1987)

Gregor, S., Benbasat, I.: Explanations from intelligent systems: theoretical foundations and implications for practise. MIS Q. **23**, 497–530 (1999)

Gronalt, M., Rauch, P.: Vendor managed inventory in wood processing industries—a case study. Sliva Fennica **42**(1), 101–114 (2008)

Grubbstrom, R.W., Huynh, T.T.T.: Multi-level, multi-stage capacity-constrained production-inventory systems in discrete time with non-zero lead times using MRP theory. Int. J. Prod. Econ. **101**(1), 53–62 (2006)

Grubbstrom, R.W., Wang, Z.P.: A stochastic model of multi-level/multi-stage capacity-constrained production-inventory systems. Int. J. Prod. Econ. **81–2**, 483–494 (2003)

Guerrero, J.M., Hang, L., Uceda, J.: Control of distributed uninterruptible power supply systems. IEEE Trans. Ind. Electron. **55**, 2845–2859 (2008)

Gumus, Z.H., Floudas, C.A.: Global optimization of nonlinear bilevel programming problems. J. Glob. Optim. **20**(1), 1–31 (2001)

Guo, X., Lu, J.: Intelligent e-government services with recommendation techniques. E-serv. Intell. Int. J. Intell. Syst. **22**(5), 401–417 (2007)

Han, J., Kamber, M.: Data Mining Concepts and Techniques. Morgan Kaufmann Publishers, Burlington (2001)

Han, J., Lu, J., Zhang, G., Ma, S.: Multi-follower tri-level decision making with uncooperative followers. The 11th International FLINS Conference, Brazil, pp. 524–529. (2014)

Han, J., Pei, J., Yin, Y.: Mining frequent patterns without candidate generation. The 2000 ACM SIGMOD International Conference Management of Data, Dallas (2000)

Hansen, P., Jaumard, B., Savard, G.: An extended branch-and-bound rules for linear bi-level programming. SIAM J. Sci. Stat. Comput. **13**, 1194–1217 (1992a)

Hansen, P., Jaumard, B., Savard, G.: New branchandbound rules for linear bilevel programming. J. Sci. Stat. Comput. **13**, 1194–1217 (1992b)

Hara, T., Fukuda, K., Kanazawa, K., Shibata, N., Hosono, K., Maejima, H., Nakagawa, M., Abe, T., Kojima, M., Fujiu, M., Takeuchi, Y., Amemiya, K., Morooka, M., Kamei, T., Nasu, H., Wang, C.M., Sakurai, K., Tokiwa, N., Waki, H., Maruyama, T., Yoshikawa, S., Higashitani, M., Pham, T.D., Fong, Y., Watanabe, T.: A 146-mm(2) 8-Gb multi-level NAND flash memory with 70-nm CMOS technology. IEEE J. Solid-State Circuits **41**(1), 161–169 (2006)

Hay, A.M., Snyman, J.A.: A multi-level optimization methodology for determining the dextrous workspaces of planar parallel manipulators. Struct. Multi. Optim. **30**(6), 422–427 (2005)

Healy, P., Kuusik, A.: Algorithms for multi-level graph planarity testing and layout. Theoret. Comput. Sci. **320**(2–3), 331–344 (2004)

Hejazi, S.R., Memariani, A., Jahanshahloo, G., Sepehri, M.M.: Linear bilevel programming solution by genetic algorithm. Comput. Oper. Res. **29**(13), 1913–1925 (2002)

Helber, S., Sahling, F.: A fix-and-optimize approach for the multi-level capacitated lot sizing problem. Int. J. Prod. Econ. **123**(2), 247–256 (2010)

Hermanns, P.B., van Thoai, N.: Global optimization algorithm for solving bilevel programming problems with quadratic lower levels. J. Ind. Manag. Optim. **6**(1), 177–196 (2010)

Herskovits, J., Leontiev, A., Dias, G., Santos, G.: Contact shape optimization: a bilevel programming approach. Struct. Multi. Optim. **20**(3), 214–221 (2000)

Hirota, G., Maheshwari, R., Lin, M.C.: Fast volume-preserving free-form deformation using multi-level optimization. Comput. Aided Des. **32**(8–9), 499–512 (2000)

Ho, S.-J., Ku, W.-Y., Jou, J.-W., Hung, M.-H., Ho, S.-Y.: Intelligent particle swarm optimization in multi-objective problems.In: Ng, W.-K., Kitsuregawa, M., Li, J., Chang, K. (eds.) Advances in Knowledge Discovery and Data Mining, vol. 3918. Lecture Notes in Computer Science, vol. 790–800. Springer, Berlin (2006)

Hobbs, B.F., Metzler, C.B., Pang, J.S.: Strategic gaming analysis for electric power systems: an MPEC approach. IEEE Trans. Power Syst. **15**(2), 638–645 (2000)

Holsapple, C.W., Whinston, A.B.: Decision Support Systems: AKnowledge-based Approach. Western Publishing Company, Minneapolis (1996)

Holweg, M., Disney, S., Holmström, J., Småros, J.: Supply chain collaboration: making sense of the strategy continuum. Eur. Manag. J. **23**(2), 170–181 (2005)

Hu, W., Li, E.Y., Li, G.Y., Zhong, Z.H.: A metamodel optimization methodology based on multi-level fuzzy clustering space reduction strategy and its applications. Comput. Ind. Eng. **55**(2), 503–532 (2008)

Hu, X., Ralph, D.: Using EPECs to model bilevel games in restructured electricity markets with locational prices. Oper. Res. **55**(5), 809–827 (2007)

Hu, X.H., Cercone, N.: Learning in relational database: a rough set approach. Comput. Intell. **11**, 323–338 (1995)

Huang, B., Yao, L., Raguraman, K.: Bi-level GA and GIS for multi-objective TSP route planning. Transp. Plann. Technol. **29**(2), 105–124 (2006)

Huang, D.Y., Wang, C.H.: Optimal multi-level thresholding using a two-stage Otsu optimization approach. Pattern Recogn. Lett. **30**(3), 275–284 (2009)

Huang, S.J., Pai, F.S.: Design and operation of burn-in test system for three-phase uninterruptible power supplies. IEEE Trans. Ind. Electron. **49**, 256–263 (2002)

Hung, Y.F., Chien, K.L.: A multi-class multi-level capacitated lot sizing model. J. Oper. Res. Soc. **51**(11), 1309–1318 (2000)

Hunold, S., Rauber, T., Runger, G.: Combining building blocks for parallel multi-level matrix multiplication. Parallel Comput. **34**(6–8), 411–426 (2008)

Hwang, C.L., Masud, A.S.: Multiple Objective Decision Making: Methods and Applications. Springer, Berlin (1979)

Hwang, C.L., Yoon, K.: Multiple Attribute Decision Making: Methods and Applications—A State of the Art Survey. Springer, Berlin (1981)

Ignizio, J.P.: Goal Programming and Extensions. Lexington Books, Lexington (1976)

Ignizio, J.P.: The determination of a subset of efficient solutions via goal programming. Comput. Oper. Res. **8**, 9–16 (1981)

Gupta, R.J.K., Kusumakar, H.S., Jayaraman, V.K., Kulkarni, B.D.: A tabu search based approach for solving a class of bilevel programming problems in chemical engineering. J. Heuristics **9** (4), 307–319 (2003)

Jain, A.S., Meeran, S.: A multi-level hybrid framework applied to the general flow-shop scheduling problem. Comput. Oper. Res. **29**(13), 1873–1901 (2002)

Jan, G.B., Marcin, S.S., Jakub, W.: A New Version of Rough Set Exploration System. Lecture Notes in Computer Science, pp. 397–404. Springer, Berlin (2002)

Janikow, C.Z.: Fuzzy decision trees: issues and methods. IEEE Trans. Syst. Man Cybern. Part B Cybern. **28**(1), 1–14 (1998)

Jelassi, M.T., Williams, K., Fidler, C.S.: The emerging role of DSS: From passive to active. Decis. Support Syst. **3**, 299–307 (1987)

Jeunet, J., Jonard, N.: Single-point stochastic search algorithms for the multi-level lot-sizing problem. Comput. Oper. Res. **32**(4), 985–1006 (2005)

Ji, X.Y., Zhen, S.: Model and algorithm for bilevel newsboy problem with fuzzy demands and discounts. Appl. Math. Comput. **172**(1), 163–174 (2006)

Jordan, C., Koppelmann, J.: Multi-level lotsizing and scheduling by batch sequencing. J. Oper. Res. Soc. **49**(11), 1212–1218 (1998)

Josefsson, M., Patriksson, M.: Sensitivity analysis of separable traffic equilibrium equilibria with application to bilevel optimization in network design. Transp. Res. Part B Methodol. **41**(1), 4–31 (2007)

Kacprzyk, J., Yager, R.R.: Management Decision Support Systems Using Fuzzy Sets and Possibility Theory. Verlag TUV Rheinland, Cologne (1985)

Kalashnikov, V., Ríos-Mercado, R.: A natural gas cash-out problem: a bilevel programming framework and a penalty function method. Optim. Eng. **7**(4), 403–420 (2006)

Kalashnikov, V.V., Rios-Mercado, R.Z.: A natural gas cash-out problem: a bilevel programming framework and a penalty function method. Optim. Eng. **7**(4), 403–420 (2006)

Keen, P.G.W., Scott Morton, M.S.: Decision Support Systems: An Organizational Perspective. Addison-Wesley Publication Company, Reading (1978)

Kennedy, J., Eberhart, R.: Particle swarm optimization. IEEE International Conference on Neural Networks, Australia, pp. 1942–1948. (1995)

Kiak, A.: Rough set theory: a data mining tool for semiconductor manufacturing. IEEE Trans. Electron. Packag. Manufact. **24**, 44–50 (2001)

Kimms, A.: A genetic algorithm for multi-level, multi-machine lot sizing and scheduling. Comput. Oper. Res. **26**(8), 829–848 (1999)

366

References

Kirchberg, M., Schewe, K.D., Tretiakov, A., Wang, B.R.: A multi-level architecture for distributed object bases. Data Knowl. Eng. 60(1), 150–184 (2007)
Kitayama, S., Yasuda, K.: A method for mixed integer programming problems by particle swarm optimization. Electr. Eng. Jpn 157(2), 40–49 (2006)
Kocvara, M.: Topology optimization with displacement constraints: a bilevel programming approach. Struct. Optim. 14(4), 256–263 (1997)
Kodiyalam, S., Sobieszczanski-Sobieski, J.: Bilevel integrated system synthesis with response surfaces. AIAA J. 38(8), 1479–1485 (2000)
Kolodneer, J.L.: Improving human decision making through case-based decision aiding. AI Magazine 12, 52–68 (1991)
Kristensen, A.R., Jorgensen, E.: Multi-level hierarchic Markov processes as a framework for herd management support. Ann. Oper. Res. 94, 69–89 (2000)
Kuhn, H.W., Tucker, A.W.: Nonlinear Programming. Uniersity of California Press, Oakland (1951)
Kuhne, T., Schreiber, D.: Can programming be liberated from the two-level style? Multi-level programming with DeepJava. Acm Sigplan Not. 42(10), 229–243 (2007)
Kuntagod, N., Mukherjee, C.: Mobile decision support system for outreach health worker. e-Health Networking Applications and Services (Healthcom), 13th IEEE International Conference on, Columbia, MO,USA, pp. 56–59. (2011)
Kuo, R.J., Huang, C.C.: Application of particle swarm optimization algorithm for solving bi-level linear programming problem. Comput. Math. Appl. 58(4), 678–685 (2009)
Labbe, M., Marcotte, P., Savard, G.: A bilevel model of taxation and its application to optimal highway pricing. Manag. Sci. 44(12), 1608–1622 (1998)
Ladeur, K.H.: The introduction of the precautionary principle into EU law: A pyrrhic victory for environmental and public health law? Decision-making under conditions of complexity in multi-level political systems. Common Mark. Law Rev. 40(6), 1455–1479 (2003)
Lai, Y.-J.: IMOST: interactive multiple objective system technique. J. Oper. Res. Soc. 46(8), 958–976 (1995)
Lai, Y.-J., Hwang, C.-L.: Fuzzy multiple objective decision making. Beckmann, M., Kunzi, H. P. (eds), Fuzzy Multiple Objective Decision Making, vol. 404. Lecture Notes in Economics and Mathematical Systems, pp. 139–262. Springer, Berlin (1994)
Lai, Y.J.: Hierarchical optimization: a satisfactory solution. Fuzzy Sets Syst. 77, 321–335 (1996)
Lam, E.Y.: Blind bi-level image restoration with iterated quadratic programming. IEEE Trans. Circuits Syst. II Express Briefs 54(1), 52–56 (2007)
Lambert, D.M., Cooper, M.C.: Issues in supply chain management. Ind. Mark. Manag. 29, 65–83 (2000)
Lasry, A., Zaric, G.S., Carter, M.W.: Multi-level resource allocation for HIV prevention: a model for developing countries. Eur. J. Oper. Res. 180(2), 786–799 (2007)
Leake, D., Wilson, D.: When experience is wrong: examining CBR for changing tasks and environments. Althoff, K., Bergmann, R., Branthing, K.L. (eds.) Case-Based Reasoning Research and Development, pp. 218–232. Springer, Berlin (1999)
Leblanc, L., Boyce, D.: A bilevel programming algorithm for exact solution of the network design problem with useroptimal flows. Transp. Res. 20, 259–265 (1986)
Lee, B.K., Kang, K.H., Lee, Y.H.: Decomposition heuristic to minimize total cost in a multi-level supply chain network. Comput. Ind. Eng. 54(4), 945–959 (2008)
Lee, C.H.: Inventec group worldwide operation. Supply Chain Management Conference- with Notebook Computers as Example, Chung Yuan Christian University, Chungli, Taiwan, pp. 71–78. (2002)
Lee, H.H.: The investment model in preventive maintenance in multi-level production systems. Int. J. Prod. Econ. 112(2), 816–828 (2008)
Leisten, R.: An LP-aggregation view on aggregation in multi-level production planning. Ann. Oper. Res. 82, 413–434 (1998)

Levitin, G., Dai, Y.S., Xie, M., Poh, K.L.: Optimizing survivability of multi-state systems with multi-level protection by multi-processor genetic algorithm. Reliab. Eng. Syst. Saf. **82**(1), 93–104 (2003)

Lew, A., Tamanaha, D.: Decision table programming and reliability.The 2nd International Conference of Software Engineering, San Francisco, pp. 345–349 (1976)

Li, C.L., Li, L.Y.: Multi-level scheduling for global optimization in grid computing. Comput. Electr. Eng. **34**(3), 202–221 (2008)

Li, C.T., Li, J., Yin, S.H., Hudson, T.D., McMillen, D.K.: Synthesized multi-level composite filter for synthetic-aperture radar image identification. Opt. Commun. **146**(1–6), 285–301 (1998)

Li, H., Jiao, Y.C., Zhang, F.S., Zhang, L.: An efficient method for linear bilevel programming problems based on the orthogonal genetic algorithm. Int. J. Innovative Comput. Inf. Control **5** (9), 2837–2846 (2009)

Li, Q.S., Liu, D.K., Fang, J.Q., Tam, C.M.: Multi-level optimal design of buildings with active control under winds using genetic algorithms. J. Wind Eng. Ind. Aerodyn. **86**(1), 65–86 (2000)

Li, Q.S., Liu, D.K., Zhang, N., Tam, C.M., Yang, L.F.: Multi-level design model and genetic algorithm for structural control system optimization. Earthq. Eng. Struct. Dyn. **30**(6), 927–942 (2001)

Li, W., Du, W.: Optimization Models and Algorithms for the Daily Working Plan of Railway Technical Station. Southwest Jiaotong University Publisher, Chengdu (2002)

Liang, L., Sheng, S.H.: The stability analysis of bi-level decision and its application. Decis. Decis. Support Syst. **2**, 63–70 (1992)

Lim, Y.I.: An optimization strategy for nonlinear simulated moving bed chromatography: Multi-level optimization procedure MLOP. Korean J. Chem. Eng. **21**(4), 836–852 (2004)

Lin, B., Ho, P.H.: Dimensioning and location planning of broadband wireless networks under multi-level cooperative relaying. IEEE Trans. Wirel. Commun. **8**(11), 5682–5691 (2009)

Lin, B., Zhu, S.: Synthetic optimization of train routing and makeup plan in a railway network. J. China Railw. Soc. **18**(1), 1–7 (1996)

Lin, L.J.: Existence theorems for bilevel problem with applications to mathematical program with equilibrium constraint and semi-infinite problem. J. Optim. Theory Appl. **137**(1), 27–40 (2008)

Lin, L.J., Shie, H.J.: Existence theorems of quasivariational inclusion problems with applications to bilevel problems and mathematical programs with equilibrium constraint. J. Optim. Theory Appl. **138**(3), 445–457 (2008)

Liou, Y.C., Yao, J.C.: Bilevel decision via variational inequalities. Comput. Math. Appl. **49**(7–8), 1243–1253 (2005)

Lisi, F.A., Malerba, D.: Inducing multi-level association rules from multiple relations. Mach. Learn. **55**(2), 175–210 (2004)

Liu, B.: Theory and Practice of Uncertain Programming. Springer, New York (2002)

Liu, G.S., Han, J.Y., Zhang, J.Z.: Exact penalty functions for convex bilevel programming problems. J. Optim. Theory Appl. **110**(3), 621–643 (2001)

Liu, G.S., Zhang, J.Z.: Decision making of transportation plan, a bilevel transportation problem approach. J. Ind. Manag. Optim. **1**(3), 305–314 (2005)

Liu, W., Butler, R., Mileham, A.R., Green, A.J.: Bilevel optimization and postbuckling of highly strained composite stiffened panels. AIAA J. **44**(11), 2562–2570 (2006)

Liu, W.L., Butler, R., Kim, H.A.: Optimization of composite stiffened panels subject to compression and lateral pressure using a bi-level approach. Struct. Multi. Optim. **36**(3), 235–245 (2008)

Lu, J.: A framework and prototype for intelligent multiple objectives group decision support systems. Ph.D. thesis, Curtin University of Technology, Australia (2000)

Lu, J., Ma, J., Zhang, G., Zhu, Y., Zeng, X., Koehl, L.: Theme-based comprehensive evaluation in new product development using fuzzy hierarchical criteria group decision-making method. IEEE Trans. Ind. Electron. **58**(6), 2236–2246 (2011)

Lu, J., Ruan, D., Wu, F., Zhang, G.: An α-fuzzy goal approximate algorithm for solving fuzzy multiple objective linear programming problems. Soft Comput. **11**(3), 259–267 (2007a)

Lu, J., Shi, C., Zhang, G.: On bilevel multi-follower decision making: general framework and solutions. Inf. Sci. **176**(11), 1607–1627 (2006)

Lu, J., Shi, C., Zhang, G., Dillon, T.: Model and extended Kuhn-Tucker approach for bilevel multi-follower decision making in a referential-uncooperative situation. J. Glob. Optim. **38**(4), 597–608 (2007b)

Lu, J., Shi, C., Zhang, G., Ruan, D.: An extended branch and bound algorithm for bilevel multi-follower decision making in a referential-uncooperative situation. Int. J. Inf. Technol. Decis. Mak. **6**(2), 371–388 (2007c)

Lu, J., Wu, F., Zhang, G.: On generalized fuzzy goal optimization for solving fuzzy multi-objective linear programming problems. J. Intell. Fuzzy Syst. **18**(1), 83–97 (2007d)

Lu, J., Zhang, G., Dillon, T.: Fuzzy multi-objective bilevel decision making by an approximation Kth-best approach. J. Multiple-Valued Log. Soft Comput. **14**(3–5), 205–232 (2008a)

Lu, J., Zhang, G., Montero, J., Garmendia, L.: Multifollower trilevel decision making models and system. IEEE Trans. Ind. Inf. **8**(4), 974–985 (2012)

Lu, J., Zhang, G., Ruan, D., Wu, F.: Multi-objective Group Decision Making: Methods, Software and Applications with Fuzzy Set Technology. Imperial College Press, London (2007e)

Lu, J., Zhang, G., Wu, F.: Team situation awareness using web-based fuzzy group decision support systems. Int. J. Comput. Intell. Syst. **1**(1), 50–59 (2008b)

Lukač, Z., Šorić, K., Rosenzweig, V.V.: Production planning problem with sequence dependent setups as a bilevel programming problem. Eur. J. Oper. Res. **187**(3), 1504–1512 (2008)

Lv, Y.B., Hu, T.S., Wang, G.M., Wan, Z.P.: A penalty function method based on Kuhn-Tucker condition for solving linear bilevel programming. Appl. Math. Comput. **188**(1), 808–813 (2007)

Lv, Y.B., Hu, T.S., Wang, G.M., Wan, Z.P.: A neural network approach for solving nonlinear bilevel programming problem. Comput. Math. Appl. **55**(12), 2823–2829 (2008)

Ma, J., Lu, J., Zhang, G.: Decider: a fuzzy multi-criteria group decision support system. Knowl. Based Syst. **23**(1), 23–31 (2010)

Maher, M.J., Zhang, X.Y., Van Vliet, D.: A bi-level programming approach for trip matrix estimation and traffic control problems with stochastic user equilibrium link flows. Transp. Res. Part B Methodol. **35**(1), 23–40 (2001)

Mao, H.I., Zeng, X.J., Leng, G., Zhai, Y.J., Keane, J.A.: Short-term and midterm load forecasting using a bilevel optimization model. IEEE Trans. Power Syst. **24**(2), 1080–1090 (2009)

Marcotte, P.: Network optimization with continuous control parameters. Transp. Sci. **17**, 181–197 (1983)

Marcotte, P.: Network design with congestion effects: A case of bilevel programming. Math. Program. **34**, 142–162 (1986)

Marcotte, P., Savard, G., Semet, F.: A bilevel programming approach to the travelling salesman problem. Oper. Res. Lett. **32**(3), 240–248 (2004)

Marcotte, P., Savard, G., Zhu, D.L.: A trust region algorithm for nonlinear bilevel programming. Oper. Res. Lett. **29**(4), 171–179 (2001)

Marcotte, P., Savard, G., Zhu, D.L.: Mathematical structure of a bilevel strategic pricing model. Eur. J. Oper. Res. **193**(2), 552–566 (2009)

Marinakis, Y., Migdalas, A., Pardalos, P.M.: A new bilevel formulation for the vehicle routing problem and a solution method using a genetic algorithm. J. Glob. Optim. **38**(4), 555–580 (2007)

Martins, A.X., de Souza, M.C., Souza, M.J.F., Toffolo, T.A.M.: GRASP with hybrid heuristic-subproblem optimization for the multi-level capacitated minimum spanning tree problem. J. Heuristics **15**(2), 133–151 (2009)

Mersha, A.G., Dempe, S.: Linear bilevel programming with upper level constraints depending on the lower level solution. Appl. Math. Comput. **180**(1), 247–254 (2006)

Miller, T., Friesz, T., Tobin, R.: Heuristic algorithms for delivered price spatially competitive network facility location problems. Ann. Oper. Res. **34**, 177–202 (1992)

Mishra, S.: Weighting method for bi-level linear fractional programming problems. Eur. J. Oper. Res. **183**(1), 296–302 (2007)

Mishra, S., Ghosh, A.: Interactive fuzzy programming approach to bi-level quadratic fractional programming problems. Ann. Oper. Res. **143**(1), 251–263 (2006)

Mitsos, A., Bollas, G.M., Barton, P.I.: Bilevel optimization formulation for parameter estimation in liquid-liquid phase equilibrium problems. Chem. Eng. Sci. **64**(3), 548–559 (2009a)

Mitsos, A., Chachuat, B., Barton, P.I.: Towards global bilevel dynamic optimization. J. Glob. Optim. **45**(1), 63–93 (2009b)

Mitsos, A., Lemonidis, P., Barton, P.I.: Global solution of bilevel programs with a nonconvex inner program. J. Glob. Optim. **42**(4), 475–513 (2008)

Mohammad Nezhad, A., Aliakbari Shandiz, R., Eshraghniaye Jahromi, A.: A particle swarm–BFGS algorithm for nonlinear programming problems. Comput. Oper. Res. **40**(4), 963–972 (2013)

Mollestad, T., Skowron, A.: A rough set framework for data mining of propositional default rules. The 9th International Symposium on Methodologies for Intelligent System. Zakopane, Poland (1996)

Montero, J., Gómez, D., Bustince, H.: On the relevance of some families of fuzzy sets. Fuzzy Sets Syst. **158**(22), 2429–2442 (2007a)

Montero, J., López, V., Gómez, D.: The role of fuzziness in decision making. Wang, P., Ruan, D., Kerre, E. (eds.) Fuzzy Logic, vol. 215. Studies in Fuzziness and Soft Computing, pp. 337–349. Springer, Berlin (2007b)

Moussouni, F., Kreuawan, S., Brisset, S., Gillon, F., Brochet, P., Nicod, L.: Multi-level design optimization using target cascading, an improvement of convergence. COMPEL Int. J. Comput. Math. Electr. Electron. Eng. **28**(5), 1162–1178 (2009)

Mumford, M.D., Antes, A.L., Caughron, J.J., Friedrich, T.L.: Charismatic, ideological, and pragmatic leadership: multi-level influences on emergence and performance. Leadersh. Q. **19**(2), 144–160 (2008)

Muu, L.D., Quy, N.V.: A global optimization method for solving convex quadratic bilevel programming problems. J. Glob. Optim. **26**(2), 199–219 (2003)

Nash, J.: Non-cooperative games. Ann. Math. **54**(2), 286–295 (1951)

Nenortaitė, J.: A particle swarm optimization approach in the construction of decision-making model. Inf. Technol. Control **36**, 158–163 (2007)

Niu, L., Lu, J., Zhang, G.: Cognition-Driven Decision Support for Business Intelligence—Models, Techniques, Systems and Applications. Springer, Berlin (2009)

Osman, M.S., Abo-Sinna, M.A., Amer, A.H., Emam, O.E.: A multi-level non-linear multi-objective decision-making under fuzziness. Appl. Math. Comput. **153**(1), 239–252 (2004)

Otto, A.: Supply chain event management: three perspectives. Int. J. Logistics Manag. **14**(2), 1–13 (2003)

Pal, B.B., Moitra, B.N.: A fuzzy goal programming procedure for solving quadratic bilevel programming problems. Int. J. Intell. Syst. **18**(5), 529–540 (2003)

Pang, J.-S., Fukushima, M.: Quasi-variational inequalities, generalized Nash equilibria, and multi-leader-follower games. Comput. Manag. Sci. **2**(1), 21–56 (2005)

Papavassilopoulos, G.: Algorithms for static Stackelberg games with linear costs and polyhedral constraints. The 21st IEEE Conference on Decisions and Control. Orlando, FL (1982)

Park, H.W., Kim, M.S., Choi, D.H.: A new decomposition method for parallel processing multi-level optimization. KSME Int. J. **16**(5), 609–618 (2002)

Parsopoulos, K.E., Vrahatis, M.N.: Recent approaches to global optimization problems through particle swarm optimization. Nat. Comput. **1**(2–3), 235–306 (2002)

Patriksson, M.: On the applicability and solution of bilevel optimization models in transportation science: a study on the existence, stability and computation of optimal solutions to stochastic mathematical programs with equilibrium constraints. Transp. Res. Part B Methodol. **42**(10), 843–860 (2008)

Patriksson, M., Rockafellar, R.T.: A mathematical model and descent algorithm for bilevel traffic management. Transp. Sci. **36**(3), 271–291 (2002)

Pawlak, Z.: Rough sets. Int. J. Inf. Comput. Sci. **11**, 341–356 (1982)

Pawlak, Z.: Rough Sets:Theoretical Aspects of Reasoning about Data. Kluwer Academic Publishers, Boston (1991)

Pawlak, Z., Slowinski, K.: Rough classification of patients after highly selective vagotomy duodenal ulcer. Int. J. Man Mach. Stud. **24**, 413–433 (1986)

Pei, S.C., Horng, J.H.: Design of FIR bilevel Laplacian-of-Gaussian filter. Signal Process. **82**(4), 677–691 (2002)

Pitakaso, R., Almeder, C., Doerner, K.F., Hartl, R.F.: Combining population-based and exact methods for multi-level capacitated lot-sizing problems. Int. J. Prod. Res. **44**(22), 4755–4771 (2006)

Pitakaso, R., Almeder, C., Doerner, K.F., Hartl, R.F.: A MAX-MIN ant system for unconstrained multi-level lot-sizing problems. Comput. Oper. Res. **34**(9), 2533–2552 (2007)

Pooch, U.W.: Translation of decision tables. ACM Comput. Surv. **6**, 125–151 (1974)

Power, D.J.: Decision Support Systems: Concepts and Resources for Managers. Greenwood Publishing Group, Westport (2002)

Power, D.J.: Specifying an expanded framework for classifying and describing decision support systems. Commun. Assoc. Inf. Syst. **13**, 158–166 (2004)

Power, D.J., Kaparthi, S.: Building web-based decision support systems. Stud. Inf. Control **11**(4), 291–302 (2002)

Pramanik, S., Roy, T.K.: Fuzzy goal programming approach to multilevel programming problems. Eur. J. Oper. Res. **176**(2), 1151–1166 (2007)

Qi, L., Teo, K.L., Yang, X.: Optimization and Control with Applications. Spinger, New York (2005)

Quaddus, M.A., Holzman, A.G.: IMOLP: an interactive method for multiple objective linear programs. IEEE Trans. Syst. Man Cybern. **16**(3), 462–468 (1986)

Quinlan, J.R.: Learning efficient classification procedures and their application to chess end-games. Michalski, R.S., Carbonell, J.G., Mitchell, T.M. (eds.) Machine Learning: An Artificial Intelligence Approach, pp. 463–482. Morgan Kaufmann, Palo Alto (1983)

Quinlan, J.R.: Induction of decision trees. Mach. Learn. **1**, 81–106 (1986)

Quinlan, J.R.: C4.5: Programs for Machine Learning. Morgan Kaufmann, San Mateo (1993a)

Quinlan, J.R.: Programs for machine learning. Mach. Learn. **16**, 235–240 (1993b)

Ragon, S.A., Gurdal, Z., Haftka, R.T., Tzong, T.J.: Bilevel design of a wing structure using response surfaces. J. Aircr. **40**(5), 985–992 (2003)

Reinartz, T., Iglezakis, I., Roth-Berghofer, T.: Review and restore for case-base maintenance. Comput. Intell. **17**(2), 214–234 (2001)

Richter, M.M.: Introduction. Lenz, M.,Bartsch-Spörl, B., Burkhard, H., Wess, S. (eds.) Case-Based Reasoning Technology: From Foundations to Applications pp. 1–15. Springer, Berlin (1998)

Rodrigues, M.T.M., Latre, L.G., Rodrigues, L.C.A.: Short-term planning and scheduling in multipurpose batch chemical plants: a multi-level approach. Comput. Chem. Eng. **24**(9–10), 2247–2258 (2000)

Rodriguez, H.M., Burdisso, R.A.: Structural-acoustic control system design by multi-level optimization. J. Acoust. Soc. Am. **104**(2), 926–936 (1998)

Roghanian, E., Aryanezhad, M.B., Sadjadi, S.J.: Integrating goal programming, Kuhn-Tucker conditions, and penalty function approaches to solve linear bi-level programming problems. Appl. Math. Comput. **195**(2), 585–590 (2008)

Roghanian, E., Sadjadi, S.J., Aryanezhad, M.B.: A probabilistic bi-level linear multi-objective programming problem to supply chain planning. Appl. Math. Comput. **188**(1), 786–800 (2007)

ROSETTA: The ROSETTA homepage. http://www.rosettaproject.org/.

Rosset, S.: Bi-level path following for cross validated solution of kernel quantile regression. J. Mach. Learn. Res. **10**, 2473–2505 (2009)

Roy, R.: Problems and methods with multiple objective functions. Math. Program. **1**, 239–266 (1971)

Ruan, G.Z., Wang, S.Y., Yamamoto, Y., Zhu, S.S.: Optimality conditions and geometric properties of a linear multilevel programming problem with dominated objective functions. J. Optim. Theory Appl. **123**(2), 409–429 (2004)

Rushton, A., Saw, R.: A methodology for logistics strategy planning. Int. J. Logistics Manag. **3**(1), 46–62 (1992)

Ryu, J.H., Dua, V., Pistikopoulos, E.N.: A bilevel programming framework for enterprise-wide process networks under uncertainty. Comput. Chem. Eng. **28**(6–7), 1121–1129 (2004)

Saaty, T.: The Analytic Hierarchy Process. McGraw-Hill, New York (1980)

Saharidis, G.K., Ierapetritou, M.G.: Resolution method for mixed integer bi-level linear problems based on decomposition technique. J. Glob. Optim. **44**(1), 29–51 (2009)

Sahin, K.H., Ciric, A.R.: A dual temperature simulated annealing approach for solving bilevel programming problems. Comput. Chem. Eng. **23**(1), 11–25 (1998)

Sakawa, M.: Fussy Sets and Interactive Mulitobjective Optimization. Plenum Press, New York (1993)

Sakawa, M., Nishizaki, I.: Interactive fuzzy programming for decentralized two-level linear programming problems. Fuzzy Sets Syst. **125**, 301–315 (2001a)

Sakawa, M., Nishizaki, I.: Interactive fuzzy programming for two-level linear fractional programming problems. Fuzzy Sets Syst. **119**, 31–40 (2001b)

Sakawa, M., Nishizaki, I.: Interactive fuzzy programming for two-level nonconvex programming problems with fuzzy parameters through genetic algorithms. Fuzzy Sets Syst. **127**, 185–197 (2002)

Sakawa, M., Nishizaki, I., Hitaka, M.: Interactive fuzzy programming for multi-level 0–1 programming problems with fuzzy parameters through genetic algorithms. Fuzzy Sets Syst. **117**(1), 95–111 (2001a)

Sakawa, M., Nishizaki, I., Uemura, Y.: Interactive fuzzy programming for multi-level linear programming problems with fuzzy parameters. Fuzzy Sets Syst. **109**(1), 3–19 (2000a)

Sakawa, M., Nishizaki, I., Uemura, Y.: Interactive fuzzy programming for two-level linear fractional programming problems with fuzzy parameters. Fuzzy Sets Syst. **115**, 93–103 (2000b)

Sakawa, M., Yauchi, K.: Interactive decision making for multiobjective nonconvex programming problems with fuzzy numbers through coevolutionary genetic algorithms. Fuzzy Sets Syst. **114**, 151–165 (2000)

Salukvadze, M.: On the extension of solutions in problems of optimization under vector valued criteria. J. Optim. Theory Appl. **13**(2), 203–217 (1974)

Sarma, K.C., Adeli, H.: Bilevel parallel genetic algorithms for optimization of large steel structures. Comput Aided Civ. Infrastruct. Eng. **16**(5), 295–304 (2001)

Sauter, V.L.: Decision Support Systems: An Applied Managerial Approach. Wiley, New York (1997)

Scaparra, M.P., Church, R.L.: A bilevel mixed-integer program for critical infrastructure protection planning. Comput. Oper. Res. **35**(6), 1905–1923 (2008)

Schlimmer, J.C., Fisher, D.A.: A case study of incremental concept induction. The 5th International Conference on Artificial Intelligence, Pheladelphia, USA (1986)

Sern, L.C.: Present and Future of Supply Chain in Information and Electronic Industry. The Supply Chain Management Conference for Electronic Industry, National Tsing Hua University Hsinchu, Taiwan (2003)

Shan, N., Ziarko, W., Hamilton, H.J., Cercone, N.: Using rough sets as tools for knowledge discovery. Paper presented at the first international conference on knowledge discovery and data mining Menlo Park, CA (1995)

Shedabale, S., Ramakrishnan, H., Russell, G., Yakovlev, A., Chattopadhyay, S.: Statistical modelling of the variation in advanced process technologies using a multi-level partitioned response surface approach. IET Circuits Devices Syst. **2**(5), 451–464 (2008)

Sherali, H.D.: A multiple leader stackelberg model and analysis. Oper. Res. **32**(2), 390 (1984)

Shi, C.: Linear Bilevel Programming Technology: Models and Algorithms. University of Technology, Sydney (2005)

Shi, C., Lu, H., Zhang, G.: An extended Kth-best approach for linear bilevel programming. Appl. Math. Comput. **164**(3), 843–855 (2005a)

Shi, C., Lu, J., Zhang, G.: An extended Kuhn-Tucker approach for linear bilevel programming. Appl. Math. Comput. **162**(1), 51–63 (2005b)

Shi, C., Lu, J., Zhang, G., Zhou, H.: An extended branch and bound algorithm for linear bilevel programming. Appl. Math. Comput. **180**(2), 529–537 (2006)

Shi, C., Zhang, G., Lu, H.: On the definition of linear bilevel programming solution. Appl. Math. Comput. **160**(1), 169–176 (2005c)

Shi, C., Zhang, G., Lu, J.: The Kth-best approach for linear bilevel multi-follower programming. J. Glob. Optim. **33**(4), 563–578 (2005d)

Shi, C., Zhou, H., Lu, J., Zhang, G., Zhang, Z.: The Kth-best approach for linear bilevel multifollower programming with partial shared variables among followers. Appl. Math. Comput. **188**(2), 1686–1698 (2007)

Shi, F., Fang, Q., Li, X., Mo, H., Huang, Y.: A bi-level programming optimization method for the layout of technical service station. J. China Railw. Soc. **25**(2), 1–4 (2003)

Shi, X., Xia, H.: Interactive bilevel multi-objective decision making. J. Oper. Res. Soc. **48**, 943–949 (1997)

Shih, H.S., Lai, Y.J., Lee, E.S.: Fuzzy approach for multilevel programming problems. Comput. Oper. Res. **23**, 73–91 (1996)

Shih, H.S., Lee, E.S.: Fuzzy multi-level minimum cost flow problems. Fuzzy Sets Syst. **107**(2), 159–176 (1999)

Shim, J.P., Warkentin, M., Courtney, J.F., Power, D.J., Sharda, R., Carlsson, C.: Past, present, and future of decision support technology. Decis. Support Syst. **33**, 111–126 (2002)

Silvent, A.S., Dojat, M., Garbay, C.: Multi-level temporal abstraction for medical scenario construction. Int. J. Adapt. Control Signal Process. **19**(5), 377–394 (2005)

Simon, H.A.: Decision making: rational, nonrational, and irrational. Educ. Adm. Q. **29**, 392–411 (1993)

Sinha, S.: A comment on Anandalingam. A mathematical programming model of decentralized multi-level systems. J. Oper. Res. Soc. **39**, 1021–1033 (1988). J. Oper. Res. Soc. **52**(5), 594–596 (2001)

Sinha, S.: Fuzzy mathematical programming applied to multi-level programming problems. Comput. Oper. Res. **30**(9), 1259–1268 (2003a)

Sinha, S.: Fuzzy programming approach to multi-level programming problems. Fuzzy Sets Syst. **136**(2), 189–202 (2003b)

Sinha, S., Sinha, S.B.: KKT transformation approach for multi-objective multi-level linear programming problems. Eur. J. Oper. Res. **143**(1), 19–31 (2002)

Sinha, S.B., Sinha, S.: A linear programming approach for linear multi-level programming problems. J. Oper. Res. Soc. **55**(3), 312–316 (2004)

Skowron, A., Polkowski, L.: Rough Sets in Knowledge Discovery. Physica Verlag, Heidelberg (1998)

Smith, J.C., Lim, C., Alptekinoglu, A.: New product introduction against a predator: a bilevel mixed-integer programming approach. Nav. Res. Logistics **56**(8), 714–729 (2009)

Sobieszczanski-Sobieski, J., Agte, J.S., Sandusky, R.R.: Bilevel integrated system synthesis. AIAA J. **38**(1), 164–172 (2000)

Sobieszczanski-Sobieski, J., Altus, T.D., Phillips, M., Sandusky, R.: Bilevel integrated system synthesis for concurrent and distributed processing. AIAA J. **41**(10), 1996–2003 (2003)

Song, H., Diolata, R., Joo, Y.H.: Photovoltaic system allocation using discrete particle swarm optimization with multi-level quantization. J. Electr. Eng. Technol. **4**(2), 185–193 (2009)

Sonia Khandelwal, A., Puri, M.C.: Bilevel time minimizing transportation problem. Discrete Optim. **5**(4), 714–723 (2008)

Sonia Puri, M.C.: Bilevel time minimizing assignment problem. Appl. Math. Comput. **183**(2), 990–999 (2006)

Spangler, W.: The role of artificial intelligence in understanding the strategic decision-making process. IEEE Trans. Knowl. Data Eng. **3**, 145–159 (1991)

St John, C.H., Pouder, R.W., Cannon, A.R.: Environmental uncertainty and product-process life cycles: a multi-level interpretation of change over time. J. Manag. Stud. **40**(2), 513–541 (2003)

Stackelberg, H.V.: The Theory of Market Economy. Oxford University Press, Oxford (1952)

Stein, O.: Bi-level Strategies in Semi-Infinite Programming. Kluwer Academic Publishers, Dordrecht (2003)

Stein, O., Still, G.: On generalized semi-infinite optimization and bilevel optimization. Eur. J. Oper. Res. **142**(3), 444–462 (2002)

Steuer, R.E.: An interactive multiple objective linear programming procedure. TIMS Stud. Manag. Sci. **6**, 225–239 (1977)

Still, G.: Linear bilevel problems: Genericity results and an efficient method for computing local minima. Math. Methods Oper. Res. **55**(3), 383–400 (2002)

Stoffel, D., Kunz, W., Gerber, S.: AND/OR reasoning graphs for determining prime implicants in multi-level combinational networks. IEICE Trans. Fundam. Electron. Commun. Comput. Sci. **E80A**(12), 2581–2588 (1997)

Sun, D.Z., Benekohal, R.F., Waller, S.T.: Bi-level programming formulation and heuristic solution approach for dynamic traffic signal optimization. Comput. Aided Civ. Infrastruct. Eng. **21**(5), 321–333 (2006)

Sun, H.J., Gao, Z.Y., Wu, H.J.: A bi-level programming model and solution algorithm for the location of logistics distribution centers. Appl. Math. Model. **32**(4), 610–616 (2008)

Taiani, F., Killijian, M.O., Fabre, J.C.: COSMOPEN: Dynamic reverse engineering on a budget. How cheap observation techniques can be used to reconstruct complex multi-level behaviour. Softw. Pract. Exp. **39**(18), 1467–1514 (2009)

Takeuchi, K., Kameda, Y., Fujimura, S., Otake, H., Hosono, K., Shiga, H., Watanabe, Y., Futatsuyama, T., Shindo, Y., Kojima, M., Iwai, M., Shirakawa, M., Ichige, M., Hatakeyama, K., Tanaka, S., Kamei, T., Fu, J.Y., Cernea, A., Li, Y., Higashitani, M., Hemink, G., Sato, S., Oowada, K., Lee, S.C., Hayashida, N., Wan, J., Lutze, J., Tsao, S., Mofidi, M., Sakurai, K., Tokiwa, N., Waki, H., Nozawa, Y., Kanazawa, K., Ohshima, S.: A 56-nm CMOS 99-mm(2) 8-Gb multi-level NAND flash memory with 10-MB/s program throughput. IEEE J. Solid-State Circuits **42**(1), 219–232 (2007)

Tan, S.X.D., Shi, C.J.R.: Balanced multi-level multi-way partitioning of analog integrated circuits for hierarchical symbolic analysis. Integr. VLSI J. **34**(1–2), 65–86 (2003)

Tang, K., Yang, J., Chen, H., Gao, S.: Improved genetic algorithm for nonlinear programming problems. J. Syst. Eng. Electron. **22**(3), 540–546 (2011)

Thiyagarajan, N., Grandhi, R.V.: Multi-level design process for 3-D preform shape optimization in metal forming. J. Mater. Process. Technol. **170**(1–2), 421–429 (2005)

Tiryaki, F.: Interactive compensatory fuzzy programming for decentralized multi-level linear programming (DMLLP) problems. Fuzzy Sets Syst. **157**(23), 3072–3090 (2006)

Tsoukalas, A., Rustem, B., Pistikopoulos, E.N.: A global optimization algorithm for generalized semi-infinite, continuous minimax with coupled constraints and bi-level problems. J. Glob. Optim. **44**(2), 235–250 (2009)

Turban, E., Aronson, J.E.: Decision Support Systems and Intelligent Systems. Pearson Education, India (1998)

Turban, E., Aronson, J.E., Liang, T.P.: Decision Support Systems and Intelligent Systems, 7th edn. Pearson Prentice Hall, New Jersey (2005)

Turban, E., Watkins, P.R.: Integrating expert systems and decision support systems. MIS Q. **10**, 121–136 (1986)

Tuy, H., Migdalas, A., Hoai-Phuong, N.T.: A novel approach to bilevel nonlinear programming. J. Glob. Optim. **38**(4), 527–554 (2007)

Uno, T., Katagir, H., Kato, K.: An evolutionary multi-agent based search method for stackelberg solutions of bilevel facility location problems. Int. J. Innov. Comput. Inf. Control 4(5), 1033–1042 (2008)

Vafaeesefat, A., Khani, A.: Head shape and winding angle optimization of composite pressure vessels based on a multi-level strategy. Appl. Compos. Mater. 14(5–6), 379–391 (2007)

Varzaghani, A., Yang, C.K.K.: A 6-GSamples/s multi-level decision feedback equalizer embedded in a 4-bit time-interleaved pipeline A/D converter. IEEE J. Solid-State Circuits 41(4), 935–944 (2006)

Voros, J.: On the relaxation of multi-level dynamic lot-sizing models. Int. J. Prod. Econ. 77(1), 53–61 (2002)

Wang, C.: Research on the model and algorithm of dynamic wagon-flow allocating in a marshalling Station. J. China Railw. Soc. 26(1), 1–6 (2004)

Wang, G.M., Wan, Z.P., Wang, X.J., Lv, Y.B.: Genetic algorithm based on simplex method for solving linear-quadratic bilevel programming problem. Comput. Math. Appl. 56(10), 2550–2555 (2008)

Wang, G.M., Wang, X.J., Wan, Z.P.: A fuzzy interactive decision making algorithm for bilevel multi-followers programming with partial shared variables among followers. Expert Syst. Appl. 36(7), 10471–10474 (2009)

Wang, G.M., Wang, X.J., Wan, Z.P., Jia, S.H.: An adaptive genetic algorithm for solving bilevel linear programming problem. Appl. Math. Mech. Engl. Ed. 28(12), 1605–1612 (2007a)

Wang, G.M., Wang, X.J., Wan, Z.P., Lv, Y.B.: A globally convergent algorithm for a class of bilevel nonlinear programming problem. Appl. Math. Comput. 188(1), 166–172 (2007b)

Wang, G.Y., He, X.: A self-learning model under uncertain condition. Chin. J. Softw. 14, 1096–1102 (2003)

Wang, G.Y., Zheng, Z., Zhang, Y.: RIDAS-a rough set based intelligent data analysis system. The First International Conference on Machine Learning and Cybernetics, Beijing, China (2002)

Wang, X.C., Baggio, P., Schrefler, B.A.: A multi-level frontal algorithm for finite element analysis and its implementation on parallel computation. Eng. Comput. 16(4), 405–427 (1999)

Wang, Y.P., Jiao, Y.C., Li, H.: An evolutionary algorithm for solving nonlinear bilevel programming based on a new constraint-handling scheme. IEEE Trans. Syst. Man Cybern. Part C Appl. Rev. 35(2), 221–232 (2005)

Watson, L.: Applying Case-based Reasoning: Techniques for Enterprise Systems. Morgan Kaufmann, San Francisco (1997)

Wen, U.-P., Hsu, S.-T.: Linear bi-level programming problems-A review. J. Oper. Res. Soc. 42, 125–133 (1991)

Werner, A.: Bilevel stochastic programming problems: Analysis and application to telecommunications. Ph.D. thesis, Norwegian University of Science and Technology, Norway (2005)

White, D., Anandalingam, G.: A penalty function approach for solving bi-Level linear programs. J. Glob. Optim. 3, 397–419 (1993)

White, D.J.: Penalty function approach to linear trilevel programming. J. Optim. Theory Appl. 93 (1), 183–197 (1997)

Wilson, D.C., Leake, D.B.: Maintaining case-based reasoners: Dimensions and directions. Comput. Intell. 17, 196–213 (2001)

Wu, W.C., Chao, T.S., Peng, W.C., Yang, W.L., Chen, J.H., Ma, M.W., Lai, C.S., Yang, T.Y., Lee, C.H., Hsieh, T.M., Liou, J.C., Chen, T.P., Chen, C.H., Lin, C.H., Chen, H.H., Ko, J.: Optimized ONO thickness for multi-level and 2-bit/cell operation for wrapped-select-gate (WSG) SONOS memory—art. no. 015004. Semicond. Sci. Technol. 23(1), 15004 (2008)

Xu, Z.K.: Deriving the properties of linear bilevel programming via a penalty function approach. J. Optim. Theory Appl. 103(2), 441–456 (1999)

Yadav, S., Xu, Y., Xue, D.: A multi-level heuristic search algorithm for production scheduling. Int. J. Prod. Res. 38(12), 2761–2785 (2000)

Yager, R.R.: Modeling prioritized multi-criteria decision making. IEEE Trans. Systems Man Cybern. Part B 34, 2386–2404 (2004)

Yang, H., Bell, M.G.H.: Transport bilevel programming problems: recent methodological advances. Transp. Res. Part B Methodol. **35**(1), 1–4 (2001)

Yang, J., Zhang, M., He, B., Yang, C.: Bi-level programming model and hybrid genetic algorithm for flow interception problem with customer choice. Comput. Math. Appl. **57**(11–12), 1985–1994 (2009)

Yang, L., Mahadevan, R., Cluett, W.R.: A bilevel optimization algorithm to identify enzymatic capacity constraints in metabolic networks. Comput. Chem. Eng. **32**(9), 2072–2085 (2008)

Yang, P.C., Wee, H.M., Yu, J.: Collaborative P&R policy for Hi-Tech industry. J. Oper. Res. Soc. **58**, 894–909 (2007a)

Yang, W., Li, L., Ma, S.: Coordinating supply chain response-time: A bi-level programming approach. Int. J. Adv. Manuf. Technol. **31**(9–10), 1034–1043 (2007b)

Yao, Y., Edmunds, T., Papageorgiou, D., Alvarez, R.: Trilevel optimization in power network defense. IEEE Trans. Syst. Man Cybern. Part C Appl. Rev. **37**(4), 712–718 (2007)

Yao, Y.Y., Yao, J.T.: Granular computing as a basis for consistent classification problems. Commun. Inst. Inf. Comput. Mach **5**, 101–106 (2002)

Yeh, W.C.: A two-stage discrete particle swarm optimization for the problem of multiple multi-level redundancy allocation in series systems. Expert Syst. Appl. **36**(5), 9192–9200 (2009)

Yi, Q., Kennedy, K.: Improving memory hierarchy performance through combined loop interchange and multi-level fusion. Int. J. High Perform. Comput. Appl. **18**(2), 237–253 (2004)

Yin, Y.F.: Genetic-algorithms-based approach for bilevel programming models. J. Transp. Eng. Asce **126**(2), 115–120 (2000)

Yip, R.K.K.: A multi-level dynamic programming method for line segment matching in axial motion stereo. Pattern Recogn. **31**(11), 1653–1668 (1998)

Yip, R.K.K., Ho, W.P.: A multi-level dynamic programming method for stereo line matching. Pattern Recogn. Lett. **19**(9), 839–855 (1998)

Shi, Y., Eberhart, R.: A modified particle swarm optimizer.IEEE International Conference on Evolutionary Computation, Anchorage, AK, USA (1998)

Yun, W.Y., Kim, J.W.: Multi-level redundancy optimization in series systems. Comput. Ind. Eng. **46**(2), 337–346 (2004)

Yun, W.Y., Song, Y.M., Kim, H.G.: Multiple multi-level redundancy allocation in series systems. Reliab. Eng. Syst. Saf. **92**(3), 308–313 (2007)

Zachary, W.: A cognitively based functional taxonomy of decision support techniques. Hum. Comput. Interact. **2**, 25–63 (1986)

Zachary, W.: A cognitively-based functional taxonomy of decision support techniques. ACM SIGCHI Bull. **19**, 72–73 (1987)

Zadeh, L.A.: Optimality and non-scalar valued performance criteria. IEEE Trans. Autom. Control **AC-8**, 59–60 (1963)

Zadeh, L.A.: Fuzzy sets. Inf. Control **8**, 338–353 (1965)

Zadeh, L.A.: The concept of a linguistic variable and its application to approximate reasoning - Part I. Inf. Sci. **8**, 199–249 (1975)

Zhang, D., Lin, G.-H.: Bilevel direct search method for leader–follower problems and application in health insurance. Comput. Oper. Res. **41**, 359–373 (2014)

Zhang, G., Dillon, T., Cai, K., Ma, J., Lu, J.: Operation properties and δ-equalities of complex fuzzy sets. Int. J. Approximate Reasoning **50**(8), 1227–1249 (2009a)

Zhang, G., Lu, J.: The definition of optimal solution and an extended Kuhn-Tucker approach for fuzzy linear bi-level programming. IEEE Comput. Intell. Bull. **2**(5), 1–7 (2005a)

Zhang, G., Lu, J.: Solution definition and extended Kuhn-Tucker approach for fuzzy linear bilevel programming. IEEE Intell. Inf. Bull. **6**(2), 1–7 (2005b)

Zhang, G., Lu, J.: Model and approach of fuzzy bi-level decision making for logistics planning problem. J. Enterp. Inf. Manag. **20**, 178–197 (2007)

Zhang, G., Lu, J.: Fuzzy bilevel programming with multiple objectives and cooperative multiple followers. J. Global Optim. **47**(3), 403–419 (2010)

Zhang, G., Lu, J., Dillon, T.: An approximation branch and bound approach for fuzzy linear bi-level decision making. The 1st International Symposium Advances in Artificial Intelligence and Applications, Wisla, Poland (2006)

Zhang, G., Lu, J., Dillon, T.: Kth-best algorithm for fuzzy bilevel programming. Paper presented at the International Conference on Intelligent Systems and Knowledge Engineering (ISKE2006), Shanghai, China (2006)

Zhang, G., Lu, J., Dillon, T.: Decentralized multi-objective bilevel decision making with fuzzy demands. Knowl. Based Syst. **20**(5), 495–507 (2007a)

Zhang, G., Lu, J., Dillon, T.: Fuzzy linear bi-level optimization: Solution concepts, approaches and applications. Wang, P., Ruan, D., Kerre, E. (eds.) Fuzzy Logic—A Spectrum of Theoretical and Practical Issues, pp. 351–379. Springer, Berling, Berlin (2007b)

Zhang, G., Lu, J., Dillon, T.: Solution concepts and an approximation Kuhu-Tucker approach for fuzzy multi-objective linear bi-level programming. Chinchuluun, A., Migdalas, A., Pardalos, P. M, Pitsoulis, L. (eds.) Pareto Optimality, Game Theory and Equilibria, pp. 307–330. Springer, Berling, Berlin (2007c)

Zhang, G., Lu, J., Gao, Y.: An algorithm for fuzzy multi-objective multi-follower partial cooperative bilevel programming. J. Intell. Fuzzy Syst. **19**(4–5), 303–319 (2008a)

Zhang, G., Lu, J., Gao, Y.: Fuzzy bilevel programming: multi-objective and multi-follower with shared variables. Int. J. Uncertain. Fuzziness Knowl. Based Syst. **16**, 105–133 (2008b)

Zhang, G., Lu, J., Montero, J., Zeng, Y.: Model, Solution concept and the Kth-best algorithm for linear tri-level programming. Inf. Sci. **180**(4), 481–492 (2010a)

Zhang, G., Lu, J., Wu, Y.: Formulation of linear programming problems with fuzzy coefficients of objective functions and constraints. Asian Inf. Sci. Life **2**, 57–68 (2003a)

Zhang, G., Ma, J., Lu, J.: Emergency management evaluation by a fuzzy multi-criteria group decision support system. Stochast. Env. Res. Risk Assess. **23**(4), 517–527 (2009b)

Zhang, G., Shi, C., Lu, J.: An extended K-best approach for referential-uncooperative bilevel multi-follower decision making. Int. J. Comput. Intell. Syst. **1**(3), 205–214 (2008c)

Zhang, G., Wu, Y., Remia, M., Lu, J.: Formulation of fuzzy linear programming problems as four-objective constrained problems. Appl. Math. Comput. **139**, 383–399 (2003b)

Zhang, G., Yoo, W.J.: V-th control by complementary hot-carrier injection for SONOS multi-level cell flash memory. IEEE Trans. Electron Devices **56**(12), 3027–3032 (2009)

Zhang, G., Zhang, G., Gao, Y., Lu, J.: Competitive strategic bidding optimization in electricity markets using bilevel programming and swarm technique. IEEE Trans. Ind. Electron. **58**(6), 2138–2146 (2011a)

Zhang, G., Zheng, Z., Lu, J., He, Q.: An algorithm for solving rule sets based bilevel decision problems. Comput. Intell. **27**, 235–259 (2010b)

Zhang, J.-R., Zhang, J., Lok, T.-M., Lyu, M.R.: A hybrid particle swarm optimization–back-propagation algorithm for feedforward neural network training. Appl. Math. Comput. **185**(2), 1026–1037 (2007d)

Zhang, K.S., Han, Z.H., Li, W.J., Song, W.P.: Bilevel adaptive weighted sum method for multidisciplinary multi-objective optimization. AIAA J. **46**(10), 2611–2622 (2008d)

Zhang, N., Ye, H., Liu, Y.T.: Decision support for choosing optimal electromagnetic loop circuit opening scheme based on analytic hierarchy process and multi-level fuzzy comprehensive evaluation. Eng. Intell. Syst. Electr. Eng. Commun. **16**(4), 183–191 (2008e)

Zhang, Q., Varshney, P.K., Wesel, R.D.: Optimal bi-level quantization of i.i.d. sensor observations for binary hypothesis testing. IEEE Trans. Inf. Theory **48**(7), 2105–2111 (2002)

Zhang, R., Lu, J., Zhang, G.: A knowledge-based multi-role decision support system for ore blending cost optimization of blast furnaces. Eur. J. Oper. Res. **215**(1), 194–203 (2011b)

Zheng, Z., Lu, J., Zhang, G., He, Q.: Rule sets based bilevel decision model and algorithm. Expert Syst. Appl. **36**(1), 18–26 (2009)

Zheng, Z., Wang, G.Y.: RRIA: A rough set and rule tree based incremental knowledge acquisition algorithm. Fundam. Inf. **59**, 299–313 (2004)

Ziarko, W., Cercone, N., Hu, X.: Rule discovery from databases with decision matrices. Paper presented at the The 9th International Symposium on Foundations of Intelligent Systems, Zakopane, Poland (1996)

Zimmermann, H.J.: Fuzzy Sets, Decision Making and Expert Systems. Kluwer Academic Publishers, Boston (1986)

Zionts, S., Wallenius, J.: An interactive multiple objective linear programming method for a class of underlying nonlinear utility functions. Manag. Sci. 29(5), 519–529 (1983)

Printed in the United States
By Bookmasters